Mechanical and Solid-Fluid Operations

The text comprehensively discusses the transport mechanism, storage, and conveying of the material, which are essential requirements for transporting solids in various process units, especially in mineral and chemical industries. It covers the properties of particles and particulate systems and focuses on their characterization and analysis.

This book:

- Presents a discussion of theoretical principles coupled with illustrative examples to help readers learn how to operate, optimize, and innovate particle processing technologies.
- Covers transport characterization of the solid-fluid operations, slurry physical properties, and properties of particles.
- Illustrates systematic and comprehensive understanding of fundamental phenomena of properties of particles and handling of particulate systems.
- Explains graphical representation of particle size, particle size measurement, and particle size distribution.
- Includes ancillary materials such as numerical problems, review questions, multiple-choice questions, and exercises at the end of each chapter.

It is primarily written for senior undergraduates, graduate students, and academic researchers in fields including chemical engineering, mechanical engineering, environmental engineering, industrial engineering, manufacturing engineering, and chemistry.

Mechanical and Solid-Fluid Operations

Characteristics and Handling of Solids

Subrata Kumar Majumder

CRC Press
Taylor & Francis Group
Boca Raton London New York

CRC Press is an imprint of the
Taylor & Francis Group, an **informa** business

Designed cover image: Subrata Kumar Majumder

First edition published 2025
by CRC Press
2385 NW Executive Center Drive, Suite 320, Boca Raton FL 33431

and by CRC Press
4 Park Square, Milton Park, Abingdon, Oxon, OX14 4RN

CRC Press is an imprint of Taylor & Francis Group, LLC

ISBN: 9781032944982 (hbk)
ISBN: 9781032944999 (pbk)
ISBN: 9781003571025 (ebk)

DOI: 10.1201/9781003571025

Typeset in Sabon
by codeMantra

I dedicate this book to my parents and beloved family.

Contents

SECTION II
Size Reduction of Solids

Preface

Particulate materials, powders, or bulk solids are used widely in all areas of the process industries, such as food processing, pharmaceutical, biotechnology, oil, chemical, mineral processing, metallurgical, detergent, power generation, paint, plastics, and cosmetics industries. These industries involve many professional scientists and engineers, such as chemical engineers, chemists, biologists, physicists, pharmacists, mineral engineers, food technologists, metallurgists, material scientists/engineers, environmental scientists/engineers, mechanical engineers, combustion engineers, and civil engineers.

The knowledge of the science and technology behind mechanical and solid-fluid operations in the process industries is valuable. Very often, the characteristics of solids and their hydrodynamic behavior in the fluid systems are ignored or overlooked at the design stage, and the results of related problems cause an inordinate number of production stoppages. However, it has been demonstrated that applying even a basic understanding of how materials behave can minimize these processing problems, resulting in less downtime and improvements in quality control and environmental emissions. So, this book will provide basics to the learners to assess the process with bulk solids and solid-fluid systems in industry, academia, and other sectors.

Most of the chemical manufacturing processes involve small solid particles at some point. Proper design and handling of these fine particles often make the difference between the success and failure of the product. Many products, such as catalysts, pigments, fertilizers, cement, ceramics, and pharmaceuticals, are manufactured in particulate forms. Mechanical and solid-fluid operations deal with the science and technology of particulate matter, a multidisciplinary field including materials science, environmental, biomedical, aerospace, agricultural, chemistry, microbiology and cell science, pharmacy, and medicine.

Therefore, this book aims to develop undergraduate students' basic science and technology skills related to the characteristics and processing of solids in the presence and absence of fluid. The idea of writing this book was conceived when I started teaching in the undergraduate solid-fluid operations program of chemical engineering at the Indian Institute of Technology Guwahati, India. Especially when I got constructive feedback from the creamy-level undergraduate students in chemical engineering on the course structure, it struck my mind to make the topics for the students more exciting and comprehensive, which drove me to make this book published and valuable to the learners and industries.

Mechanical and solid-fluid operations are extensive areas, and it is very difficult to accumulate all of the topics in a single book. Based on the contents covering a vast syllabus of most institutions/universities worldwide, this multi-volume book enables a vast syllabus coverage. This multi-volume book contains three volumes.

This first volume deals with the characteristic features of particles, particle size reduction and enlargement, the solid-solid mixing mechanism, details of types of mixing devices, and the assessment of the degree of mixing, which are essential in particle processing, especially in chemical and mineral and pharmaceutical industries. It also includes the transport mechanism, storage, and conveying of the material, which are essential requirements for transporting solids in various process units, especially in mineral and chemical industries.

This book's first section reviews the properties of particles and particulate systems and focuses on their characterization and analysis. The primary purpose is to provide the student with theoretical and practical tools to understand the behavior of particles and particulate systems. The second section deals with particle size reduction and enlargement, the laws of power consumption of size reduction, and the size reduction equipment and their principles.

The third segment of the book refers to the mixing mechanism of solids and the assessment of the degree of mixing by probability theory. This exciting subject sometimes has not been appropriately covered in other textbooks.

Many industrial processes involve solid transport and storage. This book's last section deals with the pneumatic transport of solids and their assessment, transport by conveyor and elevator systems, and storage mechanism of solids in hoppers. This book focuses on the pointwise discussion of theoretical principles coupled with ample illustrative examples, which will provide information and skills to the learners to operate, optimize, and innovate particle processing technologies.

The contents are made easy to understand and explained pointwise so that students can understand the concepts easily. The syllabus of the Graduate Aptitude Test in Engineering (GATE) (especially in India) and the syllabus of the course approved by different universities and colleges worldwide are covered in the book. After describing the concept, illustrative examples are given for each topic. Also, multiple-choice questions, objective type, and numerical problems for learner self-assessment are provided at the end of each chapter. The multiple-choice questions are made as per the Graduate Aptitude Test in Engineering (GATE) pattern.

This book will be particularly beneficial for academics of institutes or universities where mechanical and solid-fluid operations are taught. This book, however, can also be helpful as a reference book for students, teachers, and managers who study particle technology and applications.

Subrata Kumar Majumder

Acknowledgments

The author deeply acknowledges all his Ph.D. and M.Tech. students who helped him to review and make different exercises and assessment problems based on the book's contents. They suggested to include the topics in the book that are useful and suitable for the undergraduate course.

The author wishes to acknowledge the valuable suggestions and comments provided by peer reviewers.

The author also thanks the Taylor & Francis team for their editorial and publication assistance.

About the Author

Subrata Kumar Majumder is a professor in the Chemical Engineering Department at the Indian Institute of Technology Guwahati, India. He completed his Ph.D. in Chemical Engineering from the Indian Institute of Technology Kharagpur. His research interests include applied multiphase flow and reactor development, mineral processing, extraction, micronano bubble science and technology, and its applications in wastewater treatment and enhanced oil recovery. He is a Fellow of the Council of Engineering and Technology (India); a Fellow of the International Society for Research and Development, London; a Fellow of the Indian Institute of Chemical Engineers; and a Fellow of the Association of Researchers & Academicians (India). He is a member of the Indian Institute of Mineral Engineers; a life member of the Institute of Engineers (India); a member of the Asia-Pacific Chemical, Biological & Environmental Engineering Society (PCBEE); and a member of the International Association of Engineers (IAE), Japan. He was the Editor of the *Journal of Chemical Engineering Research Studies*; Guest editor of the *American Journal of Fluid Dynamics*, Scientific & Academic Publishing Co., USA; Editorial board member of the *Scientific Journal of Materials Science*; Editorial board member of the *Journal of Science and Technology*, Scientific and Academic Publishing, USA; and Advisory board member of Excelling Tech Publishers (ETP), London. He was awarded Dr. A.V. Rama Rao Best Research Award in 2022 by the Indian Institute of Chemical Engineers and the IIME Award on mineral beneficiation in 2008 from the Indian Institute of Mineral Engineers (IIME). He authored five other books and has over 120 publications in several reputed international journals. He has completed several sponsored and consultancy projects. He collaborates with Aalborg University, Denmark; University of Los Andes, Colombia; Curtin University, Australia; and Edith Cowan University, Australia. He has developed six massive open online courses (including the subject of this book) sponsored by the Government of India. He has more than 20 years of experience in teaching, research, and development.

For more details: https://www.iitg.ac.in/chemeng/faculty_profile.php?name=skm

Section I

Introduction and Characteristics of Solids

Chapter 1

Introduction

> A good scientist is a person with original ideas. A good engineer is a person who makes a design that works with as few original ideas as possible. There are no prima donnas in engineering.
>
> —*Freeman Dyson (theoretical physicist and mathematician)*

1.1 WHAT ARE MECHANICAL AND SOLID-FLUID OPERATIONS?

- The processes based on solid-solid interactions, transport of solids and fluids (gas and liquid), and solid-fluid interactions are known as mechanical and solid-fluid operations.
- The processes are generally applied in the removal of fine solids from a valuable liquid, recovery of useful solids from slurries, concentrating of sludges, sizing of the solid, transportation of solid, production of various chemicals by reaction in the presence of catalyst particles, mineral beneficiation, drying of solid, coating of a particle by polymeric agents, mixing by mechanical means, using a porous solid as a membrane for the filtration of contamination from the effluent, etc.
- The operations can be classified broadly into mechanical, electrochemical, and chemical processes, as described in Table 1.1. The basic principles of the different mechanical and solid-fluid operations are briefly described in successive sections.
- The mechanical and solid-fluid operations are essential in various industries, such as:
  - Chemical industry
  - Mineral processing, metallurgical industries
  - Detergent, power generation
  - Food processing industry
  - Pharmaceutical industry
  - Biotechnology, oil industries
  - Paint, plastics, and cosmetics industries.

DOI: 10.1201/9781003571025-2

Table 1.1 Mechanical and solid-fluid operations and their important process units (Ghosal et al., 2017)

Different Solid-Fluid Operations					
Mechanical Process		*Electrochemical Process*		*Thermochemical Process*	
Process	*Process Unit*	*Process*	*Process Unit*	*Process*	*Process Unit*
Transportation of materials: Conveying, piping, pumping, compressing, blowing (transportation of materials)	Belt conveyor, screw conveyor, pump, compressor, fan, blower, pipe, valve	Electrostatic separation (based on electrostatic force)	Cottrell precipitator	Crystallization (by a change in temperature)	Crystallizer
Size reduction: Crushing and grinding (reduction of size to separate ingredients)	Dodge crusher, blake crusher, gyratory crusher, roll crusher, ball mills and tube mills, hammer mills, burrstone mills	Magnetic separation (due to magnetic force)	Magnetic separator	Drying (removal of water moisture or moisture from another solvent by evaporation)	Dryer
Mixing (uniform distribution of materials):	Agitator, double-cone mixer, kneading machines, pugmills, colloid mills, mixing rolls, pan mixers, putty chasers	Electrodialysis (separation of ions by diffusion by imposing electromotive force)	Porous membrane	Leaching and extraction (removal of materials by dissolving them from solids)	Extractor
Mechanical separations:		Electroosmosis (based on applied electric field)	Membrane	Chromatography (separation of a gas or liquid solute by solid, adsorption)	Adsorber, disorber, chromatograph
Sieving (size-based separation)	Sieve, grizzlies, etc.				

(*Continued*)

Table 1.1 (Continued) Mechanical and solid-fluid operations and their important process units (Ghosal et al., 2017)

Different Solid-Fluid Operations					
Mechanical Process		*Electrochemical Process*		*Thermochemical Process*	
Process	*Process Unit*	*Process*	*Process Unit*	*Process*	*Process Unit*
Filtration (separation of solids from the liquid-solid mixture by a porous media or membrane by pressure difference)	Different types of filters, such as plate and frame filters, rotary drum filters, leaf filters, centrifuge, etc.	Electrophoresis (separation of colloidal particles under the influence of electrical potential)	Membrane		
Decantation (separation of two immiscible liquids)	Decanter	Ion exchange (removal of inorganic materials and its wastes from wastewater)	An ion exchanger called a deionizer		
Settling and sedimentation (separation based on centrifugal or gravitational forces)	Thickener, mineral jigs, centrifuge, and cyclone separators	Gas permeation (the penetration of a permeate (such as a liquid, gas, or vapor) through a solid)	Mini-permeameter		
Flotation (separation of materials based on hydrophobic or hydrophilic properties of solid)	Mechanical cell (agitair flotation machine, Denver cell, Steffensen flotation machine), flotation column	Pervaporation (by partial vaporization through a membrane)	Membrane		
Fluidization (solid phase is subjected to behave more or less like a fluid by the upwelling flow of gas or liquid stream)	Fluidized bed				

1.2 PROCESSES INVOLVING MECHANICAL OPERATIONS

1.2.1 Processes Involving the Transportation of Materials

- In industry, conveying the materials from one point to another is important for different applications.
- Transporting fluids is much easier, cheaper, and less troublesome than transporting solids because of their specific properties.
- The engineer prefers to transport everything in the form of liquids, solutions, or suspension as per convenience.
- The choice of equipment for transportation depends on the necessary capacity, natural shape and size of the material, and the distance through which the material has to be transported.
- Pipelines, pumps, fans, blowers, and compressors are used in every industry to transport fluids, finely divided solids in the form of slurry or solution from one point to another, whereas railways, trucks, and ropeways, ships, and conveyors transport solids not finely crushed.

1.2.2 Processes Involving Size Reduction

- The process of particle size reduction is called comminution. The primary comminution processes are crushing and grinding. Size reduction mainly occurs by compression, impact, and attrition.

1.2.3 Crushing Process

- Crushing is breaking materials into the desired size (to a specified size range) by a mechanical device.
- The types of crushing equipment used are (Prasher, 1987):
  - Based on compression mechanism: Example: jaw and cone crushers.
  - Based on impact mechanism: Example: bar blow crushers and vertical shaft impactors.

1.2.4 Grinding Process

- It is a surface generation process that produces micro or nanoparticles.
- The following factors influence the efficiency of grinding (Prasher, 1987):
  - **Type of material:** more rigid materials are suitable for finer finishes.
  - **Type of grains:** finer grains yield finer finishes.
  - **Dressing method:** appropriate dressing of wheels may make a perfect surface.

- **Feed rate:** slower feed rates yield finer finishes.
- **Machine quality:** older, worn machines are not suitable for a good product.
- **Condition of filtration:** coolant filtration removes waste that may deface the material surface.

- The usual name of the grinding machine is a grinder. The most common grinders are belts, jigs, and gear grinders; different types of mills are ball, rod, autogenous, pebble, buhrstone, tower, gap, blade mills, etc.

1.2.5 Processes Involving Mechanical Separations

- The mechanical separation processes are classified based on phase contact. The solid-solid separation processes are carried out mainly based on the solids' size, shape, density, and wetting. The solid-liquid separation processes are based on density difference and the pressure gradient. The different mechanical separation processes are briefly described as follows:

1.2.6 Screening and Classification

- Screening is a process of separating solid particles from multi-solid mixtures into multiple grades according to their sizes.
- Mining and mineral, pharmaceutical, food, plastics, and agriculture industries use this process for separating and classifying the particles into different grades.
- Classification is a process for separating mixtures of particles into two or more fractions based on the velocity with which the particles fall through a fluid medium.
- The process depends on the size of the particle. The course material (above about 250 μm) is suitable for screening as its efficiency decreases with the decreasing size of the particle.
- Finer size (below about 250 μm) is normally processed by classification.
- The different types of screens are generally used in industry and laboratory: stationary screens, vibrating screens, rotating screens, etc.

1.2.7 Filtration Process

- It is a process for separating solid particles from a solid-fluid mixture by a filter medium due to pressure differences across the medium.
- The fluid of the slurry may be liquid or gas. Fluid flows through the medium by a pressure differential across the medium.
- The pressure difference in the filtration process may be due to gravity, vacuum, or centrifugal forces.

- The filtration process can be differentiated as (Sparks and Chase, 2015):
  - **By driving force:** The filtration is governed either by the hydrostatic head (gravity) or pressure difference between upstream and downstream of the medium or by centrifugal force.
  - **By filtration mechanism:** By depositing solids (or forming a cake of a certain thickness), called cake filtration, and trapping the solids within the pores of the medium, called depth or filter-medium filtration process.
  - **By operating cycle:** The filtration may be batch or continuous, either at a constant pressure, at a constant flowrate, or in a cycle.
  - **By nature of the particle:** Filtration depends on the nature of particles, whether it is deformable under compression or not. Example: The separation of malleable materials, mainly by plate and filter press.
- Types of equipment for the filtration can be described as follows:
  - **Cross-flow filters:** The filters are designed to concentrate suspensions of fine particles or colloidal material and solutions of macromolecules. It is generally used for particles ranging from 0.1 to 5 µm (McCabe et al., 2017), (e.g., micro-screens, porous tube filters, and microstraining filters).
  - **Deep-bed filters:** The particles are separated by the bed of another particle (e.g., filtration by a bed of sand).
  - **Cake filter:** A cake acts as a filter that is formed by the accumulation of the solids on a septum (e.g., filter press, shell-and-leaf filters, and rotary vacuum filters).
  - **Syringe and capsule filters:** The filters made of polycarbonate, polypropylene, and polyethylene plastic are designed with a female luer lock at the inlet and a male slip fitting on the outlet of a syringe (Sundaram et al., 1999). It is also called a wheel filter. Syringe membrane filters are made in various pore sizes (0.2 or 0.45 µm). This type of filter is used when the filtration volume is less than 200 mL. Capsule filters work at liquid flow rates between 1 and 15 lpm. It is widely used in the laboratory.
  - **Membrane:** Different membranes are made of different kinds of materials, such as silver, cellulose, nylon, and ceramic.

1.2.8 Separation of Particulate Matter

- Particulate matter is a solid particle (particulate) or liquid droplet of size 10 µm in diameter or less suspended in the atmosphere (aerosol).
- The particulate matter is denoted by a conventional symbol as PM_d. For example, $PM_{2.5}$ is referred to as particulate matter, where the diameter of the particle is 2.5 µm or less.
- Various methods like wet scrubbing, cyclones, candle filters, or a combination of these sedimentation, filtration, and biological treatment can separate these particulate matters.

1.2.9 Decantation Process

- It is a process for the separation of immiscible liquids of different densities. This is obtained by pouring a solution from a container, leaving the residue (sediments) in the bottom of the original container. It can also be done by centrifugal force.
- It is also used to separate a liquid from an insoluble solid. Examples:
  - Decantation of wine from the potassium bitartrate crystals in red wine.
  - Decantation of olive oil is done after extracting oil and water from olives.
  - Kerosene and water are separated by decantation from their mixture.
  - Decantation is also used in the food-producing industries for the separation of soymilk base fibers (okara), casein, lactose crystals, and whey fines.
- Decantation is a particular case of settling and sedimentation.

1.2.10 Settling and Sedimentation, Thickening, Classification Process

- Settling and sedimentation is the process by which particulates are separated from the fluid by gravity or centrifugal motion.
- Settling is essential in settling crystals from the mother liquor, mining, wastewater treatment, and biological treatment.
- Thickening is a gravitational settling of solid particles that are suspended in a liquid. In contrast, classifications imply the fractionation of solid particles based on their flow rates or settling through fluids.
- The falling of particles through a fluid medium without the effect of its motion by the presence of other particles or the container walls is called free settling.
- A settling is called free settling when the particle concentration is less than 0.2% by volume in the slurry.
- The settling of particles in a thick suspension in water through which their fall is hindered by fluid viscosity and the interaction of particles is called hindered settling.
- When the solid particles are separated from the fluid by centrifugal forces acting on the particles, it is called centrifugal settling and sedimentation. Gravity settlers and centrifuges are used for the operation.

1.2.11 Flotation Process

- Flotation is a process for separating suspended particles based on particles' hydrophobic or hydrophilic nature by introducing fine gas bubbles.
- The adhesion of gas bubbles to the solid particles makes them buoyant and results in separation (Nguyen, 2003).

- Various flotation processes are froth flotation, dissolved air flotation (DAF), and induced gas flotation (IGF) (typical in the mineral processing industry).
- Flotation is done in flotation machines, which are mechanically agitated tanks, flotation columns, Jameson cells, etc.

1.2.12 Agitation and Mixing Process

- The agitation and mixing process makes the homogeneous mixing to (concentration, phase, and temperature) achieve the desired process yield by increasing mass and/or heat transfer, reaction rate, or product properties.
- It is the central feature of many chemical processes: catalytic reaction in a reactor, paints, and coatings; production of synthetic rubbers and resin, sealants and adhesives; food processing; making juice; making candy, biofuels, and ethanol; and widely used in making pharmaceuticals.
- The equipment (mixer) used for mixing depends on the nature of the materials.

1.3 BASIC FEATURES OF THE THERMOCHEMICAL PROCESSES

1.3.1 Drying Process

- It removes liquid from a solid at a temperature below the boiling point by circulating air or some other carrier gas over the solid.
- The primary difference from the evaporator is that the evaporator is used to remove water from the solution instead of solid.
- Heat is supplied to vaporize the liquid, which diffuses through various resistances (Fellows, 2009).
- Three different drying periods occur consecutively in time: preheating period, constant-rate period, and falling-rate period. The drying equipment is classified based on:
  - **Operational mode:** Batch types are generally preferred for small production requirements, whereas continuous type is used for large-scale production.
  - **Properties of the material:** It depends on the physical properties of the feed material. The application includes drying of slurry, paste, powder, crystalline solids, etc.
  - **Method of supplying heat to the material undergoing drying:** It is classified as adiabatic and non-adiabatic. In adiabatic drying, evaporation is caused by the sensible heat of a gas that comes in contact with the solid. In non-adiabatic drying, evaporation is caused by radiant heat or heat transfer through walls.

- Adiabatic dryers are spray, flash, rotary, tunnel, and tray dryers, whereas non-adiabatic are vacuum type (shelf, rotary), purge type (steam tube, drum, continuous rotary, continuous pan), and radiant type (infrared, dielectric) (Fellows, 2009).

1.3.2 Heat Exchange Process

- It is a process to transfer heat between two fluids, two solids or fluids, and solids.
- Heat is transferred from one material to the other through the surface.
- This concerns the principles that determine the accumulation and transfer of thermal energy due to temperature differences.
- It occurs by the mechanism of convection, conduction, and radiation.
- The flow of thermal energy from higher molecular kinetic energy regions to lower molecular energy regions through direct molecular collisions is called conduction (Hewitt et al., 1994).
- The transporting of heat by fluid motion is called convection.
- The emission of energy as electromagnetic waves or subatomic particles is called radiation.

1.3.3 Crystallization Process

- The formation of solid crystals precipitating from a super-saturated solution containing dissolved solids (solute), melts, vapor, or a different solid phase at a fixed temperature is called crystallization.
- The process is driven by a mass transfer of a solute from the liquid solution to a crystalline phase.
- The first step of the process is called "nucleation."
- The supersaturation can be obtained by cooling and solvent evaporation.
- Fractional crystallization is one of the most widely used methods of separating and purifying chemicals.
- Crystallization equipment is classified according to the method of generating supersaturation. Different crystallizers are tank or batch-type crystallizers, scraped surface crystallizers, circulating liquid evaporator crystallizers, circulating vacuum crystallizers, etc. (Tavare, 1995).

1.3.4 Leaching Process

- Leaching is the process of separating a solute from a solid by an insoluble solvent. Solutes are separated based on the solute's absorption capability to the solvent.
- The material (e.g., ore) is graded chemically with a solvent by dissolving the active component into the solvent (Geankoplis, 2004).
- The grading of bauxite using NaOH solution at 150°C, dissolving aluminum oxide in NaOH, extraction of silver from its ores (e.g., argentite, Ag_2S, AgCl) by the sodium cyanide, and grading gold-containing ore in potassium cyanide are some examples of leaching.

1.3.5 Calcination Process

- The calcination process is the conversion of concentrated ore into oxide by heating in the absence of air (Gilchrist, 1989).
- This process also helps to separate volatile contaminants like CO_2, SO_2, organic impurities, and moisture from the source at a fixed temperature.

1.3.6 Roasting Process

- It converts ore (usually a sulfide, below its melting temperature) into oxide in excess hot air.
- It is also known as the metallurgical process of gas-solid reactions at a temperature for purifying the metal component(s) in a blast furnace (Ghosh and Ray, 1991).
- Roasting is also referred to as frying at a fixed temperature. It is generally done in a fluidized bed. For example, roasting of coffee beans at 350°F–500°F in a fluidized bed.

1.3.7 Reaction Processes

- It is a process that leads to the formation of new molecules by rearranging or redistributing the constituent atoms.
- Chemical reactions occur by the involvement of the motion of electrons in the forming and breaking of chemical bonds, along with the evolution or absorption of energy.
- The reaction process depends on temperature, pressure, reactant mixture composition, catalyst presence, and residence time.
- There are generally four types of reactors: chemical reactor, nuclear reactor, fusion reactor, and bioreactor. Different types of solid-fluid reactions are shown in Table 1.2.
- There are three types of chemical reactors based on intermittent operation such as batch reactor, semi-batch, and flow reactors (plug flow reactor (PFR), continuous stirred tank reactor (CSTR), fixed bed reactor (FBR), moving bed reactor (MBR), fluidized bed reactor (FBR), trickle bed reactor (TBR), bubble column reactor (BCR), slurry-phase reactor (SPR)), where solid is used as a catalyst.

1.4 THE PROCESS UNITS INVOLVING ELECTROCHEMICAL PROCESS

1.4.1 Electrostatic Separation

- It is a process of separating solid particles by utilizing forces acting on charged or polarized bodies in an electric field.
- The process is essential in the production of valuable minerals and in the recovery of other useful materials.

Table 1.2 List of different types of solid-fluid-based reactions

Solid-Catalyzed Gas-Phase Reactions	*Gas-Solid Reactions*	*Gas-Liquid-Solid Processes*
• Fluid catalytic cracking • Hydrocarbon production from synthesis gas • Production of aniline, acrylonitrile, phthalic, and maleic anhydrides, etc. • Chlorination and bromination of hydrocarbons • Production of polyethylene and polypropylene. • Catalytic oxidation, etc.	• Reduction of iron oxide • Combustion of biomass and natural gas • Gasification, coking, and pyrolysis • Calcination of limestone, phosphates, aluminum hydroxide, etc. • Fluorination of uranium oxide	• **Gas-liquid-solid:** • Fluid bed catalytic cracking • Hydrotreating and hydroprocessing • Biochemical processes • Cultivation of microorganisms

1.4.2 Magnetic Separation

- It is a process of separating magnetic materials from a mixture using a magnet.
- Hematite ores (e.g., pyrolusite (MnO_2) and chromite ($FeO{\cdot}Cr_2O_3$)) are upgraded based on the magnetic properties of the ore and gangue particles.

1.4.3 Electrodialysis

- Electrodialysis is a transport process of salt ions through a selective semi-permeable membrane (cationic or anionic, called ion exchange membrane) under an electric potential (Strathmann, 1992).
- This is done in an electrodialysis cell where the salt ions move over the membranes under applied electric potential.

1.4.4 Electroosmosis

- The process by which charged particles migrate by diffusion and convective flow toward a less charged area by electroosmotic effect.
- The electroosmotic effect can significantly affect the behavior of the liquid.
- The process is applied to separate ions contained within electrolytes and solvents.
- The motions of electrolytes and solvents are driven by an applied potential across a porous material, capillary tube, membrane, or microfluidic devices (Kirby, 2010).

1.4.5 Electrophoresis

- Electrophoresis is a separation technique to separate biological molecules such as DNA, RNA, or protein based on their size and electrical charge.
- An electric current moves the molecules through a porous gel.
- The pore in the gel works like a sieve, allowing smaller molecules to move faster than larger molecules (Michov, 2022).
- Protein transport toward a positive charge is an example of this process.

1.4.6 Ion Exchange

- Ion exchange is a process by which unwanted dissolved ions (e.g., nitrate, fluoride, sulfate, and arsenic) in water and wastewater are exchanged with a similar charge by an ion-exchanged material (e.g., resin or a zeolite) (Dardeland Arden, 2008).
- The process is mainly used in water treatment. The cations are exchanged with positively charged ions available on the surface of ion-exchange material (e.g., usually sodium).
- The anions are exchanged with negatively charged ions (e.g., usually chloride).

1.4.7 Gas Permeation

- It is a fractionation of gas mixtures based on permeability to gas according to a dissolution–diffusion mechanism through a membrane.
- The gas diffusion through polymers includes random thermal movement of gas molecules in the polymer structure.
- Graphene oxide, rubbery polymer membranes, and metal-organic frameworks (MOFs) are some examples of promising materials for gas permeation due to their atomic thickness and low cross-membrane transport resistance and structural properties (Ismail et al., 2015).

1.4.8 Pervaporation

- Pervaporation is a process of separation of binary or multi-component mixtures of liquids based on partial vaporization through a polymeric or ceramic membrane.
- In the process, the feed, after heating at the operating temperature, passes through the pervaporation membrane's feed side.
- The desired component passes through the membrane and is continuously taken out in the form of vapor from the permeate side of the membrane (Strathmann, 2001).
- The partial vapor pressure gradient acts as a driving force for the process.

IMPORTANT POINTS TO REMEMBER

- The processes based on solid-solid interactions, transport of solids and fluids (gas and liquid), and solid-fluid interactions are known as mechanical and solid-fluid operations.
- The process of particle size reduction is called comminution. The primary comminution processes are crushing and grinding. Size reduction mainly occurs by compression, impact, and attrition.
- The process depends on the size of the particle. The course material (above about 250 μm) is suitable for screening as its efficiency decreases with the decreasing size of the particle. Finer size (below about 250 μm) is normally processed by classification.
- The pressure difference in the filtration process may be due to gravity, vacuum, or centrifugal forces.
- Particulate matter is a solid particle (particulate) or liquid droplet of size 10 μm in diameter or less suspended in the atmosphere (aerosol).
- Thickening is a gravitational settling of solid particles that are suspended in a liquid. In contrast, classifications imply the fractionation of solid particles based on their flow rates or settling through fluids.
- Flotation is a process for separating suspended particles based on the particles' hydrophobic or hydrophilic nature by introducing fine gas bubbles.
- The agitation and mixing process makes the homogeneous mixing to (concentration, phase, and temperature) achieve the desired process yield by increasing mass and/or heat transfer, reaction rate, or product properties.
- Three different drying periods occur consecutively in time: preheating period, constant-rate period, and falling-rate period.
- The formation of solid crystals precipitating from a super-saturated solution containing dissolved solids (solute), melts, vapor, or a different solid phase at a fixed temperature is called crystallization.
- Leaching is the process of separating a solute from a solid by an insoluble solvent. Solutes are separated based on the solute's absorption capability to the solvent.
- Roasting is the metallurgical process of gas-solid reactions at a temperature for purifying the metal component(s) in a blast furnace.
- Electrostatic separation is a process of separating solid particles by utilizing forces acting on charged or polarized bodies in an electric field.
- Electrodialysis is a transport process of salt ions through a selective semi-permeable membrane (cationic or anionic, called ion-exchange membrane) under an electric potential.
- Ion exchange is a process by which unwanted dissolved ions (e.g., nitrate, fluoride, sulfate, and arsenic) in water and wastewater are exchanged with a similar charge by an ion-exchanged material.
- Pervaporation is a process of separation of binary or multi-component mixtures of liquids based on partial vaporization through a polymeric or ceramic membrane.

SELF-TESTING QUESTIONS

I. Multiple-Choice Questions

Q1. Transporting fluids is much easier, cheaper, and less troublesome than transporting solids because of their____________________.
(a) availability
(b) process matters
(c) permission from the transport ministry
(d) specific properties

Q2. Which of the following operations deals with solids?
(a) Settling
(b) Coating
(c) Magnetic separation
(d) All the above

Q3. Which process does not involve appreciable heat and mass transfer?
(a) Leaching
(b) Crystallization
(c) Filtration
(d) Drying

Q4. Separation of mixtures of liquids by partial vaporization through a non-porous or porous membrane is called______.
(a) Ion exchange
(b) Electroosmosis
(c) Pervaporation
(d) Decantation

Q5. Breaking of materials into the desired size range by a mechanical device is called________.
(a) Extraction
(b) Crushing
(c) Leaching
(d) Filtration

Q6. The solid-liquid separation processes are suited based on which of the following options?
(a) Type of the separation vessel
(b) Density difference and the pressure gradient
(c) Availability of the materials
(d) None of these

Q7. The process of separating solid particles from multi-solid mixtures into multiple grades according to their sizes is called____________________.
(a) Leaching
(b) Extraction
(c) Screening
(d) Grinding

Q8. In ______ method, size separation is based on the low density of particles.
(a) Air separator
(b) Sieving
(c) Cyclone separator
(d) Elutriation

Q9. Which among the following size of particles can be processed by classification?
(a) $\leq 250\,\mu m$
(b) $\leq 175\,\mu m$
(c) $\leq 150\,\mu m$
(d) $\leq 225\,\mu m$

Q10. Filtration is a process for separating solid particles from a solid-fluid mixture by a filter medium due to_______________.
(a) Temperature difference
(b) Density difference
(c) Pressure differences across the medium
(d) Enthalpy difference

Q11. Decantation is a particular case of ________________.
(a) Pervaporation
(b) Settling and sedimentation
(c) Distillation
(d) Leaching

Q12. _______ implies the gravitational settling of solid particles that are suspended in a liquid.
(a) Classification
(b) Filtration
(c) Floatation
(d) Thickening

Q13. Flotation is a process for separating suspended particles based on _________.
(a) Thermal properties of particles
(b) Size of the particle
(c) Hydrophobic or hydrophilic nature of the particle
(d) Flotation device

Q14. ________ is a process of removing liquid from a solid at a temperature below the boiling point.
(a) Crystallization process
(b) Drying process
(c) Leaching process
(d) Calcination process

Q15. The material is graded chemically with a solvent by dissolving the active component from the material into the solvent and is referred to as____________.

(a) Distillation
(b) Milling
(c) Cavitation
(d) Leaching

Q16. ______________ process is the conversion of concentrated ore into oxide by heating in the absence of air.

(a) Calcination
(b) Roasting
(c) Filtration
(d) Leaching

Q17. In which process is ore converted into oxide in excess hot air?

(a) Pervaporation
(b) Grinding
(c) Roasting
(d) Calcination

Q18. __________ process helps separate solid particles by utilizing forces acting on charged or polarized bodies in an electric field.

(a) Centrifugal separation
(b) Electrochemical separation
(c) Magnetic separation
(d) Electrostatic separation

Q19. The transport process of salt ions through a selective semi-permeable membrane (cationic or anionic, called ion-exchange membrane) under an electric potential is called____________.

(a) Electroosmosis
(b) Electrodialysis
(c) Ion exchange
(d) Electrophoresis

Q20. The process by which charged particles migrate by diffusion and convective flow toward a less charged area by electroosmotic effect.

(a) Electroosmosis
(b) Electrodialysis
(c) Ion exchange
(d) Electrophoresis

Answers to multiple-choice questions

Q1. Ans. (d)
Q2. Ans. (d)
Q3. Ans. (c)
Q4. Ans. (c)
Q5. Ans. (b)
Q6. Ans. (b)
Q7. Ans. (c)
Q8. Ans. (d)
Q9. Ans. (a)
Q10. Ans. (c)
Q11. Ans. (b)
Q12. Ans. (d)
Q13. Ans. (c)
Q14. Ans. (b)
Q15. Ans. (d)
Q16. Ans. (a)
Q17. Ans. (c)
Q18. Ans. (d)
Q19. Ans. (b)
Q20. Ans. (a)

II. Objective-Type Questions

1. Give some examples of processes involving the transportation of materials.
2. Give five examples of the process which involves electrochemical phenomena.
3. What is the thermochemical process? Give some examples.
4. What is comminution? What are the main mechanisms of comminution?
5. What is the basis of the mechanical separation processes?
6. What are the different types of equipment for the filtration generally used?
7. Differentiate the settling and sedimentation, thickening, and classification process.
8. Explain the different drying periods.
9. What is crystallization?
10. What is leaching? Give three examples of leaching.
11. Is there any difference in calcination and combustion?
12. Give five examples of solid-catalyzed gas-phase reactions.
13. Explain the mechanism of ion exchange as a process.
14. What is the mechanism of gas permeation?
15. What is pervaporation? What is the driving force of pervaporation?

REFERENCES

Dardel, F. and T. V. Arden, Ion Exchangers. *Ullmann's Encyclopedia of Industrial Chemistry*, Wiley-VCH, 2008.

Fellows, P. J., (Ed.), *Food Processing Technology*, 3rd Edition, Woodhead Publishing Limited, 2009.

Geankoplis, C., *Transport Process and Separation Principles*, Prentice Hall, pp. 802–817, 2004.

Ghosal, S., S. Sanyal and S. Datta, *Introduction to Chemical Engineering*, McGraw Hill Education (India) Private Limited, 2017.

Ghosh, A. and H. S. Ray, *Principles of Extractive Metallurgy*, New Age International, 1991.
Gilchrist, J. D., *Extraction Metallurgy*, 3rd Edition, Pergamon Press, 1989.
Hewitt, G, G. Shires and T. Bott, *Process Heat Transfer*, CRC Press Inc, 1994.
Ismail, A. F., K. C. Khulbe and T. Matsuura, (Eds.), Fundamentals of gas permeation through membranes, in: *Gas Separation Membranes*, Springer, 2015. pp. 11–35.
Kirby, B. J., (Ed.), Electroosmosis, in: *Micro- and Nanoscale Fluid Mechanics: Transport in Microfluidic Devices*, Cambridge University Press, 2010. Archived from the original on 2020-11-24. Retrieved 2011-01-08. pp. 131–152.
McCabe, W. L., J. Smith and P. Harriot, *Unit Operations of Chemical Engineering*, 7th Edition, McGraw-Hill, International Edition, 2017.
Michov, B., *Electrophoresis Fundamentals: Essential Theory and Practice*, De Gruyter, 2022.
Nguyen, A., *Colloidal Science of Flotation*, CRC Press, 2003.
Prasher, C. L, *Crushing and Grinding Process Handbook*, John Wiley & Sons Ltd, 1987.
Sparks, T. and G. Chase, *Filters and Filtration Handbook*, 6th Edition, Butterworth-Heinemann, 2015.
Strathmann, H., Electrodialysis, in: W.S.W. Ho, K.K. Sirkar, (Eds.), *Membrane Handbook*, Van Nostrand Reinhold, 1992.
Strathmann, H., Membrane separation processes: Current relevance and future opportunities, *AIChE J.*, 47 (5), 1077–1087, 2001.
Sundaram, S, M. Auriemma, G. Howard Jr, H. Brandwein and F. Leo, Application of membrane filtration for removal of diminutive bioburden organisms in pharmaceutical products and processes, *PDA JPST*, 53 (4), 186–20, 1999.
Tavare, N. S., *Industrial Crystallization*, Plenum Press, 1995.

Chapter 2

Single Particle Characterization

> The scientist discovers a new type of material or energy and the engineer discovers a new use for it.
>
> —*Gordon Lindsay Glegg, British Engineer and Author*

2.1 WHAT IS A PARTICLE?

- A particle is defined as a discrete sub-portion of a substance. The definition includes solid particles, liquid droplets, or gas bubbles with physical dimensions ranging from sub-nanometer to several millimeters in size.
- The common types of materials consisting of particles are as follows:
  - powders and granules, e.g., pharmaceutical ingredients, cement, pigments
  - suspensions, e.g., vaccines
  - emulsions, e.g., milk
  - slurries, e.g., muds
  - aerosols, e.g., asthma inhalers
  - sprays, e.g., crop protection sprays.

2.2 WHY ARE PARTICLE CHARACTERISTICS/ PROPERTIES TO BE KNOWN?

Many industries routinely do particle characterization due to the following reasons:

- **Better control of product quality:** In an increasingly competitive global economy, better control of product quality delivers real economic benefits such as
  - ability to charge a higher premium for the product
  - reduce customer rejection rates and lost orders

DOI: 10.1201/9781003571025-3

  - demonstrate compliance in regulated markets.
- **Better understanding of products, ingredients, and processes:** It is required to know how particle properties affect the products, ingredients, and processes for
  - improving product performance
  - managing manufacturing and supply issues
  - optimize the efficiency of manufacturing processes
  - enhancing output or yield
  - relative economic growth.

2.3 PROPERTIES OF PARTICLE

The behavior of particulate materials is often dominated by the properties of the constituent particles. The properties of the particle affect reaction and dissolution rates, flowability and mixing, compressibility, abrasive, etc.

- The properties of the particle can be classified as:
  - particle shape
  - particle size
  - particle strength
  - porosity and microstructure
  - slurry physical properties
  - particle outline
  - zeta potential
  - particle true density
  - particle-moisture interaction.

2.3.1 Particle Shape

- Particle shape affects the efficiency of solid-fluid operation.
- Different types of particle shapes and their definitions are shown in Table 2.1.

Table 2.1 Particle shape and its definition

Shape	*Definition*
Acicular	Needle-shaped, rigid
Angular	Edgy, hard angles
Fibrous	Thread-like, non-rigid
Granular/blocky	Irregular-shaped
Spherical/cuboid/cylindrical/cone/cube	Regular-shaped

- Areas in which particle shape has an impact are as follows:
  - Reactivity and solubility
  - Powder flow and handling
  - Ceramic sinter properties
  - Abrasive efficiency
  - Texture and feel.
- The shape factors used to measure the particle shape are aspect ratio, sphericity, roundness, and particle outline.
- The projected area of bodies of different shapes such as cylinder, semi-circular, circular, and polygon is very important to assess the resistance (drag) when fluid flows over the solid surface.

2.3.2 Size of Single Particle

- The size of an irregular-shaped particle is expressed in terms of the size of an equivalent sphere based on different parameters, as shown in Figure 2.1.
- The following ways define the size of the equivalent sphere:
  - The sphere of the same volume of the particle.
  - The sphere of the same surface area of the particle.
  - The sphere of the same surface area per unit volume of the particle.
  - The sphere of the same projected area perpendicular to its direction of motion.
  - The sphere of the same projected area of the top view of the particle lying at its maximum stability. Applicable for microscopic analysis of particle size.

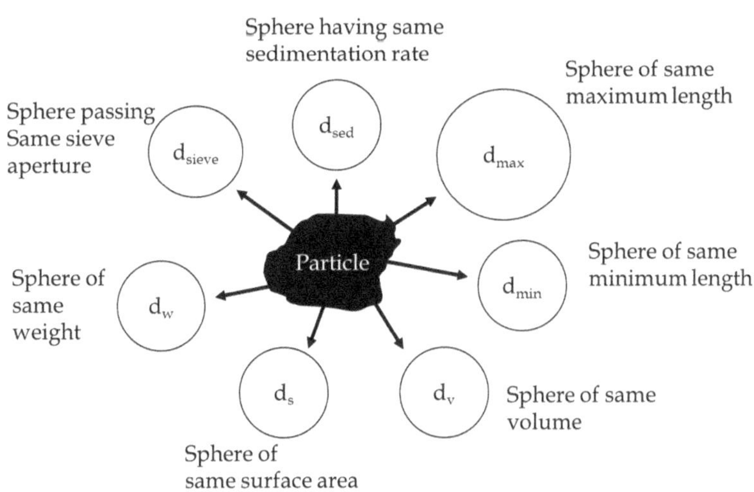

Figure 2.1 Important sizes of equivalent spheres.

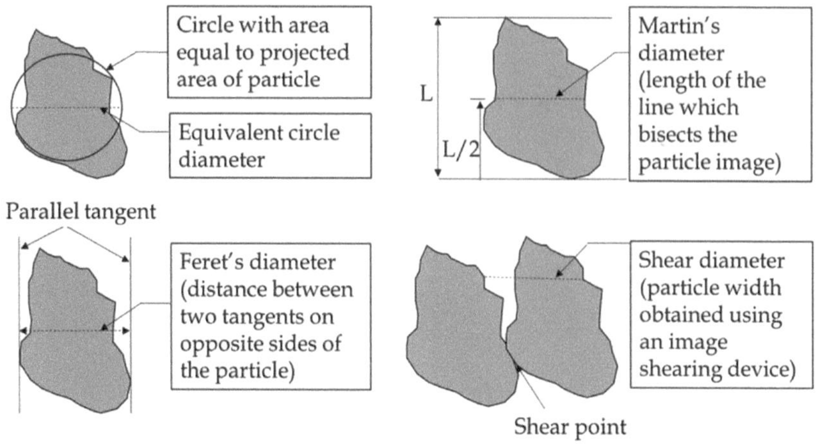

Figure 2.2 The size of a particle is based on the projection of the shape of the particle.

- The sphere passes through the same size as the square aperture through which the particle passes. Applicable for sieving of particles.
- The sphere of the same terminal velocity as that of the particle in a fluid.

- The diameters used in microscopy for assessing the size of a particle based on the projection of the shape of the particle is shown in Figure 2.2.

2.3.3 Aspect Ratio

- It is the ratio of width to height of an object.
- The aspect ratio lies within zero for a very elongated particle to unity for an equi-axed particle.
- The range of aspect ratio lies within the range: 0 < Aspect ratio < 1.

2.3.4 Sphericity

- It is referred to the degree of closeness of the shape of a particle to the perfect spherical particle. Sphericity, denoted by Φ_p, is defined as

$$\Phi_p = \frac{\text{Surface area } (A_s) \text{ of a sphere having the same volume as the particle}}{\text{Surface area of the particle } (A_p)}$$

$$= \frac{\pi^{1/3}(6V_p)^{2/3}}{A_p} \tag{2.1}$$

where V_p denotes the particle volume and A_p means the particle's surface area.

Table 2.2 Typical values of sphericity (Geldart, 1986; Kunii and Levenspiel, 1991)

Material	*Shape*	*Sphericity* (Φ_s)
Pulverized coal	Angular	$\approx$ 0.56–0.73
Crushed coal	Irregular spheroids	0.63–0.75
Activated carbon	Irregular spheroids	0.70–0.90
Cement	Angular	$\approx$ 0.59
PVC powder	Irregular spheroids	$\approx$ 0.81
Calcined alumina	Coarse spheroids	$\approx$ 0.84
Hydrated alumina	Rough spheroids	$\approx$ 0.86
Fly ash	Approximate spheres	$\approx$ 0.90
Sand	Round sand and sharp sand	0.86, 0.66 respectively
Crushed sandstone	Round and sharp	0.8–0.9
Mika flakes	Round and sharp	0.28
Fischer–Tropsch catalyst	Approximate spheres	0.58
Common salt	Cubic crystals	0.84
Crushed glass	Irregular and sharp	0.65
Silica gel	Approximate spheres	0.70–0.90
Tungsten powder	Body-centered cubic crystal	0.89
Sillimanite	Needles with square cross-sections	0.75
Wheat	Oval and spheroid	0.80

- The sphericity can also be defined as

$$\Phi_p = \frac{6V_p}{d_p A_p} \tag{2.2}$$

Example 2.1

For a spherical particle of diameter d, $V_p = \pi d^3/6$, $A_p = \pi d^2$, and so $\Phi_p = 1$.

For a cylindrical particle of diameter d (=1) and length L (= 1), $V_p = \pi d^2 L/4$, $A_p = \pi dL + \pi d^2/2$. So, $\Phi_p = 0.873$.

For a cube of width, a, $V_p = a^3$, $A_p = 6a^2$, and so $\Phi_p = 0.806$.

So, $\Phi_{\text{sphere}} > \Phi_{\text{cylinder}} > \Phi_{\text{cube}} >$ as per $1 > 0.873 > 0.806$, respectively.

- Some typical values of sphericity are listed in Table 2.2.

2.3.5 Roundness

- It is the measure of how closely the irregular shape of an object is a perfect circle.
- The degree of roundness (D_r) of a solid in a plane can be calculated based on the arithmetic mean of the roundness of the individual corners or edges in that plane, as shown in Figure 2.3.

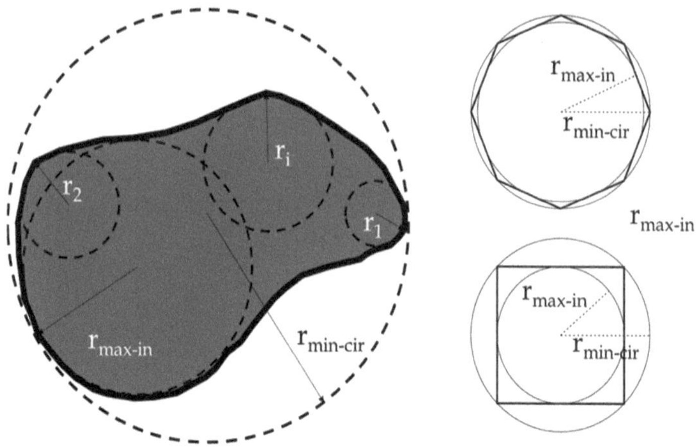

Figure 2.3 Schematic of degree of roundness.

- The "roundness" of a corner *i* can be expressed by the ratio

$$\frac{r_i}{r_{\text{max-in}}},$$

  where
  - r_i=the radius of curvature of the ith corner and
  - $r_{\text{max-in}}$=the radius of the circle, which is maximally inscribed in the plane of measurement.
- The radius of curvature is to be taken as the curvature angle to be made 10° by the arc at the center of curvature of the corner.
- Thus, the degree of roundness of a particle in a plane can be expressed as (Wadell, 1932):

$$D_r = \sum\left(\frac{r_i}{r_{\text{max-in}}}\right)/N \tag{2.3}$$

  where

  $\sum\left(\frac{r_i}{r_{\text{max-in}}}\right)$=the sum of the roundness values of the corners, and

  N=the number of corners of the particle in the given plane.
- As an example, a sphere has a roundness value of 1.0.

The shape is expressed by sphericity value, whereas a summarized expression for the characteristic of the solid is expressed by roundness. The ratio of roundness to sphericity is called an image of a solid (Wadell, 1932).

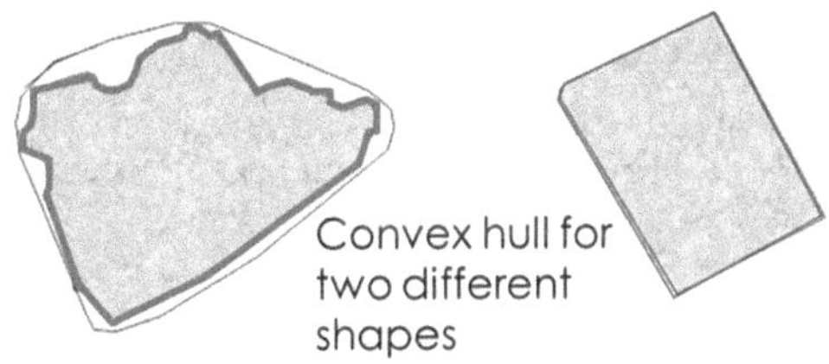

Figure 2.4 Convex hull for two different shapes.

2.3.6 Particle Outline

- The outline of a particle expresses the surface roughness of the particle. A concept of convex hull perimeter, as shown in Figure 2.4, is used for calculating particle outline. Based on the convex hull perimeter, convexity, solidity, or circularity of a particle can be defined as:
  Convexity=convex hull perimeter/actual perimeter
  Solidity=area bound by actual perimeter/area bound by convex hull perimeter
  Circularity=actual perimeter/perimeter of an equivalent area circle
  Circularity is mainly used to assess the outline of a particle.

NB: Perimeter and area are to be considered as a projected perimeter and area of 3D object.

Example 2.2

What is the convexity, solidity, and circularity of dice of the face of length 1 cm?

Solution:

Convexity=convex hull perimeter/actual perimeter=1/1=1
Solidity=area bound by actual perimeter/area bound by convex hull perimeter=1
Circularity=actual perimeter/perimeter of an equivalent area circle
Actual perimeter=4 cm
The equivalent diameter of the projection area of the circle=$\sqrt{(16\pi)}$
Therefore, circularity=$4/(\sqrt{(16\pi)}= 1/\sqrt{\pi} = 0.564$.

2.3.7 Particle Strength

- The mechanical strength of a particle is important to know for sustainable transport and storage of particles and the design of its container.
- Catalysts of a low mechanical strength may break in the reactors during transport or storage, resulting in variations in the reaction yield.
- The strength depends on the particle forces such as:

- Intramolecular forces:
  - Ionic bonding
  - Covalent bonding
- Interparticle forces
  - Ion-dipole forces
  - Dipole-dipole
  - London forces (called dispersion forces)
- Hydrogen bonding

2.3.8 Particle and Moisture Interactions

- The water or moisture content of a particle is a crucial property. The "hygroscopicity" measures a particle's ability to take up water vapor from the atmosphere.
- A porous solid (e.g., activated carbon) physically adsorbs gas on the external active surface and within the pores.
- The adsorption capacity is affected by the temperature and humidity of the atmosphere. The capacity is assessed by the isotherm of the system.
- For some solids (e.g., silica gel), the interaction at the surface depends on chemical bonds.
- The different moisture content characteristics are:
  - **Saturation moisture content (SMC):** Maximum moisture that material can retain without becoming a slurry. It is essential to know how to handle wet material.
  - **Moisture sorption/desorption:** Tendency of a material to gain or lose moisture at a constant temperature. This knowledge is important to understand the caking behavior of powder.
  - **Dust extinction moisture (DEM):** The moisture content held at which material does not emit dust.
  - **Transportable moisture limit (TML):** The limit of moisture content at which material liquefaction behavior is evaluated.

2.3.9 Zeta Potential

- A suspension of colloidal and nanoparticles under an electric field moves due to the interaction between the charged particle and the applied field.
- The electrical potential of the particle for its motion at the slipping plane (as shown in Figure 2.5) is called the zeta potential.
- The interaction effects on the electrostatic repulsion or attraction between particles under this electrical potential cause dispersion or agglomeration of the particles in the suspension.

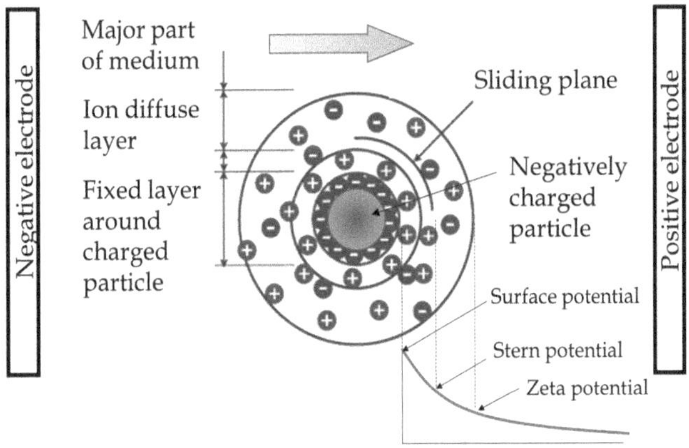

Figure 2.5 Schematic of zeta potential measurement.

- The particle's surface charge under this electrical potential depends on the voltage applied, the surface potential, rather than a charge density.
- The potential decreases exponentially with distance from the surface of the particle.
- The Smoluchowski equation estimates the zeta potential (von Smoluchowski, 1903; Hunter, 1981) as

$$\zeta = \frac{\mu_e \eta}{\varepsilon_r \varepsilon_0} \tag{2.4}$$

$$\mu_e = \frac{v}{V/L} \tag{2.5}$$

where ζ=the zeta potential, ε_r=the dielectric constant of the medium where the particle is dispersed, ε_0=the permittivity of free space, η=dynamic viscosity of the medium (Pa.s), μ_e=the electrophoretic mobility, v=speed of the particle, V=voltage, and L=the distance of electrode.

- If zeta potential is close to zero, it refers to a probable unstable aggregation of particles.
- The stability behavior of the colloid suspension based on zeta potential is as shown in Table 2.3:
- The stability behavior of the colloid suspension based on zeta potential is shown in Figure 2.6.

Table 2.3 The stability behavior of the colloid suspension based on zeta potential (Kumar and Dixit, 2017; Kim et al., 2019)

Stability Behavior of the Colloid	*Zeta Potential [mV]*
Rapid coagulation or flocculation	from 0 to ±5
Incipient instability	from ±10 to ±30
Moderate stability	from ±30 to ±40
Good stability	from ±40 to ±60
Excellent stability	more than ±61

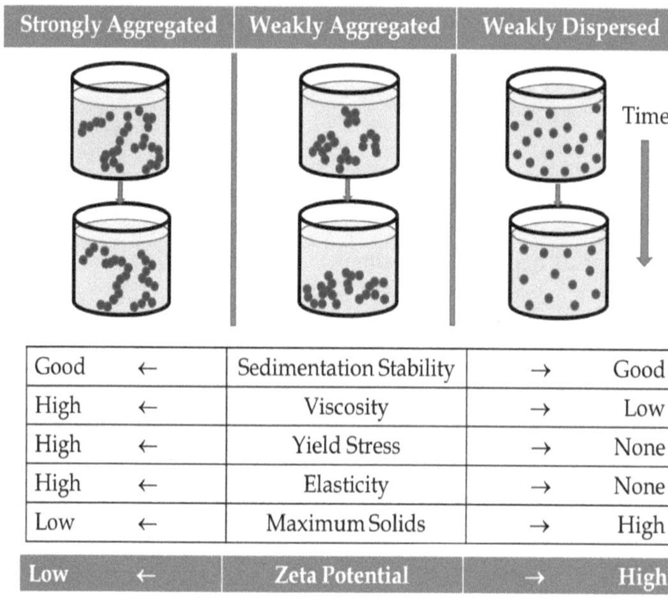

Figure 2.6 Stability behavior of the colloid suspension based on zeta potential (Fairhurst, 2013).

Example 2.3

Determine the stability behavior of the colloid in which the dielectric constant of the dispersion medium is 30 and the dynamic viscosity of the dispersion medium is 547×10^{-6} Pa.s. The electrophoretic mobility of the particles is 2.43×10^{-9} $(Vs)^{-1}$ and the permittivity of free space is 8.85×10^{-12} $F.m^{-1}$.

Solution:

Using the Smoluchowski equation:

$$\text{Zeta potential} = \frac{2.43 \times 10^{-9} \times 547 \times 10^{-6}}{30 \times 8.85 \times 10^{-12}} = 5\,\text{mV}$$

Rapid coagulation or flocculation if zeta potential ranges from 0 to $\pm 5\,\text{mV}$.

2.3.10 Particle Density

- For nonporous particles, the particle density is defined by the mass of the particle, M, divided by the volume of the particle, V_p, which can be represented by

$$\rho_p = \frac{M_p}{V_p} = \frac{\text{mass of the particle}}{\text{volume of particle}} \tag{2.6}$$

- For porous particles, as shown in Figure 2.7 with tiny pores, the density of the particle is referred to as skeletal density, which is defined by

$$\rho_p = \frac{\text{Mass of the particle } (M_p)}{\text{Enveloped volume of the particle } (V_{p,e}) \text{ as shown in Figure 2.8.}} \tag{2.7}$$

- Some particles are more easily described by length or area rather than volume.

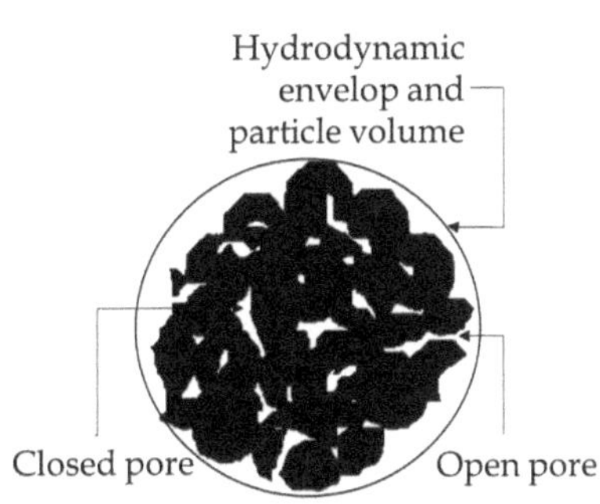

Figure 2.7 Porous particle.

- **Length density:** For something like a rod-shaped particle, one can use length density which is defined as "mass per length" and expressed as:

$$D_l = \frac{M_p}{L_p} = \frac{\text{Mass of the particle}}{\text{Length of the particle}} \tag{2.8}$$

- **Surface area density:** For something like a sheet-type particle, one can use area density which can be defined as

$$D_{sa} = \frac{M_p}{S_p} = \frac{\text{Mass of the particle}}{\text{Surface area of the particle}} \tag{2.9}$$

2.4 SLURRY PHYSICAL PROPERTIES

2.4.1 Slurry Density

- A mixture of solid and liquid is called slurry. The density of a slurry can be expressed as

$$\rho_m = \frac{100}{[c_w / \rho_s + (100 - c_w) / \rho_l]} \tag{2.10}$$

where
ρ_m = density of slurry (kg/m^3)
ρ_s = density of the solids (kg/m^3)
ρ_l = density of liquid without solids (kg/m^3)
c_w = concentration of solids by weight in the slurry (%) (also called concentration of slurry).

2.4.2 Slurry Concentration

- The concentration of slurry by weight (%) is expressed as

$$c_w = \frac{\text{Weight of dry solid}}{\text{Weight of dry solid} + \text{weight of liquid}} \times 100 \tag{2.11}$$

Example 2.4

A slurry consists of salt and brine. The slurry density is 1685 kg/m^3. Calculate the concentration of the solid by weight. Given: solid density = 2500 kg/m^3 and liquid density = 1378 kg/m^3.

Solution:

$1685 = 100/[(C_w/2500) + (100 - C_w)/1378]$
$C_w[(1/2500) - (1/1378)] = 100[(1/1685) - (1/1378)]$
$C_w = 40.59\%$

2.4.3 Slurry Viscosity

- The slurry viscosity depends on the volume fraction of particles in the slurry. The slurry viscosity can be calculated as per the following equations:

$$\mu_s = \mu_r \mu_l \tag{2.12}$$

where μ_s and μ_l are the dynamic viscosities of slurry and pure liquid, respectively. μ_r is the relative dimensionless viscosity. The relative viscosity varies with the volume fraction of solid particles in the slurry.
- For low-volume fractions of fine particles ($\varepsilon_s < 20\%$ by volume) and low shear stress (<1 kPa), the following Einstein's equation (Einstein, 1906) can be used to calculate the relative viscosity.

$$\mu_r = 1 + 2.5\varepsilon_s \tag{2.13}$$

- For higher concentrations ($20\% < \varepsilon_s < 60\%$ by volume) and low shear stress (<1 kPa) of fine particles, a modified equation based on the interactions between the solid particles is given by Thomas (1965):

$$\mu_r = 1 + 2.5\varepsilon_s + 10.05\varepsilon_s^2 + 0.00273e^{16.6\varepsilon_s} \tag{2.14}$$

- For high shear stress (>1 kPa) and higher concentrations ($\varepsilon_s > 20\%$ by volume), the relative viscosity can be predicted by the following equation given by Kitano et al., especially for polymer melts of the smooth spherical particle (Kitano et al., 1981):

$$\mu_r = \left(1 - \frac{\varepsilon_s}{0.68}\right)^{-2} \tag{2.15}$$

Note: *For particles with diameters less than 1–10 μm, colloid-chemical forces become important, causing non-Newtonian flow behavior. The result is a relative viscosity that increases as the particle size decreases but decreases to a limiting value as the shear rate is increased. For particles larger than 1–10 μm, the inertial effects due to the restoration of particle rotation after collision result in additional energy dissipation and a consequent increase in relative viscosity with increasing particle diameter.*

Example 2.5

It is seen that the transportation efficiency of the slurry through the pipe depends on the effective viscosity of the slurry. A slurry of solid concentration of 15% by volume is made to transport it for a chemical process. The liquid by which the slurry is made has a viscosity of 2 cP. Calculate the effective slurry viscosity. The shear stress of the slurry flow is less than 1 kPa.

Solution:

Volume concentration of solid $\varepsilon_s = 15\% = 0.15 < 0.20$
Liquid viscosity=2 cP=0.02 Poise
Since the volume fraction of the solid in the slurry is less than 20%, we can use Einstein's equation to calculate the effective viscosity as

$$\mu_{sl} = \mu_l \times \mu_r = \mu_l(1 + 2.5\varepsilon_s) = 0.02(1 + 2 \times 0.15) = 0.0275 \text{ Poise} = 2.75 \text{ cP}$$

2.5 TRANSPORT CHARACTERISTICS OF THE SOLID-FLUID OPERATIONS

- The particle properties affect the following transport characteristics of the solid-fluid operations:
  - Flow ability
  - Reactivity
  - Caking
  - Attrition/friability
  - Adsorption
  - Mixing and segregation behavior

2.5.1 Flowability

- Flowability refers to the ability of a particle or powder to flow. It depends on the structure and properties of the material and specific process conditions in handling the material. The flowability is assessed by the compression ratio (also called Carr's compressibility index) and Hausner ratio.
- The compression ratio is defined as (Carr, 1965)

$$\text{Compression Ratio} = \frac{\text{Tapped density} - \text{Bulk density}}{\text{Tapped density}} \tag{2.16}$$

- Hausner ratio is defined as (Hausner, 1967)

$$\text{Hausner Ratio} = \frac{\text{Tapped density}}{\text{Bulk density}} \tag{2.17}$$

- Bulk density is defined as

$$\rho_b = \frac{M}{V} \tag{2.18}$$

where M and V are the weight and volume of powder and ρ_b is the bulk density. It is measured from a sample mass of powder into a graduated cylinder and by leveling it without compacting.

Table 2.4 Powder flowability in terms of compression ratio and Hausner ratio (Turchiuli et al., 2005)

Compression Ratio (Carr's Index)	*Hausner Ratio*	*Flowability*
<10%	1.00–1.11	Excellent
11%–15%	1.12–1.18	Good
16%–20%	1.19–1.25	Average
21%–25%	1.26–1.34	Eligible
26%–31%	1.35–1.45	Poor
32%–37%	1.46–1.59	Terrible
>38%	>1.60	Can hardly flow

- The tapped density is defined as

$$\rho_t = \frac{M}{V_t} \tag{2.19}$$

where ρ_t is the tapped density, M is the powder mass, and V_t is the volume occupied after tapping it in a graduated measuring cylinder. The tapping is obtained mechanically by dropping the cylinder under its weight at a specified distance.
- The relationship among compression ratio, Hausner ratio, and powder flowability is shown in Table 2.4.
- In general, powder behavior can be split into three categories:
    - Low Carr's index and low angle of repose. Flow is good, so no future problems are likely.
    - Mid-range values for Carr's index and angle of repose. Flow can probably be improved with additives.
    - High values of Carr's index and angle of repose. Flow is extremely poor and cannot be improved sufficiently with additives.

2.5.2 Reactivity

- Reactivity of metal powder is the phenomenon of the metal elements in the powder that can form a reaction with acids, water, and oxidizing agents.
- As an example, magnesium is more reactive compared to zinc with hydrogen.

2.5.3 Caking

- Caking is a phenomenon of powders forming lumps or masses.
- Some materials change with time and lead to bulk powder caking, which makes them difficult to handle and process.

2.5.4 Attrition/Friability

- Attrition is the breakage of particles, pellets, tablets, or other bulk solid materials into smaller particles.
- The attrition affects flow, product quality, environmental and safety issues, clogging, friction, etc.
- The mechanisms that cause particle attrition are as follows:
  - by the impact on filling a bin or silo
  - by sliding the particles on one another
  - by sliding the particles onto a surface
  - by crushing due to pressure

2.5.5 Adsorption

- This is a process of adhesion of liquid or gas molecules onto the surface of a particle.
- The degree of adsorption varies with the particle size and the surface functional groups.

2.5.6 Segregation Behavior

- The segregation of particles occurs by the differences in each particle's density, size, and shape.
- Segregation based on particle size and density is common in gas-solid operations.
- Segregation of particles affects the heat and mass transfer efficiency of a solid-fluid operation.

2.5.7 Dustiness

- It is the state of being covered with dust and a quantification of dust release during a solid-fluid operation.
- It signifies the particles to become airborne in response to a mechanical or operational stimulus.
- Dustiness is affected by the particle shape, size, and inherent electrostatic forces. Dustiness increases the risk of inhalation exposure.

IMPORTANT POINTS TO REMEMBER

- A particle is defined as a discrete sub-portion of a substance.
- The shape factors used to measure the particle shape are aspect ratio, sphericity, roundness, and particle outline.
- The sphere of the same projected area of the top view of the particle lying at its maximum stability.

- The sphericity be defined as $\Phi_p = \frac{6V_p}{d_p A_p}$
- $\Phi_{sphere} > \Phi_{cylinder} > \Phi_{cube} >$ as per $1 > 0.873 > 0.806$, respectively.
- The degree of roundness of a solid in a plane can be calculated based on the arithmetic mean of the roundness of the individual corners or edges in that plane.
- The shape is expressed by sphericity value, whereas a summarized expression for the characteristic of the solid is expressed by roundness.
- The ratio of roundness to sphericity is called an image of a solid.
- Solidity is the area bound by the actual perimeter upon the area bound by the convex hull perimeter.
- The moisture content held at which the material does not emit dust is called dust extinction moisture.
- A suspension of colloidal and nanoparticles under an electric field moves due to the interaction between the charged particle and the applied field. The electrical potential of the particle for its motion is called the zeta potential.
- If zeta potential is close to zero, it refers to a probable unstable aggregation of particles.
- For particles with diameters less than 1–10 μm, colloid-chemical forces become important, causing non-Newtonian flow behavior.
- For particles larger than 1–10 μm, the inertial effects due to the restoration of particle rotation after collision result in additional energy dissipation and a consequent increase in relative viscosity with increasing particle diameter.
- The flowability is assessed by the compression ratio (also called Carr's compressibility index) and Hausner ratio.
- Low Carr's index, the low angle of repose. Flow is good, so no future problems are likely.
- The degree of adsorption varies with the particle size and the surface functional groups.

SELF-TESTING QUESTIONS

I. Multiple-Choice Questions

Q1. What is the ratio of the projected area of a sphere with diameter *d* units and a cube of side *d* units?

(a) π/4
(b) 2π/3
(c) π
(d) π/2

Q2. The compression ratio and Hausner ratio assess the ______________ of the material.
(a) Flowability
(b) Strength
(c) Sphericity
(d) Reactivity

Q3. What is the CORRECT definition of Martin's diameter?
(a) Diameter of a circle with an area equal to the projected area of the particle
(b) Distance between the two tangents on the opposite sides of the particle
(c) Length of the line that bisects the particle image
(d) Particle width obtained using an image-shearing device

Q4. If d_p is the equivalent diameter of a nonspherical particle, V_p is its volume, and S_p is its surface area, then its sphericity is defined by
(a) $\Phi = \frac{6V_p}{d_p S_p}$
(b) $\Phi = \frac{V_p}{d_p S_p}$
(c) $\Phi = \frac{6d_p S_p}{V_p}$
(d) $\Phi = \frac{d_p S_p}{V_p}$

Q5. The ratio of roundness to sphericity is called ______________.
(a) An image of a solid
(b) Shape of a solid
(c) Projected area
(d) Circularity

Q6. You have a square of side a unit and an equilateral triangle of side a unit. Which one of the following options is correct?
(a) The square has a higher circularity compared to the triangle
(b) The triangle has a higher circularity compared to the square
(c) Bothe square and triangle have the same circularity
(d) Cannot be assessed

Q7. Zeta potential close to zero refers to ______________.
(a) A probable unstable aggregation of particles
(b) Excellent stability
(c) Incipient stability
(d) Good stability

Q8. Which equation is used to find the slurry viscosity in the case of extremely low concentrations of fine particles (ϕ<20% by volume) and low shear stress (<1 kPa)?
(a) Thomas' equation
(b) Kitano et al. equation
(c) Einstein's equation
(d) All the above are applicable

Q9. The measure of degree of closeness of the shape of a particle to the perfect spherical particle is ________.
(a) Sphericity
(b) Aspect ratio
(c) Roundness
(d) Particle outline

Q10. Which of the following interparticle forces affect the particle strength?
(a) Ion-dipole interaction
(b) Covalent bonding
(c) London forces
(d) Hydrogen bonding

Q11. The size of a particle of irregular shape is defined in terms of the size of an equivalent
(a) Sphere
(b) Cylinder
(c) Cube
(d) None of these

Q12. Particle shape has a considerable impact on the performance or processing of particulate materials. In which of the following areas does particle shape has an impact?
(a) Reactivity
(b) Solubility
(c) Both (a) and (b)
(d) None of these

Q13. For the flat surface, the shape factor is equal to ________________.
(a) 1
(b) Approximately 0
(c) 1.5
(d) Infinity

Q14. Which one of the following options is correct in terms of sphericity?
(a) Cylinder<sphere<hemisphere<cube
(b) Sphere>cylinder>hemisphere>cube
(c) Hemisphere>cube>cylinder>sphere
(d) Cube>sphere>cylinder>hemisphere

Q15. Particle outline parameters can be measured using the concept known as
(a) Convex hull perimeter
(b) Color of the particle
(c) Both a and b
(d) All of the above

Q16. Which of the following parameters is not correct?
(a) Convexity=convex hull perimeter/actual perimeter
(b) Solidity=area bound by actual perimeter/area bound by convex hull perimeter
(c) Circularity=convex hull perimeter /perimeter of an equivalent area circle
(d) The most commonly used parameter is circularity

Q17. Smoluchowski equation described the relationship between
(a) Streaming potential and zeta potential
(b) Concentration and zeta potential
(c) Diffusion layer and streaming layer
(d) All of the above

Q18. For perfectly round object, the roundness value is ______.
(a) 0
(b) 1
(c) 2
(d) 0.5

Q19. What is the range of particle size based on the microscopy technique?
(a) >25 μm
(b) 10–200 μm
(c) 5–100 μm
(d) >50 μm

Q20. Match the following

(a) Acicular	p. Thread-shaped
(b) Fibrous	q. Needle-shaped
(c) Granular	r. Regular-shaped
(d) Spherical	s. Edgy-shaped

(a) a-p, b-q, c-r, d-s
(b) a-q, b-p, c-s, d-r
(c) a-s, b-r, c-q, d-p
(d) a-s, b-p, c-q, d-r

Answers to Multiple-Choice Questions

Q1. Ans. (a)
Q2. Ans. (a)
Q3. Ans. (c)
Q4. Ans. (a)
Q5. Ans. (a)
Q6. Ans. (b)
Q7. Ans. (a)
Q8. Ans. (c)
Q9. Ans. (a)
Q10. Ans. (a), (c), (d)
Q11. Ans. (a)
Q12. Ans. (c)
Q13. Ans. (b)
Q14. Ans. (b)
Q15. Ans. (a)
Q16. Ans. (b)
Q17. Ans. (a)
Q18. Ans. (b)
Q19. Ans. (c)
Q20. Ans. (b)

II. Objective-Type Questions

1. Which transport properties are affected by particle shape?
2. What are the different methods by which the size of the single particle is assessed by the size of the equivalent sphere?
3. What are the different diameters used in microscopy for assessing the size of a particle based on the projection of the shape of the particle?
4. What is sphericity? What is the sphericity of a sharp Mika flake?
5. How roundness of a particle is defined?
6. Define the convexity, solidity, and circularity with an example.
7. Why knowledge of particle strength is important?
8. What is hygroscopicity?
9. What is zeta potential? Why is zeta potential important?
10. How can zeta potential be estimated?
11. What is surface area density? How is it defined?
12. How effective does slurry viscosity depend on particle concentration?
13. What is the limit of particle size which causes non-Newtonian flow behavior?
14. How are the compression ratio and Hausner ratio related?
15. What is the significance of a low Carr's index?

III. Numerical Problems

Q1. If the bulk density of a particle is 95% of the tapped density, what will be the degree of flowability of the particle? [Ans. Excellent]

Q2. What is the sphericity (Φ_s) of a cuboid whose length, breadth, and height ratio is 1:2:3? [Ans. 0.725]

Q3. What is the sphericity of a cone having a height equal to the base diameter? [Ans. 0.778]

Q4. Consider the arc length of each corner of an octagon of side 1 cm is 0.175, and its angle made in the center of the curvature inscribed by the corner is 10°. (Radius of curvature = Arch length/angle of curvature). What will be the degree of roundness of the octagonal-shaped particle of such a condition? [Ans. $2(\sqrt{2}-1)$]

Q5. An industrial process produces a slurry with 300 kg of dry solid in 750 L water. What is the concentration of slurry by weight (in %)? [Ans. 28.57%]

Q6. In an industrial process, the concentration of slurry obtained by weight is 80%. It is also known that the slurry obtained had 50 kg of liquid. What is the weight of the dry solid (in kg) in the slurry? [Ans. 200 kg]

Q7. A slurry consists of raw magnetite in a carboxymethyl cellulose solution of 1600 kg/m^3 density. From the measurement, it was observed that the slurry weighs 2537 kg/m^3. Calculate the concentration of solids by weight. The density of magnetite is 7874 kg/m^3. [Ans. 46.35%]

Q8. An experiment showed that the density of brine was 1621.22 kg/m^3 when the brine was prepared with a salt of density 2080 kg/m^3 and a liquid of density 1281 kg/m^3. What is the concentration of salts by volume% in the brine? [Ans. 42.58%]

Q9. A slurry consists of a solution of density 1250 kg/m^3. The density of the solid is 8752 kg/m^3. From the measurement, it was observed that the slurry weighs 2674 kg/m^3. Calculate the concentration of solids in percent (%) by weight. [Ans. 62.12%]

Q10. Calculate the volume fraction of slurry (ε_s) whose solids weigh 33.62 kg. The amount of slurry to be 100 kg of density is 1937 kg/m^3. The density of slurry particles is 2420 kg/m^3. [Ans. 0.269]

Q11. Find the viscosity of the slurry in Pa·s of volume fraction 0.435 of sphericity (Φ_s>0.02) made with solid in a liquid. The viscosity of liquid at 20°C is 0.0023 Pa·s. [Ans. 0.009187 Pa·s]

Q12. A liquid-solid fluidized bed is operated at a minimum slurry velocity with 10 vol% of FCC fine catalyst of particle size 50 μm. The shear stress was found to be less than 1 kPa. What should be the approximate effective slurry viscosity (μ_s) of the slurry if the viscosity of the liquid in which the slurry is made is 0.02 Pas (μ_l). [Ans. 0.027]

Q13. Calculate the viscosity of a slurry consisting of solid (1% by weight) and brine solution. The viscosity of the brine solution is 2.0 cP, and the densities of the brine solution and solid are 1200 and 2082 kg/m^3, respectively (assume shear stresses <1 kPa). [Ans. 0.0203 Pa.s]

Q14. A slurry mixture made of solid and water has a concentration of 53% solids by weight, and the specific gravity of the solids is 4.2. What is the specific gravity of the slurry mixture? [Ans. 1.677]

Q15. Particles of a certain material can be assumed to be cube-shaped with edge lengths equal to 0.5 mm. What will be the size of the equivalent sphere (diameter) having the same surface area? [Ans. 0.691 mm]

Q16. Determine the stability behavior of the colloid in which the dielectric constant of the dispersion medium is 30 and the dynamic viscosity of the dispersion medium is 547×10^{-6} Pa.s. The electrophoretic mobility of the particles is 2.43×10^{-9} $(Vs)^{-1}$, and the permittivity of free space is 8.85×10^{-12} $F.m^{-1}$. [Ans. Zeta potential=5 mV, rapid coagulation or flocculation if zeta potential ranges from 0 to ±5 mV]

Q17. Calculate the volume fraction of the slurry mixture of limestone in water consisting of solid (60% by weight). The density of solid and liquid is given as 2430 and 1000 kg/m^3, respectively. [Ans. 0.382]

Q18. A slurry of polymeric melts of smoothed spherical particles is analyzed, and it was found that its concentration is 0.065 by volume fraction. The slurry is processed in an extrusion device to develop fibrillated tape. The performance of the process depends on the slurry viscosity. What is the slurry viscosity (in Pa.s) at the polymeric concentration of slurry whose shear stress is greater than 1 kPa? Given that the viscosity of the liquid in which the slurry is made is 0.021 Pa.s at 25°C. [Ans. 0.0256 Pa.s]

Q19. In an analysis of particle size in a sample of solids with a mass of 1 kg and a density of 5610 kg/m^3, a particle's sphericity follows the functionality $\Phi_p = 4.84V_p^{2/3}$. What is the particle's surface area per unit volume of the sample? [Ans. 5610 m^2/m^3]

Q20. A colloid suspension is prepared in a system in which the dielectric constant of the dispersion medium is 90, and the dynamic viscosity of the dispersion medium is 5.00 Pa.s. The electrophoretic mobility of the particles is 5.53×10^{-9} $(Vs)^{-1}$, and the permittivity of free space is 12.05×10^{-12} $F.m^{-1}$. What is the stability behavior of the colloidal suspension? [Ans. Zeta potential=28.05 mV, Incipient instability results if zeta potential is from ±10 to ±30 mV]

NOMENCLATURE

A_p	particle's surface area [m^2]
A_s	surface area of sphere [m^2]
b	breadth, length [m]
c_w	concentration of solids by weight in the slurry (%)
d	diameter [m]
D_l	length density [kg/m]
d_p	particle diameter [m]
D_r	degree of roundness [-]
D_{sa}	surface area density [kg/m^2]
L	distance of electrode [m]
L_p	length of particle [m]
M	powder mass [kg]
M_p	mass of the particle [kg]
N	number of corners of the particle in the given plane [-]
r_i	radius of curvature of the *i*th corner [m]
$r_{max\text{-}in}$	radius of the circle, which is maximally inscribed [m]

S_p	surface area of the particle [m²]
v	speed of the particle [m/s]
V	voltage [V]
V_p	particle volume [m³]
V_t	volume occupied after tapping [m³]

GREEK LETTERS

ζ	zeta potential
η	dynamic viscosity of the medium (Pa s)
μ_e	electrophoretic mobility [1/Vs]
μ_l	dynamic viscosity of pure liquid [Pa.s]
Φ_p	sphericity of particle [-]
μ_r	relative viscosity [-]
μ_s	dynamic viscosity of slurry [Pa.s]
ρ_t	tapped density [kg/m³]
ε_0	permittivity of free space
ε_r	dielectric constant of the medium where the particle is dispersed
ρ_l	density of liquid without solids (kg/m³)
ρ_m	density of slurry (kg/m³)
ρ_s	density of the solids (kg/m³)

SUBSCRIPTS

0	free space
e	electrostatic
i	index
l	liquid, length
m	mixture or slurry
max-in	maximally inscribed
p	particle
r	medium, roundness, relative
s	solids, slurry
sa	surface area
t	tapped
w	weight

REFERENCES

Carr, R. L., Classifying flow properties of solids, *Chem. Eng.*, 72, 69–72, 1965.

Einstein, A., Eine neue Bestimmung der Molekϋdimensionen, *Annalen der Physik*, 19, 289–306, 1906.

Fairhurst, D., An overview of the zeta potential part 3: Uses and applications, *American Pharmaceutical Review*, 2013. https://www.americanpharmaceuticalreview.com/Featured-Articles/139288-An-Overview-of-the-Zeta-Potential-Part-3-Uses-and-Applications/

Geldart, D. (Ed.), *Gas Fluidization Technology*, Wiley Interscience, 1986.

Hausner, H. H., *Friction Conditions in a Mass of Metal Powder*, Polytechnic Inst. of Brooklyn. Univ. of California, 1967.

Hunter, R. J., *Zeta Potential in Colloid Science*, Academic Press, 1981.

Kim, I. J., J. G. Park, Y. H. Han, S. Y. Kim and J. F. Shackelford, Wet foam stability from colloidal suspension to porous ceramics: A review, *J. Korean Ceram. Soc.*, 56 (3), 211–232, 2019.

Kitano, T., T. Kataoka, and T. Shirota, An empirical equation of the relative viscosity of polymer melts filled with various inorganic fillers, *Rheol. Acta*, 20, 207–209, 1981.

Kumar, A. and C. K. Dixit, Methods for characterization of nanoparticles, in: S. Nimesh, R. Chan, N. Gupta (Eds.), *Advances in Nanomedicine for the Delivery of Therapeutic Nucleic Acids*, Woodhead Publishing, 2017. pp. 43–58.

Kunii, D. and O. Levenspiel, *Fluidization Engineering*, 2nd Edition, Butterworth-Heinemann, 1991.

von Smoluchowski, M., Contribution to the theory of electro-osmosis and related phenomena, *Bull. Int. Acad. Sci. Cracovie*, 3, 184–199, 1903.

Thomas, D. G., Transport characteristics of suspension, VIII: A note on the viscosity of Newtonian suspensions of uniform spherical particles, *J. Colloid Sci.*, 20, 267–277, 1965.

Turchiuli, C., M. Fuchs, M. Bohin, M. E. Cuvelier, C. Ordonnaud, M. N. Peyrat-Maillard and E. Dumoulin, Oil encapsulation by spray drying and fluidized bed agglomeration, *Innovative Food Sci. Emerg. Technol.*, 6, 29–35, 2005.

Wadell, H. A., Volume, shape and roundness of rock particles, *J. Geol.*, 40, 443–451, 1932.

Chapter 3

Particle Size and Its Distribution

Science can amuse and fascinate us all, but it is engineering that changes the world.

—*Isaac Asimov, American writer, professor of biochemistry*

3.1 MEAN SIZE OF PARTICLES

- In various applications of soli-fluid operations, it is essential to know the mean particle size to analyze the performance of the functions.
- Many definitions of the mean particle size include arithmetic, geometric, quadratic, and harmonic, but these do not have special significance. In practice, the mean of particle sizes in a mixture of different sizes is assessed based on the following:
  - Mean particle size based on volume or mass
  - Mean particle size based on surface

3.1.1 The Mean Size of Particle Based on Volume or Mass

- There are two forms of expressing the mean size based on the volume or mass as follows: volume-mean diameter and mean volume diameter.

3.1.1.1 Volume-mean diameter

- The definition of volume-mean diameter (d_v), or the mass mean diameter, is as follows:

$$d_v = \frac{\sum x_i\, d_{p,i}}{\sum x_i} = \sum x_i\, d_{p,i};\ x_i = n_i k_1 d^3_{p,i}\, \rho_s;\ k_1 = \frac{1}{6}\pi \tag{3.1}$$

DOI: 10.1201/9781003571025-4

Or in integral forms, it is defined as follows:

$$d_v = \frac{\int_{d_{p,i}}^{d_{p,j}} f(d_p)(d_p)dd_p}{\int_{d_{p,i}}^{d_{p,j}} f(d_p)dd_p} = \frac{\int_0^1 d_p dx}{\int_0^1 dx} = \int_0^1 d_p dx \tag{3.2}$$

- Instead of mass fraction, it can be expressed in terms of particle numbers as follows:

$$d_v = \frac{\rho_s k_1 \sum (n_i d_{p,i}^4)}{\rho_s k_1 \sum (n_i d_{p,i}^3)} = \frac{\sum (n_i d_{p,i}^4)}{\sum (n_i d_{p,i}^3)} \tag{3.3}$$

3.1.1.2 Mean volume diameter

- The mean volume diameter, denoted by (d_v'), is defined as follows:

$$d_v' = \sqrt[3]{\left(\frac{\sum (n_i d_{p,i}^3)}{\sum n_i}\right)} \tag{3.4}$$

Or, in terms of mass fraction, it can be expressed as follows:

$$d_v' = \sqrt[3]{\left(\frac{\sum x_i}{\sum (x_i / d_{p,i}^3)}\right)} = \sqrt[3]{\left(\frac{1}{\sum (x_i / d_{p,i}^3)}\right)} \tag{3.5}$$

Example 3.1

In a laboratory-scale operation of grinding of particles, the following results are obtained after sieve analysis:

Mass of particles, gm	3	8	10	6	2
The mean size of particles, μm	450	580	300	200	75

Find the volume-mean diameter and mean volume diameter of the particles in μm.

Solution:

As per definition, $d_v = \frac{\sum x_i d_{p,i}}{\sum x_i} = \sum x_i d_{p,i}$

Mass of Particles (gm)	*The Mean Size of Particles (μm) (d_{pi})*	*Mass Fraction Retained (x_i)*	$x_i\, d_{pi}$	$x_i / d_{p,i}^3$
5	450	5/50=0.1	45	1.10×10^{-09}
15	580	15/50=0.3	174	1.54×10^{-09}
20	300	20/50=0.4	120	1.48×10^{-08}
6	200	6/50=0.12	24	1.50×10^{-08}
4	75	4/50=0.08	6	1.90×10^{-07}
Sum of the mass of all particles=50 g			$\sum x_i d_{p,i} = 369$	$\sum (x_i / d_{p,i}^3) = 2.22 \times 10^{-07}$

Therefore, $d_v = \sum x_i d_{p,i} = 369\ \mu\text{m}$

$$d_v' = \sqrt[3]{\left(\frac{\sum x_i}{\sum (x_i / d_{p,i}^3)}\right)} = \sqrt[3]{\left(\frac{1}{\sum (x_i / d_{p,i}^3)}\right)} = \sqrt[3]{\left(\frac{1}{2.22 \times 10^{-7}}\right)} = 165.13\ \mu\text{m}$$

3.1.2 Mean Sizes Based on the Surface

- In this case, there are two forms of expressing the mean size based on the surface as follows: the surface-mean diameter and the mean surface diameter.
- **The surface-mean diameter:** the surface-mean diameter is defined as follows:

$$d_s = \frac{\sum \left[(n_i d_{p,i}) S_i\right]}{\sum (n_i S_i)} = \frac{\sum (n_i k_2 d_{p,i}^3)}{\sum (n_i k_2 d_{p,i}^2)} = \frac{\sum (n_i d_{p,i}^3)}{\sum (n_i d_{p,i}^2)} \tag{3.6}$$

S_i is the surface of particle i, $k_2 = \pi$.

- Or, in terms of mass fraction (x_i), it is expressed as follows:

$$d_s = \frac{\sum x_i}{\sum \left(\frac{x_i}{d_{p,i}}\right)} = \frac{1}{\sum \left(\frac{x_i}{d_{p,i}}\right)} = \frac{1}{\int_0^1 \frac{dx}{d_{pi}}} \tag{3.7}$$

- The surface-mean diameter, d_s is also called the **Sauter mean diameter** or **volume to surface-mean diameter.**
- **The mean surface diameter:** the mean surface diameter denoted by d_s' is defined as follows:

$$d_s' = \sqrt{\left(\frac{\sum (n_i d_{p,i}^2)}{\sum n_i}\right)} \tag{3.8}$$

- Or in terms of mass fraction, it is defined as follows:

$$d_s' = \sqrt{\left(\frac{\sum x_i / d_{p,i}}{\sum (x_i / d_{p,i}^3)}\right)} \tag{3.9}$$

Example 3.2

In a particle size reduction operation, it was found that 325 g of samples yield the product size of the operation as follows after its analysis:

Mass of particles (gm)	50	90	100	50	20	10	5
The mean size of particles (μm)	500	425	355	300	250	212	180

Find the surface-mean diameter and mean surface diameter of the particles in μm.

Solution:

Mass of Particles (gm)	*The Mean Size of Particles (μm) (d_{pi})*	*Mass Fraction Retained (x_i)*	$\frac{x_i}{d_{p,i}}$	$x_i/d_{p,i}^3$
50	500	0.15	3.08×10^{-4}	1.23×10^{-9}
90	425	0.28	6.52×10^{-4}	3.61×10^{-9}
100	355	0.31	8.67×10^{-4}	6.88×10^{-9}
50	300	0.15	5.13×10^{-4}	5.70×10^{-9}
20	250	0.06	2.46×10^{-4}	3.94×10^{-9}
10	212	0.03	1.45×10^{-4}	3.23×10^{-9}
5	180	0.02	8.55×10^{-4}	2.64×10^{-9}
Total mass = 325 g			$\sum x_i / d_{p,i} = 2.46 \times 10^{-3}$	$\sum (x_i / d_{p,i}^3) = 2.72 \times 10^{-08}$

Therefore,

$$d_s = \frac{1}{\sum\left(\frac{x_i}{d_{p,i}}\right)} = 355.16\ \mu\text{m}; \quad d_s' = \sqrt{\left(\frac{\sum x_i / d_{p,i}}{\sum (x_i / d_{p,i}^3)}\right)} = 321.62\ \mu\text{m}$$

3.2 GENERAL DEFINITION OF THE MEAN PARTICLE SIZE IN A SAMPLE

In a sample of different sizes, the particle is expressed as a mean particle size for analysis of fluidization performance. The various means are defined in several standard documents (ISO 9276-2:2001; ASTME 799-03).

- A general definition of mean diameter can be expressed as follows (ASTME 799):

$$d_{p,mn} = \left[\frac{\int_0^{\infty} N(d_p) d_p^m \, dd_p}{\int_0^{\infty} N(d_p) d_p^n \, dd_p}\right]^{1/(m-n)} \tag{3.10}$$

where N=number function of particles of size d_p, and m and n are the coefficients.

- In terms of a finite number of discrete size classes, the mean diameter can be defined as follows:

$$d_{p,mn} = \left[\frac{\sum_{i=1}^{\infty} N_i d_{pi}^m}{\sum_{i=1}^{\infty} N_i d_{pi}^n}\right]^{1/(m-n)} \tag{3.11}$$

where N_i is the number of particles within the ith discrete classes of size.

- Different types of "mean" and their respective applications are given in Table 3.1.
- The Sauter mean diameter is commonly used as it characterizes several essential processes where the distribution of the size of the dispersed phase is wide in range.

Table 3.1 Definition of mean particle size (Majumder, 2016)

Mean Diameter	*Name*	*Definition*	*Field of Application*
d_{10}	Arithmetic/linear mean	$d_{p,10} = \frac{1}{N}\sum_{i=1}^{N} N_i d_{pi}$	Evaporation
d_{20}	Surface mean	$d_{p,20} = \left[\frac{1}{N}\left(\sum_{i=1}^{N} N_i d_{pi}^2\right)\right]^{1/2}$	Absorption
d_{30}	Volume mean	$d_{p,30} = \left[\frac{1}{N}\left(\sum_{i=1}^{N} N_i d_{pi}^3\right)\right]^{1/3}$	Hydrology
d_{21}	Surface diameter mean	$d_{p,21} = \sum_{i=1}^{\infty} N_i d_{pi}^2 / \sum_{i=1}^{\infty} N_i d_{pi}$	Adsorption
d_{31}	Volume diameter mean	$d_{p,31} = \left[\sum_{i=1}^{\infty} N_i d_{pi}^3 / \sum_{i=1}^{\infty} N_i d_{pi}\right]^{1/2}$	Evaporation and molecular diffusion

(Continued)

Table 3.1 Definition of mean particle size (Majumder, 2016)

Mean Diameter	*Name*	*Definition*	*Field of Application*
d_{32}	Sauter mean	$d_{p,32} = \sum_{i=1}^{N} (N_i d_{pi}^3) / \sum_{i=1}^{N} (N_i d_{pi}^2)$	The efficiency of mass transfer and multiphase reaction
d_{43}	De Brouckere	$d_{p,43} = \sum_{i=1}^{N} (N_i d_{pi}^4) / \sum_{i=1}^{N} (N_i d_{pi}^3)$	Combustion equilibrium, Spray system

Example 3.3

In an industrial operation, a catalytic cracking is carried out with a suitable catalyst of density 2550 kg/m³. The particle size is analyzed and found that within a certain range of the particle size, the number of the particles varies as follows.

Find the Sauter mean particle diameter.

Size range (μm)	−710+300	−300+180	−180+90	−90+38	Pan
Number of particles in the range	9.45×10^{11}	1.23×10^{13}	1.28×10^{14}	1.30×10^{15}	4.88×10^{15}

Solution:

The Sauter mean particle diameter is given as follows:

$$d_{p,32} = \sum_{i=1}^{N} \left(N_i d_{pi}^3\right) / \sum_{i=1}^{N} \left(N_i d_{pi}^2\right)$$

To find this, we have to proceed as follows:

Size range (μm)	$d_{pi\,avg}$	N_i	$N_i d_{pi}^3$	$N_i d_{pi}^2$
−710+300	(710+300/2)=505	9.45×10^{11}	121.75	2.41×10^5
−300+180	240	1.23×10^{13}	170.45	7.10×10^5
−180+90	135	1.28×10^{14}	316.56	2.34×10^6
−90+38	64	1.30×10^{15}	340.91	5.33×10^6
Pan	38	4.88×10^{15}	267.86	7.05×10^6
			$\sum_{i=1}^{N}(N_i d_{pi}^3) = 1217.54$	$\sum_{i=1}^{N}(N_i d_{pi}^2) = 1.57\times10^7$

Thus, $d_{p,32} = \sum_{i=1}^{N} (N_i d_{pi}^3) / \sum_{i=1}^{N} (N_i d_{pi}^2) = 77.69\ \mu\text{m}$

3.3 PARTICLE SIZE DISTRIBUTION (PSD)

- The population of particles in a particle mixture based on its size is described by a PSD.
- The distribution is expressed by a mathematical function that defines the relative amounts of particles in a sample, typically by number frequency, mass, or volume as a function of particle size. PSD is also termed gradation.

3.3.1 The Significances of the PSD

- The physical and chemical properties in a particulate system are related to particle size and distribution.
- PSD significantly affects the efficiency of processes depending on particles.
- It affects the load bearing and strength of the material.
- The reactivity of solids as catalysts participating in chemical reactions depends on the PSD.
- Colloid properties such as rheology and surface area are related to the PSD.
- It affects the stability of slurry/suspension/emulsion.
- Prevention of aggregation of fine particles can be accomplished by creating an interparticle steric energy barrier by controlling PSD.
- A uniform PSD ensures
  - a uniform dissolution rate,
  - a uniform sedimentation rate, and
  - minimizes stratification when are stored or transported, or processed for fluidization operation.

3.4 GRAPHICAL REPRESENTATION OF PSD

- The distribution is represented graphically as a frequency distribution curve or cumulative (undersize) distribution curve as a function of the particle size.
- In the graph, the y-axis is defined either by number count, surface area, mass, or volume (sometimes by percentage), and the x-axis is defined by particle diameter in μm as shown in Figure 3.1.
- PSDs are expressed either by frequency distribution curves or cumulative curves.
- For naturally occurring materials, the curve will generally have a single peak.
- For mixtures of particles, there may be as many peaks as components in the mix.
- The curve may have more peaks if the particles are formed by crushing larger particles.

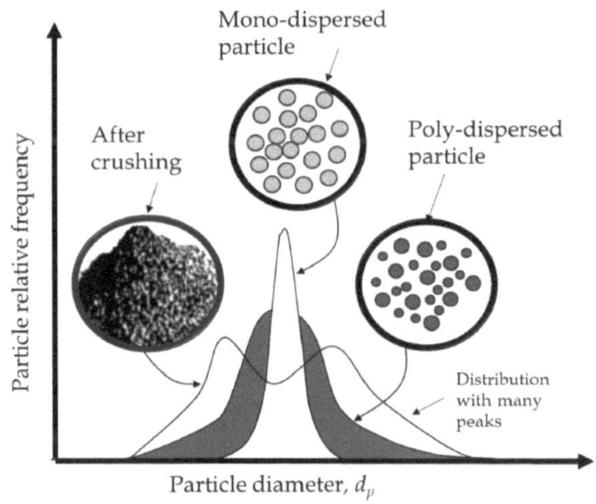

Figure 3.1 Particle size distribution (PSD).

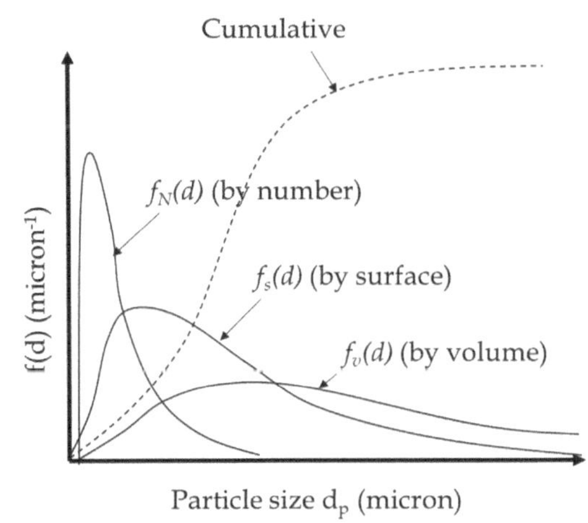

Figure 3.2 Representation of PSD.

- The cumulative distribution (F) is the integral of the frequency distribution (dF/dd_p). The frequency distribution (dF/dd_p) is often written as $f(d_p)$, as shown in Figure 3.2.
- A PSD can be represented in different ways with respect to the weighting of individual particles. The weighting mechanism depends on the measuring principle being used. The distributions can be made by either number, surface, mass, volume, or intensity.
- **Number weighted distributions:** a counting technique such as image analysis will give a number weighted distribution where each particle is given equal weighting irrespective of its size. This is most often

useful where knowing the absolute number of particles is important – in foreign particle detection for example – or where high resolution (particle by particle) is required.

- **Volume weighted distributions:** static light scattering techniques such as laser diffraction will give a volume weighted distribution. In this study, the contribution of each particle in the distribution relates to the volume of that particle (equivalent to mass if the density is uniform), i.e., the relative contribution will be proportional to $(\text{size})^3$. This is often extremely useful from a commercial perspective as the distribution represents the composition of the sample in terms of its volume/mass.
- The mass distribution and the volume distribution (where particle density does not vary with the size) are the same.
- The distributions by mass, number, and surface differ from each other (where particle density does not vary with the size).
- **Intensity weighted distributions:** dynamic light scattering techniques will give an intensity weighted distribution, where the contribution of each particle in the distribution relates to the intensity of light scattered by the particle. As per Rayleigh approximation, the relative contribution for very small particles will be proportional to $(\text{size})^6$.
- If in a sample, the total number of particles, N, the entire surface area, S, full mass m, and total volume V are constant, one can express the interrelationship of the distributions based on surface, volume, mass, and frequency as follows:

$$f_s(d_p) = k_s d_p^2 f_N(d_p) \tag{3.12}$$

$$f_V(d_p) = k_V d_p^3 f_N(d_p) \tag{3.13}$$

$$f_m(d_p) = k_m \rho_p d_p^3 f_N(d_p) \tag{3.14}$$

where

$$k_S = \frac{\alpha_S N}{S} \tag{3.15}$$

$$k_V = \frac{\alpha_V N}{V} \tag{3.16}$$

$$k_m = \frac{\alpha_m \rho_p N}{V} \tag{3.17}$$

where α_s, α_v, and α_m are the factors relating the proportionality factors of the particle to its surface, volume, and mass, respectively.

- The constants k_S, k_V, and k_m may be found by using the fact that

$$\int_0^\infty f(d_p)\,dd_p = 1 \tag{3.18}$$

3.4.1 Common Methods of Expressing Size Distributions

- Arithmetic normal distribution

$$f(d_p) = \frac{dF}{dd_p} = \frac{1}{\sigma\sqrt{2\pi}} \exp\left[-\frac{(d_{pi} - \bar{d}_{pi})^2}{2\sigma^2} \right] \tag{3.19}$$

where σ is the standard deviation, which is defined as follows:

$$\sigma^2 = \frac{1}{N} \sum_i \left(d_{pi} - \bar{d}_{pi} \right)^2 \tag{3.20}$$

where $\bar{d}_{pi}$ is the mean of the particle diameter, which is expressed as follows:

$$\bar{d}_{pi} = \frac{1}{N} \sum_i d_{pi} \tag{3.21}$$

- Log-normal distribution

$$f(d_p) = \frac{dF}{dz} = \frac{1}{(d_{pi} - d_{pi,\min})\sqrt{2\pi\sigma^2}} \exp\left[-\frac{[\ln(d_{pi} - d_{pi,\min}) - \bar{d}_{pi}]^2}{2\sigma_z^2} \right] \tag{3.22}$$

Example 3.4

In an operation of particle size reduction, the particle sizes (in mm) are obtained as follows: 1.25, 1.5, 1.8, 2.12, 2.5, 3, 3.55, 4.25, 5, 6, 7. Based on this information, find the normal and log-normal distribution function.

Solution:

$$\bar{d}_{pi} = \frac{1}{N} \sum_i d_{pi} = 3.45; \quad \sigma^2 = \frac{1}{N} \sum_i \left(d_{pi} - \bar{d}_{pi} \right)^2 = 3.31$$

Therefore

$$f(d_p)_{\text{Normal}} = \frac{1}{\sigma\sqrt{2\pi}} \exp\left[-\frac{(d_{pi} - \bar{d}_{pi})^2}{2\sigma^2} \right] = 0.22 \exp\left[-\frac{(d_{pi} - 3.45)^2}{6.62} \right]$$

$$f(d_p) = \frac{1}{(d_{pi} - d_{pi,\min})\sqrt{2\pi\sigma^2}} \exp\left[-\frac{[\ln(d_{pi} - d_{pi,\min}) - \bar{d}_{pi}]^2}{2\sigma_z^2} \right]$$

$$= \frac{0.22}{(d_{pi} - 1.25)} \exp\left[-\frac{[\ln(d_{pi} - 1.25) - 3.45]^2}{6.62} \right]$$

3.5 PARTICLE SIZE MEASUREMENT

The particle size can be measured and analyzed by the following methods:

- **Sedimentation:** by examining the settling of particles by terminal velocity based on the particle size.
- **Permeametry technique:** based on the pressure drop and flow rate through the particulate bed.
- **Centrifuge:** by sedimentation within the range of the particle size 0.01–40 μm.
- **Electrical sensing:** by suspension in an electrolyte within the range of the particle size 1–240 μm.
- **Laser diffraction:** by angular light scattering within the range of the particle size 0.04–2000 μm.
- **Light scattering:** by light scattering within the range of the particle size 0.003–3 μm.
- **Optical method:** by microscope within the range of the particle size 0.5–x μm.
- **Scanning electron:** by microscope with a resolution of 10 nm.

3.5.1 Sedimentation Technique

- The method is based on the terminal velocity of a particle. The particle's terminal velocity increases with an increase in the size of the particle. As per Stoke's law

$$u_t = \left[\frac{g d_p^2}{18\mu}(\rho_p - \rho_f)\right] \tag{3.23}$$

$$d_p = \left[\frac{18\mu}{(\rho_s - \rho_f)g}\frac{h}{t}\right]^{1/2} \tag{3.24}$$

where d_p = equivalent mean diameter of the particle.

- The size distribution is measured by settling a homogeneous suspension in a cylinder (typically 5.50 cm diameter and 550 mL volume) (Andreasen apparatus), as shown in Figure 3.3.
- The level change (h) of the suspension is noted down with a time interval, and a number of samples (~10 mL each) at the same time interval by sampling tube (graduated pipette) will be taken out. Then, the sample will be dried and measured.

  In this way, the size of the particle will be calculated using Equation (3.6), and mass fraction (x_i) will be calculated based on the total mass dried and measured.

  Based on this mass fraction and the respective size of the particle, the size distribution of the particle mixture can be drawn.

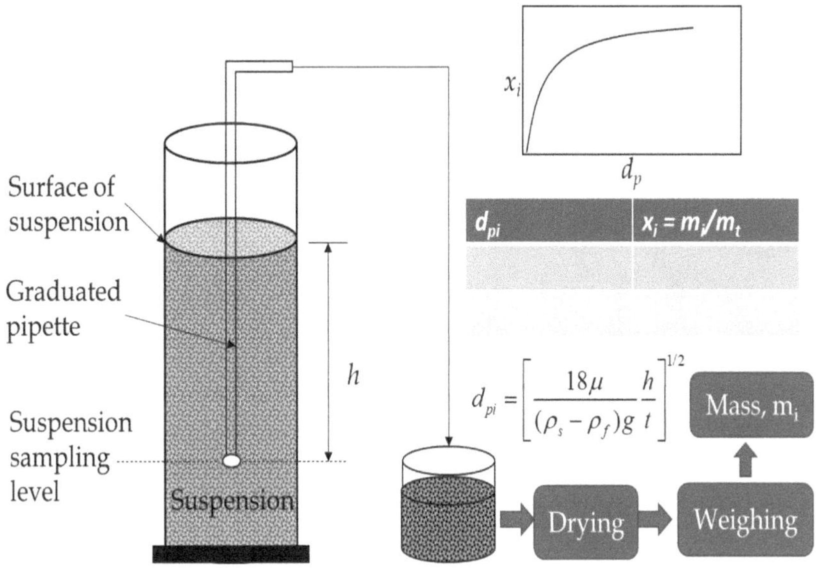

Figure 3.3 Measuring the size by settling a homogeneous suspension (1%–2% by vol.) in a cylinder (Andreasen apparatus).

- The sedimenting behavior of the particle assesses the PSD based on its terminal velocity in the following ways:
  - **Incremental method:** it is assessed by determining the change of the concentration or density of the suspension with a fixed time at fixed depths.
  - **Cumulative method:** it is assessed by measuring the rate at which the particle is settling out of suspension at a fixed level.
- The advantage is that it can measure a wide range of sizes with considerable accuracy and reproducibility.
- The disadvantage is that high concentrated suspension can deviate from accuracy due to interactive effects between particles.

3.5.2 Permeametry Method

This is a method of size analysis based on fluid flow through a packed bed, as shown in Figure 3.4.

- When the fluid flows laminarly through the pores of a randomly packed bed of particles, it exhibits a frictional pressure drop, which calculates the particle size.
- The Carman–Kozeny (C–K) equation (McCabe et al., 2017) is used to calculate the diameter of the particle once the frictional pressure drop

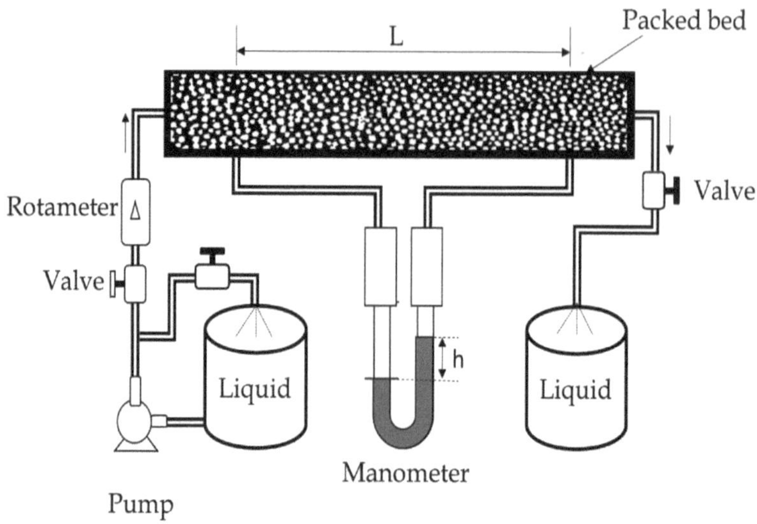

Figure 3.4 Fuid flow through a packed bed.

is known from the experiment for laminar flow through a randomly packed bed of uniformly sized spheres of diameter, d_p. The C–K equation is as follows:

$$\frac{\Delta P}{L} = 180\frac{(1-\varepsilon_f)^2}{\varepsilon_f^3}\frac{\mu_f u}{d_p^2} \tag{3.25}$$

$$\varepsilon = \frac{\text{Total volume of solids}}{\text{Total volume of bed}} = \frac{M_s / \rho_s}{A_b L} \tag{3.26}$$

- The C–K is valid only for laminar flow if $\mathrm{Re}_p = \dfrac{\rho_f u d_p}{(1-\varepsilon_f)\mu_f} < 1.0$.
- From Equation (3.27), the diameter of the particle can be calculated as follows:

$$d_p = \sqrt{180\frac{(1-\varepsilon_f)^2}{\varepsilon_f^3}\frac{\mu_f u}{\Delta P / L}} \tag{3.27}$$

where ΔP=the pressure drop across the bed, ε=the packed bed void fraction, L=the length of the bed, μ=the fluid viscosity, u=the superficial fluid velocity, d_p=Particle mean diameter, M_s=mass of solid, and ρ_s=density of solid.

Example 3.5

A bed of 750 g crushed sand of different sizes is used to measure the average size of the particle in a laboratory setup of a packed bed of length 50 cm and diameter 5 cm, as shown in Figure 3.4. The absolute density of the sand is 1682 kg/m^3. Tap water of viscosity 0.012 poise is allowed to flow through the packed bed at a flow rate of 10 L/h through a rotameter. The pressure drop reading of a manometer connected to the packed bed is found to be 2.7 cm of mercury. Estimate the average particle size of the sand used in the packed bed.

Solution:

C–K equation: $\dfrac{\Delta P}{L} = 180\dfrac{(1-\varepsilon)^2}{\varepsilon^3}\dfrac{\mu u}{d_p^2}$

Cross-sectional area of packed bed $(A_{\text{bed}}) = \dfrac{\pi}{4} d_{\text{bed}}^2 = 0.002\text{m}^2$

Void fraction, $\varepsilon = \dfrac{\text{Total volume of solids}}{\text{Total volume of bed}} = \dfrac{M_s/\rho_s}{A_b H}$

$$= (0.75/1682)/(0.002 \times 0.50) = 0.45$$

$P = \rho_w g h = 13500 \times 9.8 \times (2 \times 10^{-2}) = 3572.10\,\text{N/m}^2$
Flow rate of water = 100 lph = $2.77 \times 10^{-5}\,\text{m}^3/\text{s}$
Velocity of water = volumetric flow rate/cross-sectional area = 0.0014 m/s
From the C–K equation,

$$d_p = \sqrt{180\frac{(1-\varepsilon)^2}{\varepsilon^3}\frac{\mu u}{\Delta P/L}} = \sqrt{180\frac{(1-0.45)^2}{0.45^3}\frac{0.012(0.0014)}{3572.10/0.50}}$$

$= 0.0012\text{ m} = 1.2\text{mm}$

Therefore, the average estimated particle size is 3.7 mm (diameter).

Based on this particle size, $\text{Re}_p = \dfrac{\rho_f u d_p}{(1-\varepsilon_f)\mu} = \dfrac{997(0.014)(0.0012)}{(1-0.45)(0.012)} =$ $0.250 < 1.0$ (checked for validity of C–K equation).

3.5.3 Electrical Sensing Zone Analysis (Coulter Counter)

- This method works based on the principle of change of the electric resistance by particles in an electrical field between two electrodes, as shown in Figure 3.5.
- The number and size of particles suspended in an electrolyte are determined by allowing the particles to pass through an orifice that is immersed as an electrode.
- As the particle passes through the orifice, it generates voltage pulses (due to the changes in electric resistance) whose amplitude is proportional to the volume of the particles (Clark et al., 2019).

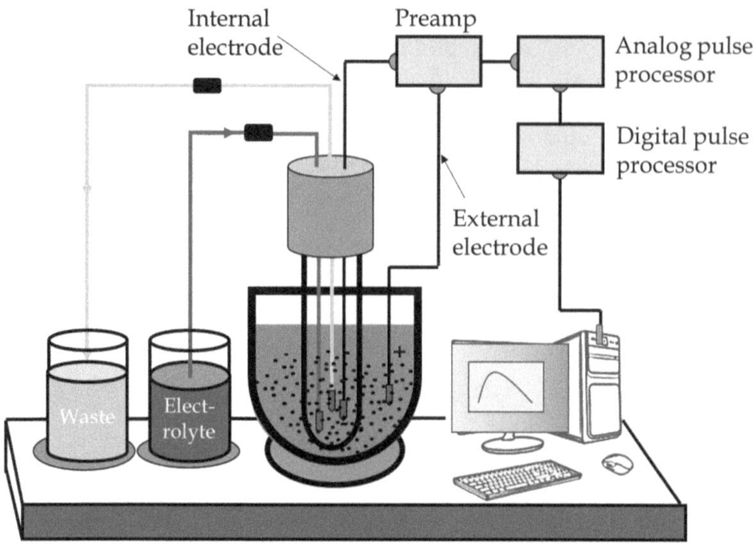

Figure 3.5 Electrical sensing zone (ESZ) analysis (Coulter counter) (Clark et al., 2019).

- The size distribution of the particle is then determined from the pulses by amplifying, sizing, and counting them.
- This method determines particles' size from submicron to a visible extent (~400–1600 μm) depending on the system properties.
- The advantage is that it helps in particle size analysis of powder, suspension, or emulsion.

3.5.4 Laser Light Scattering Particle Size Analysis

- According to the principle of laser light scattering (Mie scattering), a laser beam passes through a dispersed particulate sample of different particle sizes. The large particles in the sample scatter light at small angles, and small particles scatter light at large angles relative to the laser beam (Kurtz, 2009), as shown in Figure 3.6.
- The angular scattering intensity data are then analyzed to calculate the particles' size based on the scattering pattern. According to the Mie theory of light scattering, the diffraction angles are inversely related to the particle size.
- Mie theory is about the absorption and scattering of plane electromagnetic waves by uniform isotropic particles (e.g., sphere, infinite cylinder) in a uniform, isotropic dielectric infinite medium.
- The particle resonances that are generated by the scattering of light depend on the variation of the size, which is defined as follows (Kurtz, 2009):

$$\chi = \frac{2\pi r n_m}{\lambda} \tag{3.28}$$

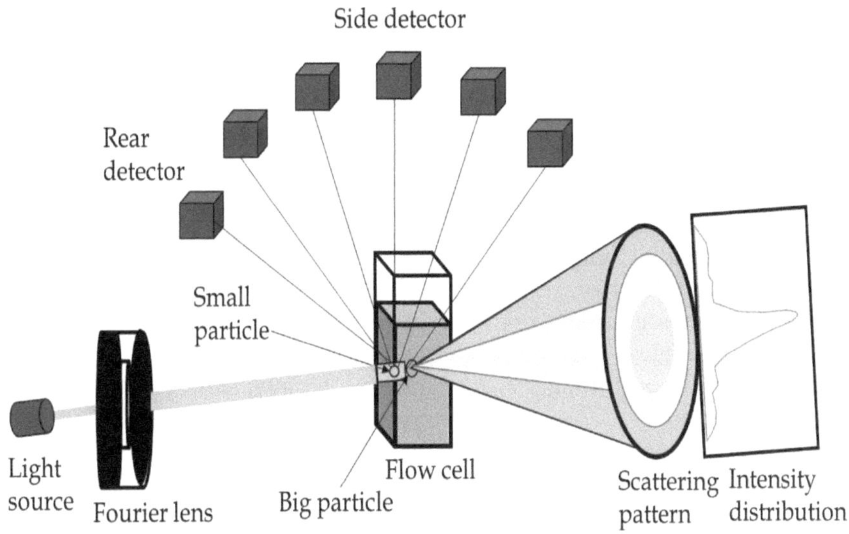

Figure 3.6 Laser light scattering particle size analysis system.

where r is the radius of a particle and λ/n_m represents the wavelength in a surrounding medium of refractive index n_m.

- The particle size is considered as a volume equivalent sphere diameter.
- Laser Light Scattering Particle Size Analysis can also be used to characterize PSDs from nanometers to millimeters.
- The advantages of the techniques are as follows:
  - Nonintrusive, a low-power laser beam can be used.
  - Fast, more straightforward measurement and its analysis.
  - Precise and wide ranges in size.
  - The measurement can be done without calibration.
  - The working principle based on the physical properties of the system and more accuracy.
- The disadvantages are as follows:
  - It is expensive.
  - The basic output form is volume measurement, assuming spherical particles.
  - A difference in refractive indices between particles and suspending medium may cause deviation of scattering resonance.

3.5.5 Microscopy

- The optical microscope also measures particle sizes up to 5 µm.
- The electron microscope may be used for size analysis below 5 µm.

- One can use an image analysis system assembled with the optical or electron microscope to assess the size distribution and shape of the particles.
- Martin's diameter, Feret's diameter, shear diameter, projected area diameters, etc. are used to calculate the particles' size based on the particles' projected image.
- Advantages of manually operated optical microscopy are as follows:
  - Inexpensive compared to other types.
  - Can individually examine particles.
  - It can detect aggregates of 2-D shape, color, etc.
  - Can examine small sample sizes.
- Disadvantages of manually operated optical microscopy are as follows:
  - Time consuming, high operator fatigue, and few particles examined.
  - Very low throughput.
  - No information on 3-D shape.
  - A certain amount of subjectivity is associated with sizing operator bias.
- Advantages of automatic and image analysis microscopes are as follows:
  - It is faster and easy to operate.
  - Bias-free operation.
- Disadvantages of automatic and image analysis microscopes are as follows:
  - It is costly.
  - It is automatic and semiautomatic, which is very erotic for agglomerates.
- Advantages of transmission and scanning electron microscopy are as follows:
  - Particles can be examined separately for each particle.
  - It has provision for visual means to see submicron particle.
  - Particles' shape can be assessed.
- Disadvantages of transmission and scanning electron microscopy are as follows:
  - It is very expensive.
  - It takes a long time for sample preparation.
  - Difficult to prepare emulsions.
  - Not suitable for high capacity.

Particle size analysis by different equipment or devices as per the particle size is shown in Table 3.2.

Table 3.2 Particle size analysis by different equipment or devices as per the particle size (Merkus, 2009)

Equipment/Device	*Particle Size (μm)*
Centrifuge/Sedimentor	0.01–40
Electrical sensor (Coulter counter)	1–240
Laser diffraction	0.04–2000
Light scattering	0.003–3
Optical microscope	0.5–3
Scanning electron microscope	0.5–*x*

3.6 DEFINITION OF POWDER AND GRANULAR MATERIAL

- Powders and granular are fine, loose dry solids composed of fine particles ranging from 100 to 10 mm in diameter, which can flow freely under the force applied.
- *Powders* are a particular subclass of granular materials, although *powder* and *granular* are sometimes used to distinguish separate material classes.
- It is produced by grinding, crushing, or disintegrating a solid substance.
- It has complex physical and chemical properties such as dustiness, cohesiveness, hygroscopicity, plasticity, stickiness, hardness, aeration capability, coating tendency, compressibility, flow attritability, combustibility, explosivity, reactivity, and frictional properties.
- Its flowability and reactivity as a catalyst suit the fluidized bed's chemical process.
- A good powder has a uniform PSD.
- If the PSD is not uniform, the powder can be segregated according to the different particle sizes, which may result in inaccurate dosing or inconsistent performance.

3.7 POWDER CLASSIFICATION

3.7.1 Brown and Richards' (1970) Classification

- According to Brown and Richards (1970) and Nedderman (1992), five categories of powder are classified, which are defined based on the particle size as shown in Table 3.3

3.7.2 Geldart's (1973) Powder Classification

- Geldart (1973) categorized powder based on the particle size and the relative density between the fluid phase and the solid particles. He identified four groups such as C-, A-, B-, and D-type powders, based on which the fluidization character can be distinctly observed, as shown in Table 3.4 and Figure 3.7

Table 3.3 Powder classification as per Richards and Brown (1970) and Nedderman (1992)

Classified as	*Particle Size Range*
Ultrafine powder	0.1–1.0 μm
Superfine powder	1–10 μm
Granular powder	10–100 μm
Granular solid	0.1–3.0 mm
Broken solid	3–10 mm

Table 3.4 Powder classification as per Geldart (1973)

Type of Powder	*Size Range of Particle (μm)*	*Example*
C (cohesive)	0–30	Flour
A (aeratable)	30–100	Milk powder and FCC catalyst
B (bubbling)	100–1000	Sand and coal for the reduction reaction
D (spoutable)	>1000	Roasting coffee beans, drying grains or peas, roasting metal ores, and drying paddy

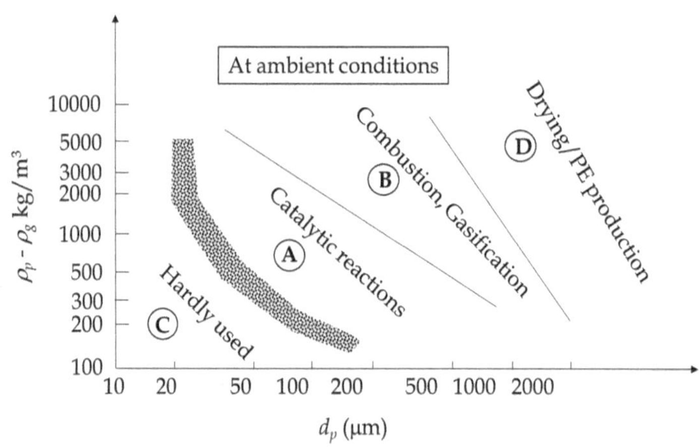

Figure 3.7 Geldart's powder classification map. C, cohesive (flour, fly ash); A, aeratable (FCC catalyst); B, bubbles (500-μm sand); and D, spoutable (wheat, 2000-μm polyethylene pellets).

IMPORTANT POINTS TO REMEMBER

- The surface-mean diameter is also called the **Sauter mean diameter** or **volume to surface-mean diameter,** which is defined as $d_s = \dfrac{\sum (n_i d_{p,i}^3)}{\sum (n_i d_{p,i}^2)} = \dfrac{1}{\sum \left(\dfrac{x_i}{d_{p,i}} \right)} = \dfrac{1}{\int_0^1 \dfrac{dx}{d_{pi}}}.$
- In terms of a finite number of discrete size classes, the mean diameter can be defined as $d_{p,mn} = \left[\dfrac{\sum_{i=1}^{\infty} N_i d_{pi}^m}{\sum_{i=1}^{\infty} N_i d_{pi}^n} \right]^{1/(m-n)}$.
- PSD is also known as gradation.
- For naturally occurring materials, the curve will generally have a single peak.
- The curve may have more peaks if the particles are formed by crushing larger particles.
- The mass distribution and the volume distribution (where particle density does not vary with size) are the same.
- As per Rayleigh approximation, the relative contribution for very small particles will be proportional to $(\text{size})^6$.
- The size distribution is measured by settling a homogeneous suspension in an apparatus (typically 5.50 cm diameter and 550 mL volume), which is called the Andreasen apparatus.
- The C–K equation is used to calculate the diameter of the particle once the frictional pressure drop is known from the experiment for laminar flow through a randomly packed bed of uniformly sized spheres of diameter.
- The size distribution of the particle is determined from the pulses by amplifying, sizing, and counting them in a Coulter counter.
- The large particles in the sample scatter light at small angles, and small particles scatter light at large angles relative to the laser beam in a laser light scattering particle size analysis system.
- Martin's diameter, Feret's diameter, shear diameter, projected area diameters, etc., are used to calculate the particles' size based on the particles' projected image.
- Powders and granular are fine, loose, dry solids composed of fine particles ranging from 100 to 10 mm in diameter, which can flow freely under the force applied.
- Geldart (1973) categorized powder based on the particle size and the relative density between the fluid phase and the solid particles.

SELF-TESTING QUESTIONS

I. Multiple-Choice Questions

Q1. For PSD, it is essential to know________________.
(a) The mean size of the particles
(b) Frequency distribution
(c) Particle diameter
(d) Particle volume

Q2. Which of the following means is not applicable for getting the mean size of the particle?
(a) Arithmetic mean
(b) Harmonic mean
(c) Root mean square
(d) Quadratic mean

Q3. Which one of the following techniques does not fall under particle size measurement?
(a) Sieving
(b) Laser diffraction
(c) Permeametry
(d) Electrostatic precipitation

Q4. Match the following based on the particle size analyzer as per the particle size.

Particle Size (μm)	*Equipment/Device*
(i) 0.5–3	(A) Light scattering
(ii) 0.01–40	(B) Optical microscope
(iii) 1–240	(C) Sedimentor
(iv) 0.003–3	(D) Electrical sensor

(a) (i) – (B), (ii) – (D), (iii) – (C), (iv) – (A)
(b) (i) – (A), (ii) – (C), (iii) – (D), (iv) – (B)
(c) (i) – (B), (ii) – (C), (iii) – (D), (iv) – (A)
(d) (i) – (A), (ii) – (D), (iii) – (C), (iv) – (B)

Q5. The Sauter mean diameter is defined as_____
(a) $d_s = 1/\sum x_i/d_{pi}$
(b) $d_s = \sum x_i/d_{pi}$
(c) $d_s = 1/\sum x_i d_{pi}$
(d) $d_s = 1/\sum d_{pi}/x_i$

Q6. Which of the following statement is true?
(a) Mass and volume distribution are always the same
(b) Cumulative distribution is the derivative of the frequency distribution

(c) The distribution by mass, number, and surface are not the same
(d) None of the above

Q7. For a laser diffraction, the diffraction angle is
(a) Proportional to the particle size
(b) Proportional to the square root of the particle size
(c) Inversely proportional to the particle size
(d) Inversely proportional to the square root of particle size

Q8. What is the correct description of permeametry?
(a) This method follows the sedimentation rate of a sample of particles in a liquid.
(b) This is a method of size analysis based on fluid flow through a packed bed.
(c) Particles are suspended in a dilute electrolyte drawn through a tiny orifice with a voltage applied.
(d) None of these.

Q9. Which of the given values of Re_p is applicable to Stokes' law?
(a) $Re_p > 0.8$
(b) $Re_p > 0.3$
(c) $Re_p < 0.3$
(d) None of these

Q10. Which of the following statements is TRUE?
(a) If light passes through a suspension, the diffraction angle is proportional to the particle size.
(b) If light passes through a suspension, the diffraction angle does not depend on particle size.
(c) If light passes through a suspension, the diffraction angle is inversely proportional to the particle size.
(d) None of these.

Q11. Which of the following statements is the basis of electrozone sensing?
(a) The amplitude of the pulse can be related to the volume of the particle passing the orifice.
(b) The pulse amplitude can be related to the electrical conductivity of the particle passing the orifice.
(c) The pulse amplitude can be related to the density of the particle passing the orifice.
(d) The pulse amplitude can be related to the thermal conductivity of the particle passing the orifice.

Q12. Which of the following is an INCORRECT match according to Richards and Brown's categorization of powder (1970)?
(a) 0.1–1.0 μm (ultrafine powder)
(b) 1–10 μm (granular powder)
(c) 10–100 μm (granular powder)
(d) 3.0–10 mm (broken solid)

Q13. Which of the following statements about laser diffraction are true?
- (a) The diffraction angle is directly proportional to particle size.
- (b) The diffraction angle is inversely proportional to particle size.
- (c) The technique usually takes less than 3 minutes to give measurements.
- (d) The method uses a very high-power laser beam.

Q14. Which of the following options are CORRECT?
- (a) In sedimentation, particles are assumed to be accelerated toward gravity.
- (b) Large particles can be measured undersized as they create turbulence, affecting the forces acting on a single particle.
- (c) The particle is assumed to be in equilibrium during the sedimentation process.
- (d) Sedimentation can be done for all the solutions irrespective of the concentration of the solution.

Q15. Which of the following statements is/are correct?
- (a) Permeametry refers to the method of size analysis based on fluid flow through a packed bed.
- (b) In electrozone sensing, the particle size should be in the range of 0.3–1000 μm.
- (c) In the sedimentation process, particles can be soluble in the suspending liquid.
- (d) If Stoke's law is assumed to apply, then, Re_p should be less than 0.3.

Q16. If the particle size is between 10 and 100 μm, then according to Richards and Brown's powder classification, it is considered as_______.
- (a) Ultrafine powder
- (b) Granular solid
- (c) Granular powder
- (d) Superfine powder

Q17. Sedimentation is applicable for condition_________________.
- (a) When Re<0.3
- (b) When Re>0.3
- (c) Particles typically less than 50 μm in diameter
- (d) Particles typically more than 50 μm in diameter

Q18. Find the correct statement.
- (a) For naturally occurring materials, the curve will generally have a single peak.
- (b) For mixtures of particles, there will be only one peak regardless of the components in the mix.
- (c) The curve may have more peaks if the particles are formed by crushing larger particles.
- (d) None of these.

Q19. Match the column

Methods of Particle Size Measurement	*Advantages*
1. Sieving	a. Permanent record photograph
2. Manual optical microscopy	b. Visual means to see submicron specimens
3. Transmission and scanning electron microscopy	c. Equipment required is relatively simple and inexpensive
4. Sedimentation	d. Wide size distribution

(a) 1-d; 2-a; 3-b; 4-c
(b) 1-a; 2-b; 3-c; 4-d
(c) 1-b; 2-c; 3-d; 4-a
(d) 1-c; 2-d; 3-a; 4-c

Q20. Among the following, which statements are true about PSD?
(a) For naturally occurring materials, the curve will generally have a single peak.
(b) For a mixture of particles, there may be as many peaks as the components in the mix.
(c) The curve may have more peaks if the particles are formed by crushing larger particles.
(d) All are correct.

Answer to multiple-choice questions

Q1. Ans. (a)
Q2. Ans. (c)
Q3. Ans. (d)
Q4. Ans. (c)
Q5. Ans. (a)
Q6. Ans. (c)
Q7. Ans. (c)
Q8. Ans. (b)
Q9. Ans. (c)
Q10. Ans. (c)
Q11. Ans. (a)
Q12. Ans. (b)
Q13. Ans. (b) and (c)
Q14. Ans. (b) and (c)
Q15. Ans. (a), (b), and (d).
Q16. Ans. (c)
Q17. Ans. (a) and (c)
Q18. Ans. (a) and (c).
Q19. Ans. (a)
Q20. Ans. (d)

II. Objective Type Questions

1. Define the mean size of a particle as per volume or mass.
2. Define the mean size of the particle based on surface area.
3. What is Sauter mean diameter? Where is it commonly used?
4. Are the mean surface diameter and surface-mean diameter the same?
5. Write the general definition of mean diameter.
6. What are the specific fields in which the different mean particle diameter is used?

7. What is PSD? Why is it important to know?
8. How PSD can be graphically expressed? Show the distributions by mass, number, and surface differ from each other.
9. Write one common model equation expressing the PSD.
10. What is log-normal distribution? How it is mathematically expressed?
11. Briefly explain the different techniques of measuring the particle size.
12. Which method is used to calculate the particle size in porous media?
13. Write a short note on the Coulter counter to analyze the particle size.
14. Explain the principle of laser light scattering particle size analysis method.
15. Classify the powder based on the particle size as per Richards and Brown's (1970) classification and Geldart (1973) classification.

III. Numerical Questions

Q1. In particle analysis, the following results are obtained:

Mass of particles (gm)	2	5	7	4	1
The mean size of particles (μm)	350	240	200	150	100

Find the volume-surface-mean diameter of the particles in μm. [Ans. 193.5 μm]

Q2. Size analysis was carried out by a series of sieves in a laboratory experiment. The mass fraction (x_i) as per particle diameter (d_{pi}) based on screen opening diameter (d_{pi}) is given in the below table.

x_i	0.20	0.42	0.35	0.17
d_{pi} (mm)	0.47	0.68	1.73	3.92

What is the mass mean diameter? [Ans. 1.6515 mm]

Q3. What is the total number of particles of a uniform particulate sample consisting of spherical particles of size 2 mm and density 2650 kg/m^3 if the sample mass is 20 kg? [Ans. 18017554]

Q4. Finely divided clay is used as a catalyst in the petroleum industry. It has a density of 1.2 g/cc and a sphericity of 0.5. The size analysis is as follows:

Average diameter, $d_{pi\,avg}$ (cm)	0.0252	0.0178	0.0126	0.0089	0.0038
Mass fraction, x_i (g/g)	0.088	0.178	0.293	0.194	0.247

What is the specific surface area of the clay material in cm^2/g? [Ans. 1235.438 cm^2/g]

Q5. What is the volume-surface-mean diameter in μm for the following particulate material?

Size range (μm)	−710+300	−300+180	−180+90	−90+38	Pan
Mass of particles in the range (gm)	30	35	65	70	55

[Ans. 79.11 μm]

Q6. A gas flows through a packed bed of 2 cm depth and a porosity of 0.5 at a superficial gas velocity of 1.5 cm/s. The pressure drop reading of a manometer connected to the unit is 2 cm of water. What is the particle size of the packed bed material? $\rho_{gas}=1.2\,kg/m^3$ and $\mu_{gas}=1.8\times10^{-5}$ Pa.s. Use the C–K equation. [Ans. 1.394 mm]

Q7. Find the surface-mean diameter for the following characteristics of particulate material.

Size range (μm)	−704+352	−352+176	−176+88	−88+44	−44−0
Mass of particles in the range (gm)	25	37.5	62.5	75	30

[Ans. 61.40 μm]

Q8. A particle *X* with a diameter *d* μm settles with a terminal velocity of *U* in oil of specific gravity 0.8 and viscosity of 30 cP under Stokes law. What is the terminal velocity of another particle, *Y*, with a diameter half of *X*, that settles in the same oil? [Ans. *U*/4]

Q9. Oil droplets having a diameter of 20 μm are to be settled from the air at an air temperature of 311 K at 101.3 kPa pressure. The density of oil is 900 kg/m^3. Given that at 311 K, the density and viscosity of air are 1.137 kg/m^3 and 1.90×10^{-5} Pa.s, respectively. If the droplet fall was in the Stokes law region, what is the terminal velocity of the droplet? [Ans. 0.01 m/s]

Q10. Find the terminal velocity for - 0.175 mm+0.145 mm of particles ($\rho=1300\,kg/m^3$) falling in water ($\rho=1000\,kg/m^3$, $\mu=1$ cP). (Assume the particles are falling in Stokes law region) [Ans. 0.0042 m/s].

Q11. A spherical particle is flowing in water. In gravitational settling, the diameter of the particle is 0.15 mm, and the density of the particle is 3500 kg/m^3. What is the settling velocity of the particle in Stokes' Regime? (Take temperature to room temperature.) The viscosity of water at room temperature is 8.891×10^{-4} Pa.s. [Ans. 0.34 m/s]

Q12. A spherical particle of the density 2800 kg/m^3 is settling in a liquid of density 2000 kg/m^3 and viscosity 0.0015 Pa.s. In the settling, the particle settles at a settling velocity of 0.05 m/s in the Stokes' Regime. Find the diameter of the spherical particle (in mm). [Ans. 0.415 mm]

Q13. In a permeameter, the volume of the bed is known to be 350 cm^3. The mass of the solid is 250 g, and its density is 4000 kg/m^3. What is the value of the bed void fraction? [Ans. 0.18]

Q14. You are performing a sedimentation experiment to find the equivalent mean diameter of a particle. The liquid used in the process is water, and the density of the particle is 3200 kg/m^3. The level observed at time 50 seconds was 15 mm. What is the equivalent mean diameter of the particle? The viscosity of water at room temperature is 8.891×10^{-4} Pa.s. [Ans. 14.92 μm]

Q16. The sedimentation process is being used to determine the particle size of sand. Determine the average diameter of a particle of sand if it is falling in the liquid with a terminal velocity of 0.28 m/s. The densities of the liquid and sand are 1000 and 1540 kg/m^3, respectively. The viscosity of water is 9×10^{-4} Pa/s. [Ans. 0.93 mm]

Q17. The size analysis of a powdered material on the mass basis is a straight line whose equation is given by the relation $d = 150x + 3$, where x is the mass fraction of the sample and d is particle diameter in μm. What is the volume-mean diameter of the sample? [Ans. 78 μm]

Q18. Calculate the distance traveled by a solid particle of diameter 2 mm in 2 seconds during sedimentation. Sedimentation takes place in water. Assume that the density of solid particles to be 1200 kg/m^3 and the density of water to be 1000 kg/m^3. Also, assume that the particle travels in water with a constant velocity. The viscosity of water is 10^{-3} Pa/s. [Ans. 0.872 m]

Q19. The relationship between mass fraction and size of the particle is obtained during size analysis of powder as $d = 80x + 3$, where d (in μm) is the particle size and x is a mass fraction. Find the surface-mean diameter of the sample. [Ans. 36.412 μm]

Q20. The terminal velocity of the iron ball, with a radius of 4 mm, falling in a liquid tank is 5 m/s. Find the viscosity of liquid if the density of the liquid is 1.5 g/cm^3 and the density of iron is 6.5 g/cm^3. Consider that gravitational acceleration is 10 m/s^2. [Ans. 0.0355 Pa.s]

Q21. The size analysis on a mass basis is shown by a straight line from 0% mass at a particle size of 1 μm to 1% mass at a particle size of 51 μm. Calculate the surface-mean diameter of the particles constituting the system. [Ans. 12.72 μm]

Q22. The terminal velocity of a copper ball of radius 2 mm falling through a tank of oil is 6.5 m/s. Compute the oil's viscosity if the oil's density is 1.5×10^3 kg/m^3 and copper's density is 8.9×10^3 kg/m^3. Consider that gravitational acceleration is 10 m/s^2. [Ans. 0.040 Pa.s]

Q23. A fluid of viscosity 0.01 poise flows through a packed bed of cross-sectional area 0.002 m^2 and depth 1.50 m, which is packed with 6.0 kg of solid particles of density 2500 kg/m^3. Calculate the packed bed void fraction. [Ans. 0.80]

Q24. A fluid of viscosity 0.01 Pa.s flows at 20°C at a flow rate of 0.10 m/s through a packed bed of cross-sectional area 0.002 m^2 and of length 1.50 m, which is packed with 5.0 kg of solid particles of density 2500 kg/m^3. The pressure drop across the bed is 1500 Pa per unit length. Find the particle mean diameter that is used in the packed bed. [Ans. 6.7 mm]

Q25. A group of particles of density 1500 kg/m^3 settled down by sedimentation in a liquid of viscosity 0.01 kg/m.s and density 1000 kg/m^3 at the bottom from the liquid surface at a distance of 3 m within 2 seconds. What would be particle diameter during the sedimentation process? Assume gravitational acceleration is 10 m/s^2. [Ans. 7.35 mm]

Q26. A solid particle of density 1250 kg/m^3 falls through a liquid (density = 800 kg/m^3) and viscosity of 8.9×10^{-4} Pa.s at a terminal velocity of 5 m/s. Calculate the particle size (diameter in mm). Gravitational acceleration is 9.81 m/s^2. [Ans. 5.72 mm]

Q27. Find out the volume-mean diameter of a particle population consisting of an equal number of particles with diameters 1, 2, 3, and 4 μm. [Ans. 3.54 μm]

Q28. Size analysis was carried out on a sample of gravel. The data for mass fraction and average particle diameter of the fraction are given in the below table:

x_i	0.2	0.4	0.4
d_{pi} (mm)	5	10	20

For the data given, find the Sauter mean diameter (d_s) of the sample (in mm). [Ans. 10 mm]

Q29. The diameter of steel particles in a sample ranges from 50 to 150 μm. The number of particles of diameter x in the sample is proportional to $1/(50+x)$. What is the average diameter of the particle in microns? [Ans. 94.3 μm]

Q30. For the particle size analysis, the following values are given:

Mass of particles (g)	3	6	8	5	2
The mean size of particles (μm)	450	330	250	140	110

Calculate the volume-surface-mean diameter of the particles in μm. [Ans. 216.71 μm]

Q31. A size analysis of different samples of a particle mixture was carried out in a laboratory. The mass fractions (x_i) of the particle as per average particle diameter d_{pi} in each sample were found to be as (x_i, d_{pi}, mm) = (0.2, 5); (0.4, 10); (0.4, 20). Find the mass mean diameter of the particle in the mixture. [Ans. 13 mm]

Q32. The sedimentation process is used to determine sand's particle size. Determine the average diameter (in mm) of a particle of sand if it is falling in a liquid with a terminal velocity of 0.28 m/s. The densities of the liquid and sand are 1000 and 1540 kg/m^3, respectively. The viscosity of the liquid is 9×10^{-4} Pa.s. [Ans. 0.93 mm]

Q33. A permeate test is supposed to be conducted to estimate the average particle size in a packed bed of volume 1 m^3. The mass of the packed solid with a density of 10 kg/m^3 used for the test is 5.5 kg. Find the void fraction in the packed bed. [Ans. 0.55]

Q34. In a sample of catalyst particles, it is screened and seen that the number of particles is distributed based on the size of the range 0.2–1 mm as per the following:

Diameter of particle (mm)	0.2	0.4	0.6	0.8	1
No of particles	20	120	240	140	20
$f(x)$ = Log-normal distribution function					

[Ans. (i) the volume-mean diameter = 0.74 mm, (ii) the surface-mean diameter of the sample = 0.70 mm, (iii) the mean surface diameter = 0.40 mm, and (iv) the log-normal distribution function based on the particle size of the sample: $f(x)$: 6.68, 1.80, 1.67, 2.15, 3.12.

NOMENCLATURE

ΔP	pressure drop [N/m^2]
A_b	bed cross-sectional area [m^2]
d_p	particle diameter [m]
$d_{p,mn}$	mean particle diameter [m]
d_s	surface-mean diameter [m]
d'_s	mean surface diameter [m]
d_v	volume-mean diameter [m]
d'_v	mean volume diameter [m]
F, f	distribution function [-]
g	gravitational acceleration [m/s^2]

h height [m]
L length [m]
m coefficient [-]
M_s mass of solid [kg]
n coefficient [-]
n number of particle [-]
n_m refractive index [-]
N number function of particles of size d_p [-]
r radius of a particle [m]
t time [s]
u velocity of fluid [m/s]
u_t terminal velocity [m/s]
x_i mass fraction [-]

GREEK LETTERS

ρ_p particle density [kg/m^3]
ρ_s density of solid [kg/m^3]
ρ_f fluid density [kg/m^3]
μ_f viscosity of fluid [Pa.s]
ε_f void fraction [-]
λ wavelength [m]

SUBSCRIPTS

b bed
f fluid
i component
m material
mn coefficient for mean
p particle
s solid, surface
t terminal
v volume

REFERENCES

Brown, R. L. and J. C. Richards, *Principles of Powder Mechanics: Essays on the Packing and Flow of Powders and Bulk Solids*, Pergamon Press, 1970.

Clark, D. L., D. A. Geeson and R. J. Hanrahan, (Eds.), Particle size distribution analysis, in: *The Plutonium Handbook*, 2nd Edition, Vol. 6, Chapter 45.6, American Nuclear Society, 2019. pp. 3323–3351.

Geldart, D., Types of gas fluidization, *Powder Technol.*, 7, 285–292, 1973.

Kurtz, S. M., *UHMWPE Biomaterials Handbook*, 2nd Edition, Academic Press, 2009.

Majumder, S. K., *Hydrodynamics and Transport Processes of Inverse Bubbly Flow*, 1st Edition, Elsevier, 2016.

McCabe, W. L., J. Smith and P. Harriot, *Unit Operations of Chemical Engineering*, 7th Edition, McGraw-Hill, International Edition, 2017.

Merkus, H., *Particle Size Measurements - Fundamentals, Practice, Quality*, Springer, 2009.

Nedderman, R. M., *Statics and Kinematics of Granular Materials*, Cambridge University Press, 1992.

Section II

Size Reduction of Solids

Chapter 4

Mechanism of Size Reduction

Scientists study the world as it is; engineers create the world that has never been.

—*Theodore von Karman, Hungarian-American mathematician, aerospace engineer, and physicist*

4.1 IMPORTANCE OF PARTICLE-SIZE REDUCTION

- Particle size is reduced to a specific size and shape to increase the surface area to enhance chemical reactions or to separate valuable minerals from its ores.
- The primary purposes of the size reduction are as follows:
  - To obtain a workable shape and size for a specific process.
  - To produce powders.
  - To enhance reactivity.
  - To enhance the separation of unwanted ingredient.
  - To improve the uniformity of mixing.
  - To improve the flow ability. The flow of powder during the compression of tablets is effective if the particle size is small.
  - To increase the efficiency of drying.
  - To improve physical stability and decrease sedimentation by reducing the particle size.
  - Increasing the dissolution rate by increasing the surface area is done by size reduction.
  - To improve adsorption rate by producing more surface.
- In the mineral, metallurgical, pharmaceutical, and food industries, nanotechnology size reduction is one of the vital processes.

DOI: 10.1201/9781003571025-6

4.2 FACTORS THAT AFFECT THE SIZE-REDUCTION BEHAVIOR OF THE SOLID MATERIALS

- For designing a process, selecting equipment, and looking for ways to increase efficiency, designers should understand the size-reduction behavior of the solid materials in their operations. To do so, one needs to consider the following properties:
  - Particle-size distribution in the feed
  - Particle shape
  - Bulk density
  - Flowability
  - Cohesiveness and adhesiveness
  - Corrosivity and composition
  - Moisture content
  - Hardness, brittleness, and friability
  - Moisture content
  - Fibrous morphology
  - Abrasiveness
  - Stickiness
  - Elasticity, plasticity, and ductility
  - Dust explosion characteristics
  - Temperature sensitivity
  - Toxicity
  - Oil and fat content
  - Reactivity or release of gases
  - Explosiveness

4.3 BASIC DEFINITION RELATED TO SIZE REDUCTION

- **Comminution:** The operation of reducing raw ore size to the specific size required for mechanical separation or metallurgical processing of the materials.
- **Crushing (dry):** The operation of size reduction by which the material of definite size converts into large fragments.
- **Grinding (wet):** The process of making materials into fine particles.

4.4 MECHANISM OF PARTICLE-SIZE REDUCTION

- Different mechanisms of particle-size reduction are shown in Figure 4.1. Common mechanisms of particle-size reduction are as follows:

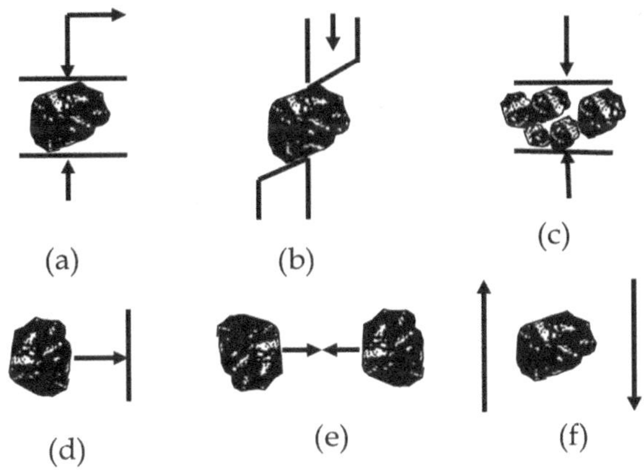

Figure 4.1 Different mechanisms of particle-size reduction. (a) Crushing or attrition between two solid surfaces, (b) shearing, (c) compression crushing in a particle bed, (d) impact at one solid surface, (e) impact between particles, and (f) shear action of the surrounding medium.

- **Compression:** Particle breakage happens by two rigid forces by the application of pressure. Compressive forces are used for the coarse crushing of hard materials. Coarse crushing results in a size reduction of about 3 mm.
- **Impact:** Impact occurs when a substance is more or less immobile and is struck by a fast-moving object or when a moving particle collides with a stationary surface. Particle breaks by a single rigid force (e.g., hammer). The particle-particle impact causes breakage and shattering. Abrasion happens during particle-wall and particle-particle impacts. There are two variations of impact, namely:
    i. **Gravity impact:** Material dropped onto a hard surface, such as a steel plate, is an example of gravity impact.
    ii. **Dynamic impact:** Material dropping before a moving hammer (both objects in motion) refers to dynamic impact.
- **Attrition:** In the case of attrition, the pressure exerted on the material by which the surfaces move relative to each other results in shear forces that break the particles.
- **Shear:** Shear happens when the particle is compressed between two edges of the hard surfaces, moving tangentially (e.g., tearing, cleaving, cutting, or shredding).
- **Nonmechanical:** Thermal shock, explosive shattering, and electrohydraulic may cause the breakage of solids by introducing nonmechanical energy.

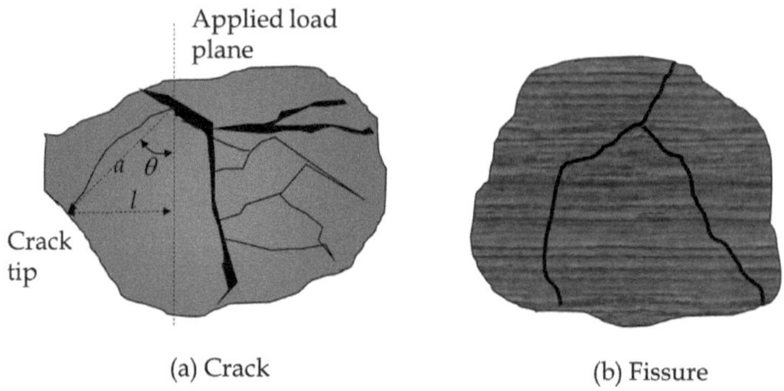

Figure 4.2 Crack and fissures of solid.

4.5 FRACTURE MECHANISM TO YIELD SIZE REDUCTION

- When applied to the particles, stress propagates cracks or in materials, fractures (as shown in Figure 4.2a) at a certain degree of deformation.
- If the deformation of the material is initially proportional to the applied stress, and the fracture occurs suddenly, the material is said to be brittle-elastic (Erdogan, 2000).
- At lower stresses, the particle deformation is linear, elastic, and reversible; whereas, as higher stresses are experienced, the material strength is exceeded locally, and cracks are triggered.
- The cracks grow extremely fast and lead to the destruction of the particle. The process consists of two parts, first opening up any small fissures (Figure 4.2b), which are already present and second forming a new surface.

4.5.1 Stress Intensity for the Propagation of the Crack

- Irwin (1957) stated that there are three independent modes of stress applied on the material to propagate the crack as follows:
  - **Mode I:** Opening mode (tensile stress normal to the plane of the crack),
  - **Mode II:** Sliding mode (shear stress acting parallel to the plane of the crack and perpendicular to the crack front), and
  - **Mode III:** Tearing mode (shear stress acting parallel to the plane of the crack and parallel to the crack front).
- According to Irwin (1957) and his coworkers, the presence of a crack modifies the stress field in its immediate location, which can be expressed as functional form as follows:

$$\tau_{ij} = \left(\frac{K_I}{\sqrt{2\pi l}}\right) f_{ij}(\theta) \tag{4.1}$$

where τ_{ij} are the Cauchy stresses, l is the distance between the crack tip and load or applied stress plane (Figure 4.2a), θ is the angle of crack tip with respect to the crack plane, and $f_{ij}(\theta)$ are functions that depend on the crack geometry and loading conditions, and the quantity is dimensionless.

- The parameter K_I is called the stress intensity factor, which depends on the geometry of the material other than the center-cracked infinite plate.
- The stress intensity factor (K_I) can be expressed in units of Pa√m. The stress intensity factor is related to the energy term as follows:

$$K_I = \tau\sqrt{\pi a} \tag{4.2}$$

where τ is called applied stress and a = crack length.

- Since the stress intensity factor differs for different geometries, a correction factor f_c (dimensionless) as a geometric shape factor is introduced to characterize the geometry. Also, the correction factor depends on the crack length and width of the material. Thus

$$K_I = f_c\tau\sqrt{\pi a} \tag{4.3}$$

Therefore, after the substitution of K_I from Equation (4.3) into Equation (4.1), one can express

$$\tau_{ij} = \left(\frac{f_c\tau\sqrt{\pi a}}{\sqrt{2\pi l}}\right) f_{ij}(\theta) = \left(\frac{f_c\tau\sqrt{a}}{\sqrt{2}\sqrt{l}}\right) f_{ij}(\theta) \tag{4.4}$$

which can be interpreted as follows:

$$\text{Stress field} \propto \sqrt{(a/l)} \tag{4.5}$$

Example 4.1

A material is to be cracked of maximum 3 cm length under a stress of 40 MPa. The cracked tip is 2 cm from the applied stress plane. The stress field follows $\tau_{ij} = 2.95\times10^7 f_{ij}(\theta)$ under a certain geometrical shape. Find the stress intensity factor and its correction factor?

Solution:

$$\tau_{ij} = \left(\frac{K_I}{\sqrt{2\pi l}}\right) f_{ij}(\theta) = 2.95\times10^7 f_{ij}(\theta) \text{ which implies,}$$

$$K_I = 2.95 \times 10^7 \sqrt{2\pi l}$$

Also, from Equation (4.3),

$$K_I = f_c \tau \sqrt{\pi a}$$

Therefore, comparing these above equations,

$$f_c \tau \sqrt{a} = 2.95 \times 10^7 \sqrt{2\pi l}$$

$$f_c = 2.95 \times 10^7 \sqrt{2l}/(\tau \sqrt{a}) = (2.95 \times 10^7 \sqrt{20.02})/(40 \times 10^6 \sqrt{0.03}) = 0.85$$

4.5.2 The Energy Required to Result in the Crack and Crack Criterion

- Irwin (1957) and his colleagues suggested a method of calculating the amount of energy available for a crack in terms of the asymptotic stress and displacement fields around a crack front in a linearly elastic solid. This asymptotic expression for the stress field in Mode I loading is related to the stress intensity factor K_I.
- The energy requirement for propagating geometrically similar cracks depends on the material's internal structure and is proportional to the surface area.
- Minimum crack energy must be applied to the particle for the crack to occur.
- Any materials tend to exhibit yield stress, above which they deform plastically. This means that there is always a zone around the tip of a crack in a metal where plastic deformation occurs (called plastic zone).
- When the plastic zone at the crack's tip is small relative to the crack length, the stress state at the crack tip results from elastic forces within the material and can be characterized using a critical parameter.
- All large lumps of material contain cracks. The size reduction occurs as a result of crack initiation, as shown in Figure 4.2, and crack propagation occurs at specific criteria.
- Griffith (1921), during World War I, developed fracture mechanics to explain the failure of brittle materials. According to Griffith, the product of the square root of the crack length (a) and the stress at fracture (τ_f) is nearly constant, which is expressed by Equation (4.6):

$$\tau_f \sqrt{a} = C\,(\text{constant}) \tag{4.6}$$

- The growth of a crack, the extension of the surfaces on either side of the crack, requires an increase in the surface energy. Griffith (1921) developed an expression for the constant C in terms of the surface energy of the crack by solving the elasticity problem of a finite crack in an elastic plate, which can be expressed as follows:

$$C = \sqrt{\frac{2Y\gamma}{\pi}} \tag{4.7}$$

where Y is the Young's modulus of the material and γ is the surface energy (J/m^2) of the material. Young's modulus of different materials can be obtained from Kinloch (1987) or https://www.engineeringtoolbox.com/young-modulus-d_417.html. For surface energy of the material, you can follow: https://www.twi-global.com/technical-knowledge/faqs/faq-what-are-the-typical-values-of-surface-energy-for-materials-and-adhesives.

Example 4.2

A copper plate is found fractured under a compression with a maximum cracked length of 12 cm. The surface energy of the material is 1.360 J/m^2. Young Modulus of the material is 117 GPa. Calculate the stress at which the fracture happened.

Solution:

According to Griffith, the product of the square root of the crack length (a) and the stress at fracture (τ_f) is nearly constant, which is expressed by Equation (4.6) as follows:

$$\tau_f\sqrt{a} = C\,(\text{constant})$$

where $C = \sqrt{\frac{2Y\gamma}{\pi}}$

Therefore, $\tau_f\sqrt{a} = \sqrt{\frac{2Y\gamma}{\pi}}$ which implies, $\tau_f = \sqrt{\frac{2Y\gamma}{\pi a}}$

$$= \sqrt{\frac{2\times 117\times 10^9\times 1.360}{\pi(0.12)}} = 918.78\ \text{kPa}$$

- For the case of a thin rectangular plate with a crack perpendicular to the load, the energy release rate (E in J/m^2) can be expressed as follows:

$$E = \frac{\pi\tau^2 a}{Y} \tag{4.8}$$

- Irwin (1957) stated, "The stored elastic strain energy is released when the crack grows. This is the thermodynamic driving force for fracture. The dissipated energy includes plastic dissipation and surface energy.

Table 4.1 The surface fracture energy per unit surface area for different types of material (Richardson et al., 2003)

Material	*Surface Fracture Energy Per Unit Surface Area (J/m²)*	*Remarks*
Brittle material (e.g., glass)	1–10	The surface energy term (γ) dominates the plastic dissipation
For polymers close to the glass transition temperature (e.g., plastic)	$10–10^3$	—
Ductile materials (e.g., metals like steel)	$10^3–10^5$	The plastic dissipation term dominates

The dissipated energy provides the thermodynamic resistance to fracture." Therefore, the total energy release rate is as follows:

$$E = 2\gamma + E_p \tag{4.9}$$

where γ is the surface energy and E_p is the sum of plastic dissipation and dissipation from other sources per unit area of crack growth.

- If the stress at the fracture is τ_f, the energy release rate will be as follows:

$$E_c = \frac{\pi \tau_f^2 a}{Y} \tag{4.10}$$

- If $E \geq E_c$, the crack begins to propagate. At lower values of E, elastic deformation occurs without fracture, and the energy input is entirely ineffective in achieving size reduction (Bernotat and Schönert, 2000).
- Orders of magnitude of the surface fracture energy per unit surface area for different types of material are given in Table 4.1.

Example 4.3

A thin rectangular aluminum sheet is broken under an applied stress of 200 MPa, which yields a crack length of 10 cm perpendicular to the load. When the crack begins to propagate under the stress, what would be the stress at the fracture at this crack propagation condition?

Solution:

The energy release rate $E = \frac{\pi \tau^2 a}{Y} = \frac{\pi (200)^2 a}{Y}$ J/m². When the crack begins to propagate under the stress at the fracture is τ_f, the energy release rate = $E_c = \frac{\pi \tau_f^2 a}{Y} = E = \frac{\pi (200)^2 a}{Y}$ which yield stress at the fracture is 200 MPa.

IMPORTANT POINTS TO REMEMBER

- Particle size is reduced to a specific size and shape to increase the surface area to enhance chemical reactions or to separate valuable minerals.
- Common mechanisms of size reduction are compression, impact, attrition, shear, thermal shock, explosive shattering, and electrohydraulic.
- A material is said to be brittle-elastic if the deformation of the material is initially proportional to the applied stress, and the fracture occurs suddenly.
- According to Griffith, the product of the square root of the crack length (a) and the stress at fracture (τ_f) is nearly constant.
- The stored elastic strain energy is released when the crack grows. This is the thermodynamic driving force for fracture.
- The energy requirement for propagating geometrically similar cracks depends on the material's internal structure and is proportional to the surface area.

SELF-TESTING QUESTIONS

I. Multiple-Choice Questions

Q1. Particle-size reduction results in ________________________.
- (a) Increase in the surface area
- (b) To enhance chemical reactions
- (c) To enhance the separation of valuable minerals from its ores
- (d) All of these

Q2. Which one of the following is correct for the reduced size of particle in a system?
- (a) Increasing dissolution rate
- (b) Decreasing surface area
- (c) Increasing sedimentation rate
- (d) Decreasing the efficiency of drying

Q3. Material dropping before a moving hammer (both objects in motion) refers to ________________________.
- (a) Dynamic impact
- (b) Gravity impact
- (c) Compression
- (d) Attrition

Q4. If the deformation of the material is initially proportional to the applied stress, and the fracture occurs suddenly, the material is said to be ________________________.

(a) Plasma
(b) Brittle-elastic
(c) Ductile-plastic
(d) None of these

Q5. According to Griffith, the product of the square root of the crack length (a) and the stress at fracture (τ_f) is ________________________.

(a) Increasing with surface area
(b) Decreasing with surface area
(c) Nearly constant
(d) Dependent on material density

Q6. According to Griffith, the square root of the crack length (a) is related to the stress at fracture as

(a) The square root of the crack length (a) is inversely proportional the stress at fracture
(b) The square root of the crack length (a) is directly proportional the stress at fracture
(c) The square root of the crack length (a) is independent of the stress at fracture
(d) The square root of the crack length (a) is square root of the stress at fracture

Q7. Which of the following is/are purpose of particle-size reduction?

(a) To obtain powders
(b) Obtain specific shape & size
(c) Increase reactivity
(d) Decrease reactivity

Q8. Which of the following size reduction mechanism particle disintegrate by two rigid forces?

(a) Impact
(b) Attrition
(c) Compression
(d) Shear

Q9. The energy requirement for propagating geometrically similar crack is proportional to particle diameter as________________________.

(a) d_p^3
(b) d_p^2
(c) d_p
(d) Constant

Q10. Which of the following is not a correct justification as to why particle-size reduction is done?

(a) To increase surface area available for chemical reaction
(b) To decrease reactivity

(c) To liberate valuable minerals held within particles
(d) To obtain powders

Q11. Which of the following options are not correctly matched?
(a) The capacity of a particle for storing energy is proportional to its surface area ($\propto d^2$)
(b) The energy requirement of a particle for propagating geometrically similar cracks is proportional to its surface area ($\propto d^2$)
(c) The energy available per unit crack area increases linearly with particle size ($\propto d$)
(d) The capacity of a particle for storing energy is proportional to its surface area ($\propto d^3$)

Q12. Which one of the following options is correct about the mechanism of attrition?
(a) Particle disintegration by two rigid forces
(b) Particle breaks by a single force
(c) Arising from particles scrapping between two surfaces
(d) Produced when the particle is compressed between two edges of hard surfaces

Q13. In fracture mechanism, the energy available per unit crack area is proportional to
(a) Volume
(b) Surface area
(c) Particle size
(d) None of these

Q14. Which of the following mechanisms for reducing the size of particle where particle gets disintegrated by two rigid forces?
(a) Compression
(b) Impact
(c) Attrition
(d) Shear

Q15. Match the following

A. Attrition	P. Particle is compressed between the two edges
B. Compression	Q. Size reduction by rubbing between two solid surfaces
C. Shearing	R. Crushing of particle between two solid surfaces
D. Impact	S. Breaking of particle by single rigid force

(a) A-P, B-Q, C-R, D-S
(b) A-S, B-P, C-R, D-Q
(c) A-Q, B-R, C-P, D-S
(d) A-Q, B-P, C-R, D-S

Q16. The size reduction mechanism where the particle is compressed between two edges of the hard surfaces is called ______________.

(a) Compression
(b) Impact
(c) Shear
(d) Attrition

Q17. In fracture mechanism, for the case of a thin rectangular plate with a crack perpendicular to the load, the energy release rate (E) can be expressed as ______________.

(a) $E = \dfrac{\pi\tau^2 a^2}{Y}$

(b) $E = \dfrac{\pi\tau^2 Y}{a}$

(c) $E = \dfrac{\tau^2 a}{Y}$

(d) $F = \dfrac{\pi\tau^2 a}{Y}$

Q18. At lower values of the energy release rate during crack of a metal under a load perpendicular to the crack, which one of the folowing options is correct?

(a) Elastic deformation occurs with fracture
(b) Elastic deformation occurs without fracture
(c) No deformation occurs
(d) None of the above

Q19. Lower values of critical value of energy required to propagate a crack in a material implies

(a) Elastic deformation occurs without fracture
(b) Elastic deformation occurs with fracture
(c) Energy input is completely ineffective in achieving size reduction
(d) Energy input is completely effective in achieving size reduction

Q20. Order of magnitude of surface fracture energy of glass per unit surface area are

(a) 1–10 J/m^2
(b) 10–10^3 J/m^2
(c) 10^3–10^5 J/m^2
(d) None of the above

Solution to multiple-choice questions

Q1. Ans. (d)
Q2. Ans. (a)
Q3. Ans. (a)
Q4. Ans. (b)
Q5. Ans. (c)
Q6. Ans. (a)
Q7. Ans. (a), (b), and (c)
Q8. Ans. (c)
Q9. Ans. (b)
Q10. Ans. (b)
Q11. Ans. (a)
Q12. Ans. (c)
Q13. Ans. (c)
Q14. Ans. (a)
Q15. Ans. (c)
Q16. Ans. (c)
Q17. Ans. (d)
Q18. Ans. (b)
Q19. Ans. (a) and (c)
Q20. Ans. (a)

II. Objective Type Questions

1. Why size reduction of particles is important?
2. Explain the different factors that affect the size reduction.
3. What are the basic definitions of the comminution?
4. Explain briefly the different mechanisms of the size reduction.
5. How does the fracture mechanism work for size reduction?
6. Explain the stress intensity for the propagation of the crack according to Irwin's (1957) principle.
7. How does crack length (a) related to the stress at fracture?
8. What is a plastic zone?
9. How is a growth of a crack related to surface energy?
10. What is the thermodynamic driving force for fracture?

III. Numerical Questions

Q1. The critical energy required to reduce the size of a particle of glass is 100 J/m^2. If the crack length is 2 mm and Young's modulus of glass is 60 GPa, what shear stress (in MPa) must be applied? [Ans. 100 J/m^2].

Q2. Find the critical value of energy (in J/m^2) required to propagate a crack in a material of Young's modulus 2×10^{11} Pa if the length of the crack is formed under stress of 40.00 MPa at its fracture tip is 10 cm. [Ans. 2513.27 J/m^2]

Q3. Find the critical value of energy (in J/m^2) required to propagate a crack in a material of Young's modulus 120×10^{11} Pa if the length of the crack is formed under a stress of 125.0 MPa at its fracture tip is 7 cm. [Ans. 286.34 J/m^2]

Q4. At the same critical elastic deformation condition, what amount of stress is to be applied to the same material to get the crack double its initial value? [Ans. $\tau_2 = 0.7071\ \tau_1$]

Q5. Find the critical energy required per unit area (in J/m^2) of a material of Young's modulus 2000 Pa, beyond which the material loses its elastic deformation if the crack length is 15 cm and the stress applied at the fracture tip is 180 Pa. [Ans. 7.63 J/m^2]

Q6. A stress of 500 Pa is applied to a material of unit volume to propagate a crack of length 5.0 cm on its surface. The geometry of the material resulted in a correction factor of 0.1 for this crack. What would be the stress intensity factor for the fracture of the material? [Ans. 19.82 $Pa\sqrt{m}$]

NOMENCLATURE

a	crack length [m]
E_c	energy release rate at fracture [J/m^2]
f_c	geometric shape factor correction factor [-]
$f_{ij}(\theta)$	functions [-]
K_I	stress intensity factor [$Pa\sqrt{m}$]
l	distance between the crack tip and load or applied stress plane [m]
E_p	the sum of plastic and other dissipations and /m^2 crack growth [J/m^2]

GREEK LETTERS

γ	surface energy [J/m^2]
τ	applied stress [N/m^2]
τ_f	stress at fracture [N/m^2]
τ_{ij}	the Cauchy stresses [N/m^2]
θ	angle of the crack tip with respect to the crack plane [Radian]
Y	Young's modulus of the material [Pa]

SUBSCRIPTS

c	correction, critical
f	fracture
I	intensity
ij	direction
p	plastic

REFERENCES

Bernotat, S. and K. Schönert, Size Reduction. *Ullmann's Encyclopedia of Industrial Chemistry*, Wiley-VCH Verlag GmbH & Co. KGaA, 2000.

Erdogan, E., Fracture mechanics, *Int. J. Solids Struct.*, 37, 171–183, 2000.

Griffith, A. A., The phenomena of rupture and flow in solids, *Philos. Trans. R Soc. London, A*, 221 (582–593), 163–198, 1921.

Irwin, G., Analysis of stresses and strains near the end of a crack traversing a plate, *J. Appl. Mech.*, 24, 361–364, 1957.

Kinloch, A. J., *Adhesion and Adhesives, Science and Technology*, Chapman & Hall, 1987.

Richardson, J. F., J. H. Harker and J. R. Backhurst, *Chemical Engineering, Vol. 2, Particle Technology and Separation Processes, A volume in Chemical Engineering Series*, 5th Edition, Elsevier, 2003.

Chapter 5

General Machines for Size Reduction

> There is nothing I believe more strongly than getting young people interested in science and engineering, for a better tomorrow, for all humankind.
>
> – *Bill Nye (science educator, former mechanical engineer)*

5.1 SIZE REDUCTION EQUIPMENT

- Different types of size reduction equipment are available as shown in Figure 5.1, and each has its size reduction method.
- The choice of equipment for a given size reduction operation is influenced by the nature of the material, quantity, and size of the material to be handled.
- The more essential properties of the material apart from its size are:
  - Hardness
  - Crushing strength
  - Soapiness
  - Structure
  - Friability
  - Explosiveness
  - Moisture content
  - Stickiness
- **Criteria for selecting size-reducing equipment:** The following are the criteria used in choosing the right type of equipment.
  - **Production requirements:** include output size, shape, and the required capacity.
  - **Material characteristics:** include material specification, feed (input) size, friability, and abrasiveness.
  - **Operational considerations:** include power demand, equipment availability (hours/annum), availability and cost of replaceable parts, reduction ratio, maintenance requirements, needed

DOI: 10.1201/9781003571025-7

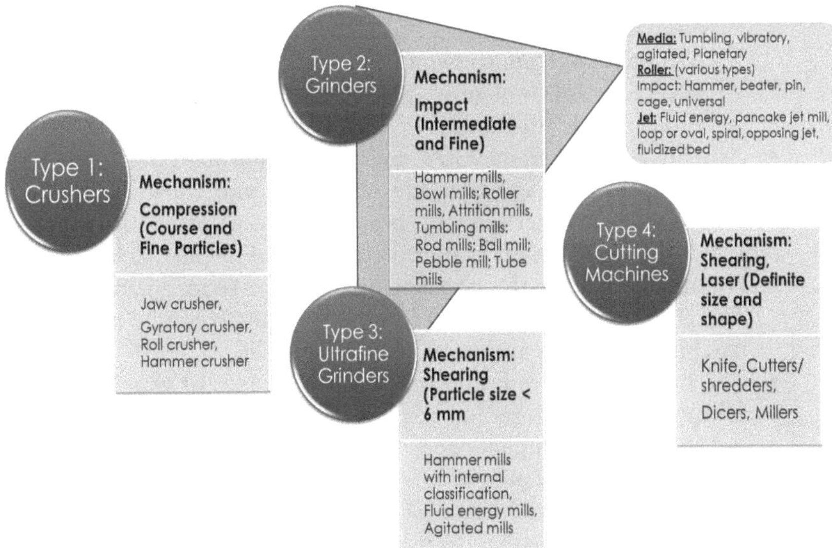

Figure 5.1 Broad classifications of size reduction equipment.

workforce, approachability of parts for maintenance, availability of spares, safety, and environment.
  - **Capital cost:** The capital cost of the equipment and the total cost of the size reduction are also important for the selection of equipment.
- The size-reducing equipment can broadly be classified into four types described in the following section.

5.2 CLASSIFICATION OF SIZE REDUCTION EQUIPMENT

- The size reduction equipment can be broadly classified into four types, as shown in Figure 5.1, which are described subsequently.
- **Type 1: Crushers (applied to coarse particles):**
  - The crusher is distinguished as primary crushing (up to 100–300 mm), secondary crushing (25–100 mm), and pulverization (5–25 mm) (Richardson et al., 2003).
  - Depending on the shape of the crushing member, crushers are divided into five classes such as (a) jaw crusher, (b) cone crusher for primary crushing, (c) cone crusher for secondary crushing and pulverization, (d) smooth-roll crusher, (e) toothed-roll crusher, (f) hammer crusher.

  - The mechanism and feed and product size ranges of crushers are as follows:
    - **Mechanism: Compression**
    - **Feed size: 1500–40 mm**
    - **Product size: 50–5 mm**
  - A primary crusher operates on run-of-mine material, accepting anything from the mine face and breaking into 150–250 mm.
  - A secondary crusher reduces these lumps to 6 mm.
- **Type 2: Grinders (applied to intermediate and fine particles):**
  - For medium duty, the grinders are used to reduce it to powder. The grinders mostly used are hammer mills, rolling compression machines, attrition mills, and tumbling mills.
  - The mechanism, feed, and product size ranges of grinders are as follows:
    - **Mechanism: Impact**
    - **Feed size: 50–2 mm**
    - **Product size: 5–0.1 mm**
  - Grinders reduce crushed feed to powders.
  - The product from an intermediate grinder might pass a 40-mesh screen.
  - Most of the product from a fine grinder would pass a 200-mesh screen with a 74-μm opening.
- **Type 3: Ultrafine grinders (applied to ultrafine particles):**
  - For ultrafine duty, the grinders are used to reduce it to ultrafine powder. The grinders used most are hammer mills with internal classification, fluid energy, and agitated mills.
  - The mechanism, feed, and product size ranges of ultrafine grinders are as follows:
    - **Mechanism: Shearing**
    - **Feed size: <6 mm**
    - **Product size: down to 0.01 μm**
  - Mills that are used to reduce solids to fine particles of average size 1–20 μm are called ultrafine grinders.
  - Ultrafine grinding can be done on a dry and wet basis.
    - **Dry basis:** high-speed hammer mills with internal and external classification.
    - **Wet basis:** fluid energy or jet, colloid, and agitated mills.
- **Type 4: Cutting machines (applied to definite size and shape of particle as per design)**
  - The equipment reduces the particle to a fixed designated size. These requirements are met by devices that cut, chop, or tear the feed into a product with the desired characteristics. The machines include rotary knife cutters and granulators. The most used cutting machines are knife cutters, dicers, and millers.

- The mechanism and feed and product size ranges of cutting machines are as follows:
  - **Mechanism:** Shearing, laser
  - **Definite size and shape**
  - **The product size will be 2–10 mm in length.**
- The machines reduce the particle to a fixed required size.
- These requirements are met by devices that cut, chop, or tear the feed into a product with the desired characteristics.

5.3 OVERVIEW OF VARIOUS SIZE REDUCTION EQUIPMENT

- The following sections describe some important size-reducing equipments' principles, applications, advantages, and disadvantages.

5.3.1 Jaw Crusher

5.3.1.1 Working principle

- The jaw crusher works based on mechanical pressure by reciprocating the two jaws, of which one is fixed and the other is swinging at an angle of 20°–30° with the fixed jaw in a horizontal plane.
- The jaws open or close at 250–400 times per minute.
- It is driven by an eccentric, applying an excellent compression force to particles caught between the jaws.
- Based on the position of the pivoting of the swing jaw, the crushers are classified as:
  - **Blake crusher:** The swing jaw is fixed at the upper position, as shown in Figure 5.2.
  - **Dodge crusher:** The swing jaw is fixed at the lower position.
  - **Universal crusher:** The swing jaw is fixed at an intermediate position.

5.3.1.2 The capacity of jaw crushers

- The theoretical mass capacity (Q_m) of a jaw crusher can be calculated as per the following equation.

$$Q_m = \frac{\rho_p A w N_j (1-\varepsilon)}{60} \text{ kg/s,} \tag{5.1}$$

where ρ_p=density of material (kg/m^3), w=jaw width (m), A=area of swing (m^2)=gap of the opening area of jaws at the product outlet (or called jaw throw)×width of jaw, N_j=number of swings per minute (min^{-1}), ε=porosity of particles (-).

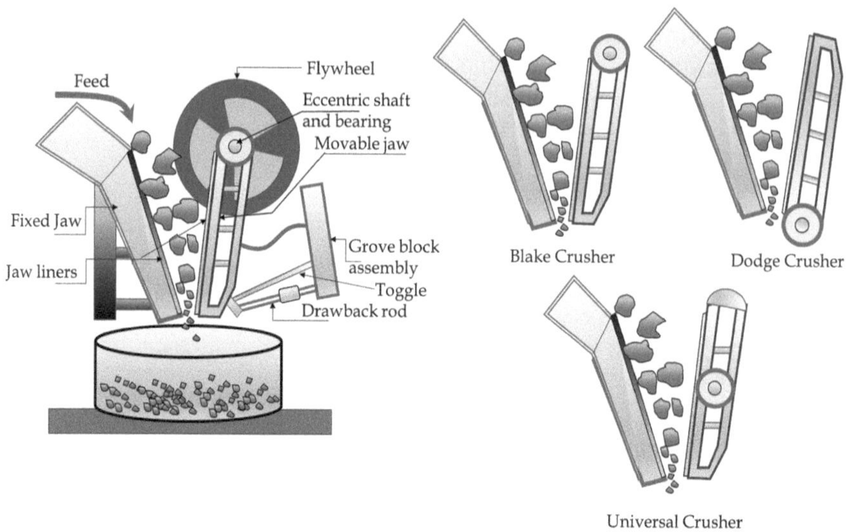

Figure 5.2 Schematic of a typical jaw crusher.

Example 5.1

A jaw crusher is used to reduce the size of the material of density of 2,700 kg/m^3 and porosity of 0.26. The crusher jaw width is 0.21 m. The gap between the opening area of the jaws is 0.10 m. If the jaw crusher is operating at a speed of 250 rpm, what is the capacity of the jaw crusher?

Solution:

ρ_p = density of material = 2,700 kg/m^3, W_j = jaw width = 0.21 m, gap of jaw at the outlet = 0.10 m, A = area of swing = jaw width × gap = 0.021 m^2, N_j = number of swings per minute = 250 rpm, ε = porosity of particles = 0.26 (-).

The theoretical mass capacity (Q_m) of a jaw crusher can be calculated as per Equation (5.1) as

$$Q_m = \frac{\rho_p A W_j N_j (1-\varepsilon)}{60}\ \text{kg/s} = \frac{2700 \times 0.021 \times 0.21 \times 250 \times (1-0.26)}{60}$$

$$= 37.21\ \text{kg/s} = 133.95\ \text{t/h}$$

- Because of the complex nature of the jaw and gyratory crusher, Broman (1984) developed simple models for optimizing the capacity of the jaw and gyratory crusher. The volumetric capacity of the jaw crusher is expressed as (Wills and Finch, 2016):

$$Q_v = \frac{bS_{\text{OS}}skN\cot(a)}{60}\ \text{m}^3/\text{s}, \tag{5.2}$$

where Q_v is volumetric throughput (capacity) (m^3/s); b is inner width of crusher (m); S_{os} is open side setting (m); s is jaw throw (m); a is the angle of nip; k is a material constant, normally having values 1.5–2.0; N is the speed of crusher (rpm).

Example 5.2

A bauxite ore of density 3,200 kg/m³ and porosity 0.18 is crushed by a jaw crusher. The inner width of the crusher is 0.16 m with an open side setting of 0.11 m and jaw throw of 0.15 m. The angle of the nip is found to be 15°. The material constant is to be considered as 1.7. If the jaw crusher is operating at a speed of 260 rpm, what is the capacity of the jaw crusher in m³/h as per Broman's (1984) correlation?

Solution:

Given that B is the inner width of crusher (m) = 0.16 m; S is open side setting (m) = 0.11 m; s is jaw throw (m) = 0.15 m; a is the angle of nip = 15°; k is a material constant = 1.5; N is the speed of crusher = 260 rpm.

Therefore, as per the equation, the volumetric capacity would be

$$Q_v = \frac{bS_{os}skN\cot(a)}{60}\ \mathrm{m^3/s}$$

$$= \frac{(0.16)(0.11)(0.15)(1.7)(260)\cot(15\times\pi/180)}{60}$$

$$= 0.073\ \mathrm{m^3/s} = 0.073\times 3600\ \mathrm{m^3/h} = 261.29\ \mathrm{m^3/h}$$

- As per Sastri (1994), the capacity of a jaw crusher ($Q_{S,m}$) can be calculated from the equation:

$$Q_{S,m} = 60N_{op}w\left(S+\frac{T}{2}\right)aK_1K_2K_3\rho_p,\ \mathrm{kg/h}, \tag{5.3}$$

where N_{op} is the operating speed of the crusher in rpm or strokes/min, w is called the width of jaws (in meters), and S is the closed side setting as shown in Figure 5.2. The parameter T is the throw. The parameter a is called the advance of material in the crusher per stroke (in m), which is defined as:

$$\begin{aligned} a &= \frac{h_vT}{G-(S+T)} \quad \text{for} \quad N \le N_c \\ &= \frac{450g}{N^2} \quad \text{for} \quad N \ge N \end{aligned} \tag{5.4}$$

- The parameter h_v (in m) is called vertical depth between jaws

$$\begin{aligned} h_v &= 3.25G^{1.15} && \text{for} \quad G \leq 0.25\text{ m} \\ &= 0.21 + 1.8G && \text{for} \quad G \geq 0.25\text{ m} \end{aligned} \tag{5.5}$$

- The parameters K_1, K_3, and K_3 are related to feed size (d_F) to gap (G) of crusher ratio, throw (T) of crusher to gap (G) of crusher ratio, and the nature of the material, respectively. The variation of K_1 with d_F/G ratio can be expressed as

$$K_1 = 0.85 - (d_F/G)^{2.5}, \tag{5.6}$$

where

$$\left.\begin{aligned} d_F &= 0.8d_{F,\max} && \text{for unscalped feed} \\ &= 0.8d_{F,\max} + 0.2S_c && \text{for scalped feed} \end{aligned}\right|_{d_{F,\max}=0.85G} \tag{5.7}$$

- The parameter S_c, is called the opening of the scalping screen (in meters). The variation of K_2 with T/G ratio can be expressed as

$$K_2 = 1.92 \times 10^{-6.5T/G} \tag{5.8}$$

- Also, the throw of the crusher T should satisfy:

$$T = \frac{0.037G}{K_2} \tag{5.9}$$

- It is suggested that the value of K_3 would be 0.6 for softer materials and 1.0 for harder materials.
- For speeds below a specific value, the capacity varies directly with the speed of the machine, and above this speed, the capacity varies inversely with the speed. The transition at which this occurs is defined (Rose and English, 1967) as the critical speed (N_c, in rpm), and it is evident that the maximum capacity of a machine will be at the critical speed. The critical speed can be calculated by balancing the advance of material in the crusher per stroke at low and high speeds of the crusher, i.e.,

$$\frac{h_v T}{G - (S + T)} = \frac{450g}{N_c^2}, \tag{5.10}$$

which yields

$$N_c = 66.4\left(\frac{G - (S + T)}{h_v T}\right)^{0.5} \tag{5.11}$$

- Normal operating speed (in rpm) can be correlated with the gap of the jaw crusher which can be expressed as

$$N_{op} = 280 \times 10^{-0.175G^2} \tag{5.12}$$

Example 5.3

A jaw crusher is of width (w) 10 cm, the closed side setting (S) is 15 mm, and the gap (G) of the crusher jaws is 26 cm. The crusher is to be used for crushing a hard material (grossular garnet), which is screened by an unscalped screen with an opening size of 0.44 mm. Calculate the throw of the crusher for harder materials using Sastri's (1994) concept. What should be the critical speed of the crusher? What is the operating speed of the crusher? What is the capacity ($Q_{S,m}$) of the jaw crusher? The density of the material is 3,560 kg/m³.

Solution:

The parameters are given as:
$G = 0.26$ m; $w = 0.10$ m; $S = 0.015$ m;

$$
\begin{aligned}
h_v &= 3.25G^{1.15} && \text{for} \quad G \le 0.25\ \text{m} \\
&= 0.21 + 1.8G && \text{for} \quad G \ge 0.25\ \text{m} = 0.21 + 1.8(0.26) = 0.678
\end{aligned}
$$

For scalped feed,

$$d_{F,\max} = 0.85G = 0.85(0.26) = 0.221\ \text{m}$$

$$
\begin{aligned}
d_F &= 0.8d_{F,\max} && \text{for unscalped feed} = 0.8(0.221) = 0.1768 \\
&= 0.8d_{F,\max} + 0.2S_c && \text{for scalped feed}
\end{aligned}
$$

$$K_1 = 0.85 - (d_F/G)^{2.5} = 0.85 - (0.1768/0.26)^{2.5} = 0.469$$

Assume that throw of the crusher, $T = 0.0325$ m

$$K_2 = 1.92 \times 10^{-6.5T/G} = 1.92 \times 10^{-6.5(0.0325)/(0.26)} = 0.296$$

Check whether the following equation is satisfied or not for assumed $T = 0.0325$ m.
To do this:

$$T = \frac{0.037G}{K_2} = \frac{0.037(0.26)}{0.296} = 0.0325 \approx 0.0325\ \text{(Checked)}$$

Therefore, the throw of the crusher is 0.021 m.

The theoretical speed of the crusher

$$N_c = 66.4\left(\frac{G-(S+T)}{h_v T}\right)^{0.5} = 66.4\left(\frac{0.26-(0.015+0.0325)}{(0.678)(0.0325)}\right)^{0.5}$$

$$= 206.20 \text{ rpm}$$

Operating speed of the crusher

$$N_{op} = 280\times 10^{-0.175G^2} = 280\times 10^{-0.175(0.26)^2} = 272.48 \text{ rpm} > N_c$$

$$a = \frac{h_v T}{G-(S+T)} \qquad \text{for } N \le N_c$$

$$= \frac{450g}{N^2} = \frac{450(9.81)}{272.48^2} = 0.059 \qquad \text{for } N \ge N_c$$

Therefore,

$$Q_{S,m} = 60N_{op}w\left(S+\frac{T}{2}\right)aK_1K_2K_3\rho_p, \text{ kg/h}$$

$$= 60(272.48)(0.10)\left(0.015+\frac{0.0325}{2}\right)$$

$$(0.059)(0.469)(0.296)(1.0)(3560) = 1489.69 \text{ kg/h}$$

5.3.1.3 *Advantages of jaw crusher*

- The main advantages of jaw crusher are:
  - simple structure and low cost
  - convenient to inspect, repair, and maintain
  - reliable work
  - small machine body
  - easy device
  - the ores with water and high viscosity are not easy to jam

5.3.1.4 *Disadvantages of jaw crusher*

- The main disadvantages of jaw crusher are:
  - low productivity
  - lower power consumption
  - bigger vibration
  - small crushing ratio
  - uneven particle size

5.3.1.5 *Applications*

- Jaw crushers are generally used for rock quarries, mining, construction and demolition recycling, construction aggregates, road and railway construction, metallurgy, and the chemical industry.

5.3.2 Smoothed-Roll Crusher

5.3.2.1 *Working principle*

- Two heavy smoothed-face metal rolls rotating on parallel horizontal axes. These are the functional elements of the smooth-roll crusher, as shown in Figure 5.3.
- Particles of feed caught between the rolls are broken in compression and drop out below. The rolls turn toward each other at the same speed, ranging from 50 to 300 rpm.
- Through two opposite rotating grinding rollers, with friction and gravity, the material is beaten into the crushing chamber, making it squeeze, grind, and break. Finally, the material is discharged from the discharging port.
- Crushing forces in a roll crusher can be used as:
    - Figure 5.3 shows the geometry for smooth-surfaced roll crushers for calculating maximum feed size.
    - When the rolls grip a particle, one can write:

$$C \sin(\theta/2) = \mu \cos(\theta/2) \tag{5.13}$$

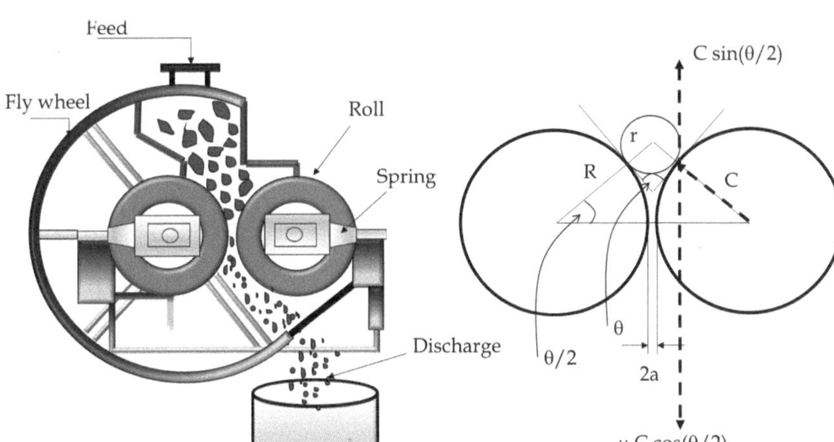

Figure 5.3 Schematic of smoothed-roll crusher.

Therefore,

$$\tan(\theta/2) = \mu \tag{5.14}$$

- Typically, μ=0.2–0.3, so, θ should be less than 25° to prevent particle slippage in the rolls.
- The angle of nip also establishes the maximum particle size according to

$$\cos(\theta/2) = \frac{R_r + a}{R_r + r_p} = \frac{d_r + 2a}{d_r + d_p} = \frac{\text{diameter of roll} + \text{gap between two rolls}}{\text{diameter of roll} + \text{diameter of particle}} \tag{5.15}$$

  where R_r=roll's radius, r_p=particle radius, θ=angle of nip, μ=coefficient of friction, C=compressive force, $2a$=roll's gap.
- The limiting size ($d_{p,\max}$) of particles that the rolls can nip depends on the coefficient of friction between the particle and the roll surface, but in most cases, it can be estimated from the relation (McCabe et al., 2017)

$$d_{p,\max} = 0.04R_r + a, \tag{5.16}$$

  where R_r=roll's radius and a=half the gap width between the rolls.
- The maximum size of the product is approximately equal to $2a$.

5.3.2.2 *The capacity of a smooth-roll crusher*

- The capacity depends on:
  - Type of material
  - The size of the feed
  - The size of the finished product
  - The width of the rolls
  - The rotating speed of the rolls
  - Uniformity of feed of the material
- The mass capacity ($Q_{W\text{-}N,m}$) of the roll crusher is directly proportional to its width w, diameter d_r of roll, and the speed of the revolution of rolls.
- Under continuous and steady feeding conditions, the capacity is given by Wills and Napier-Munn (2006):

$$Q_{W-N,m} = \pi d_r w \omega L \rho_p \text{ kg/s}, \tag{5.17}$$

where d_r=diameter of roll (m), w=width of the roll, ω=speed of roll, rps, L=roll's gap (m)=$2a$, ρ_p=density of the particle or material (kg/m^3).

Example 5.4

A roll crusher with crushing rolls, of 0.5 m in diameter, is set so that the crushing surfaces are 10 mm apart and the angle of nip is 31°, what is the maximum size of particle that should be fed to the rolls? Under continuous and steady feeding conditions, as per (Wills and Napier-Munn, 2006), if the actual capacity of the machine is 20% of the theoretical, calculate the throughput in kg/s when running at 1.0 Hz if the working face of the rolls is 0.25 m long and the bulk density of the feed is 2,700 kg/m^3.

Solution:

$$\cos(\theta/2) = \frac{R_r + a}{R_r + r_p} = \frac{d_r + 2a}{d_r + d_p}$$

$$= \frac{\text{diameter of roll} + \text{gap between two rolls}}{\text{diameter of roll} + \text{diameter of particle}}$$

As per the case,

$$\cos(\theta/2) = \frac{R_r + a}{R_r + r_p} \Rightarrow \cos[(31/2)\times\pi/180]$$

$$= \frac{0.25 + 0.005}{0.25 + r_p} \Rightarrow r_p = 0.014 \text{ m}$$

Under continuous and steady feeding conditions, the capacity is given by Wills and Napier-Munn (2006) as:

$$Q = \pi d_r w \omega L \rho_p = \pi \times 0.5 \times 0.25 \times 1.0 \times 0.02 \times 2700$$

$$= 21.21 \text{ kg/s} = 76.34 \text{ ton/h}.$$

5.3.2.3 *Advantages of smooth-roll crusher*

- It is widely used for crushing coal or large ore, with Mohs hardness below 9. Mohs hardness scale is a standard for measuring rock hardness. It uses a diamond drill to scratch the surface of tested rocks (Woodcock and Mason, 1987).
- Energy saving and environmental protection.

5.3.2.4 *Disadvantages of smooth-roll crusher*

- The production capacity of the roll crusher is low.
- The abrasion of the roller surface increases the gap between the two rollers, resulting in uneven product size, which requires frequent repairs.
- Leaking of raw materials reduces the actual crushing effect.

5.3.2.5 *Applications*

- Roll crushers have long been used in various applications, including run-of-mine (ROM) feeds, coal, Trona (trisodium hydrogen dicarbonate dihydrate), salt, glass, and other friable minerals in the mining and recycling industries.

5.3.3 Hammer Crusher/Mill

5.3.3.1 *Working principle*

- The hammer crusher/mill is essentially a steel drum mainly composed of a transmission device, rotary disk (150–450 mm in diameter), lining plate, grate screen, and shell on which hammers (typically 4–8 hammers mounted on each disk) are mounted as shown in Figure 5.4.
- The crusher mainly breaks materials by the impact utilizing kinetic energy from the high-speed hammers.
- The impacts occur at high speed on the crushing plate and grate screen. The material smaller than the gap of the grate screen is discharged, whereas the material with a larger particle size is bounced back to the lining plate and crushed by the additional impact of the hammers.

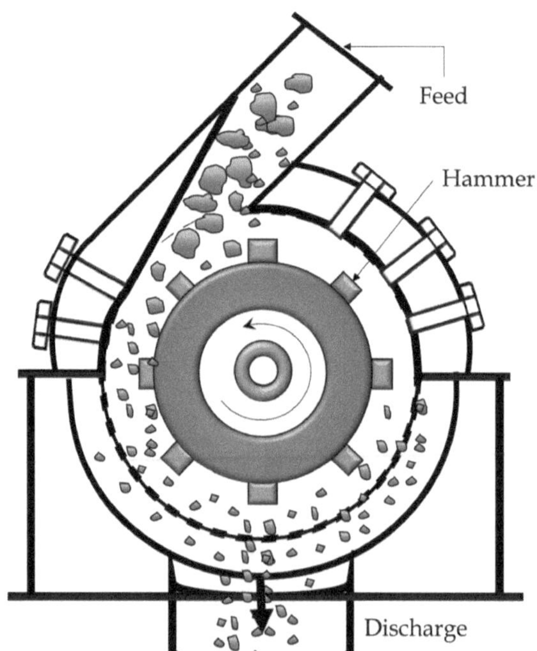

Figure 5.4 Schematic of hammer crusher/mill.

- The materials also impact each other during the whole crushing process in the crusher.
- Hammer mills crush materials into two stages.
  - Size reduction, which occurs by dynamic impact.
  - Sizing occurs by attrition and shear in the second zone where a small clearance between the hammer and the screen bar produces a particle range of 15–50 μm.
- Hammer crushers can be divided into the following types:
  - Based on rotor amount: single-rotor hammer crusher and double-rotor hammer crusher.
  - Based on rotor rotation direction: reversible and irreversible hammer crusher.
  - Based on hammer rows: single-row hammer crusher and multi-row hammer crusher.
- The rotor is spun at high speed, around 110 m/s inside the drum, while the material is fed into a feed hopper.
- The product size will be 25 mm to 20 mesh (0.833 mm) in particle size.
- The capacity of commercial hammer mills ranges from 60 to 240 kg of solid per kilowatt hour of energy consumed.

5.3.3.2 *The capacity of the hammer mill*

- The maximum crushing capacity of a mill is decided by two factors, i.e., the maximum hold-up, $H_{u,\max}$ in the mill and the mean residence time t through the mill.
- The maximum crushing capacity of the mill $Q_{\max}$ can be calculated by the relationship expressed in terms of those factors (Endoh, 1992):

$$F_{\max} = H_{u,\max} / t \tag{5.18}$$

$$H_{u,\max} = 39.5 \times 10^{-2} \rho_b N_r r l S_r^{1/2} \tag{5.19}$$

$$t = 3.98 \times 10^3 \tau \left[0.693 - \left(\frac{d_v}{D_e} \right)^n \right] / \left(d_v^{1.3} \omega^2 L^{-2/3} \right), \tag{5.20}$$

where N_r=the number of hammers, r=distance from the top of a hammer to the center of the rotational axis of the mill [m], l=the width of a hammer in the rotational direction [m], S_r=the opening area ratio of the screen (%), t=the mean residence time [h], d_v=the sphere equivalent diameter of products (mm), D_e=the sphere equivalent diameter of feed [mm], ω=the rotational speed of the hammers [rad/s], L=the length of the mill [m], τ=the breakage characteristic time [h], and n=the distribution constant of feed. Some typical values are given in Table 5.1.

Table 5.1 The distribution constant of feed (Endoh, 1992) based on feed size

Material	*D_e (mm)*	*n*	*Density (kg/m³)*
Charcoal	9.3	1.3	2,780
	4.3	1.0	
	3.5	0.76	
Rubber	3.3	6.0	999
Coal	3.4	1.19	1,240
Sand	1.06	1.9	2,610
	17.9	10.0	
	15.4	10.0	

Example 5.5

A hammer mill is used to reduce the size of a slab of charcoal of mean size 60 mm whose bulk density is 2,550 kg/m³ to a product mean size of 20 mm. The design specification of the hammer mill is given as follows. What is the mean residence time [h] and the maximum crushing capacity of the hammer mill?

Given: N_r=the number of hammers=52; r=the distance from the top of a hammer to the center of the rotational axis of the mill [m]=0.34 m; l=the width of a hammer in the rotating direction=0.123 m; S_r=the opening area ratio of screen (%)=41.7%; d_v=the sphere equivalent diameter of products (mm); D_e=the sphere equivalent diameter of feed=9.3 mm; ω=the rotating speed of the hammers [rad/s]=178 rad/s; L=the length of mill [m]=0.79 m; τ=the breakage characteristic time [h]=0.1 h; n=the distribution constant of feed, n=1.3; ρ_b=bulk density of material=2.55 g/cc.

Solution:

$$H_{u,\max} = 39.5\times10^{-2}\rho_b N_r r l S_r^{1/2} = 39.5\times10^{-2}(2550)(52)$$

$$(0.34)(0.123)(0.417)^{1/2} = 1414.47 \text{ kg}$$

$$t = 3.98\times10^{3}\tau\left[0.693-\left(\frac{d_v}{D_e}\right)^{n}\right]/\left(d_v^{1.3}\omega^{2}L^{-2/3}\right)$$

$$= 3.98\times10^{3}(0.1)\left[0.693-\left(\frac{0.020}{0.093}\right)^{n}\right]/\left(0.020^{1.3}\times1.78^{2}\times0.79^{-2/3}\right)$$

$$= 0.97 \text{ h}$$

Therefore, maximum capacity

$$F_{\max} = H_{u,\max}/t = 1414.47/0.97 = 1462.12 \text{ kg/h}$$

5.3.3.3 *Advantages of hammer mill*

- It works without the need for a closed-circuit crushing system.
- It works with a minimum of fines due to self-classification.
- It has the capacity of a high reduction ratio.
- Comparatively reasonable energy requirements.

5.3.3.4 *Disadvantages of hammer mill*

- It is unsuitable for fine grinding of hard and abrasive material due to excessive wear.
- Not suitable for low-melting sticky or plastic-like material due to heat generation in the mill head due to mill fouling.
- Chocking of the mill at a specific operating condition leads to damage.
- Clogging of the screen may reduce the efficiency of the mill.

5.3.3.5 *Application*

- It is widely used in cement manufacturing, chemical, electric power, metallurgy, and pharmaceutical industries to process wet or dry granulations and disperse powder mixtures.

5.3.4 Attrition Mills (Rolling Compression Mill)

5.3.4.1 *Working principle*

- The materials, especially grains, are rubbed between the grooved flat faces of rotating circular roughened horizontal or vertical axes disks with a pin, as shown in Figure 5.5.
- There are two types of attrition mills: single disk-runner mills and double disk-runner mills.
- The most common type of rolling compression machine is a pulverizer.
- In a single-runner mill, one plate is stationary and fixed with the body of the mill while the other is a rotating disk, whereas, in a double-runner machine, both disks are driven at high speed in opposite directions.
- Material is fed through the opening between the plates and is reduced by crushing and shear.
- It is used for blending and feather curling, unrelated to size reduction.
- Disks of a single-runner mill turn at 350–700 r/min, whereas disks of a double-runner mill turn at 1200–7000 r/min.
- The capacity of attrition mills ranges from 0.5 to 8 tons/h for producing particles of size 200-mesh screen.
- The energy required depends on the nature of the feed, and the degree of reduction typically ranges between 8 and 80 kWh per ton of product.
- A comparison between a single disk-runner mill and a double disk-runner mill is shown in Table 5.2.

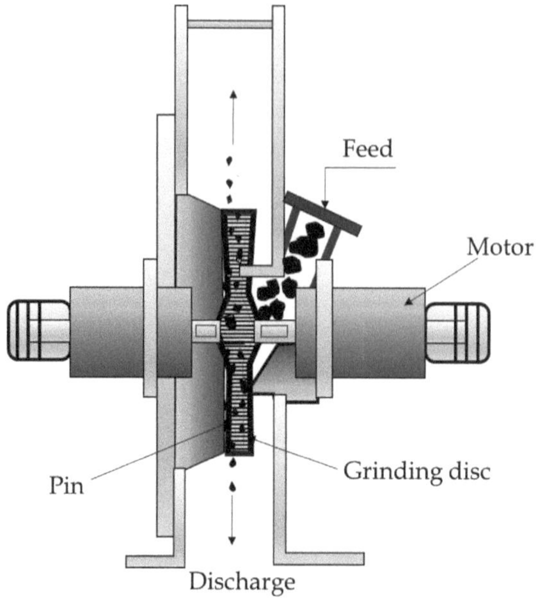

Figure 5.5 Schematic of attrition mill.

Table 5.2 Comparison between single disk-runner and double disk-runner mill

Parameter Runner	*Single Disk*	*Double Disk-Runner*
Disk diameter	250–1,400 mm	250–1,400 mm
Speed	350–700 rpm	1,200–7,000 rpm
Capacity	0.5–8 ton/h	0.5–8 ton/h
Product size	Passed through 200 mesh, i.e., 0.074 mm	Passed through 200 mesh, i.e., 0.074 mm
Power consumption	8–80 kWh per ton of product	8–80 kWh per ton of product

5.3.4.2 Advantages of attrition mill

- Used for grinding soft materials.
- Spacing between the plates is adjustable.
- The disks are driven at high speed in opposite directions for high-capacity operation.

5.3.4.3 Disadvantages of attrition mill

- The performance is reduced due to overfeeding.
- Heat is generated significantly during milling.
- Use in spice grinding is limited.

5.3.4.4 *Application*

- The mills are mainly used to reduce limestone, cement clinker, and coal.
- The mill is also used to reduce the size of solids like clay and talc, wood, starch, insecticide powders, and carnauba wax.

5.3.5 Tumbling Mills

5.3.5.1 *Working principle*

- The material is processed for its size reduction in a cylindrical shell that is turning about its horizontal axis filled with a grinding medium like metal rods in a rod mill, lengths of chain or balls of stainless steel, rubber, or wood in a ball mill, flint pebbles or porcelain or zircon spheres in a pebble mill as shown in Figure 5.6.
- The shell is usually a steel line with carbon steel plate, porcelain, silica rock, or rubber.
- The mills grind the material by impact and attrition.
- This speed is usually 50%–75% of the critical speed.

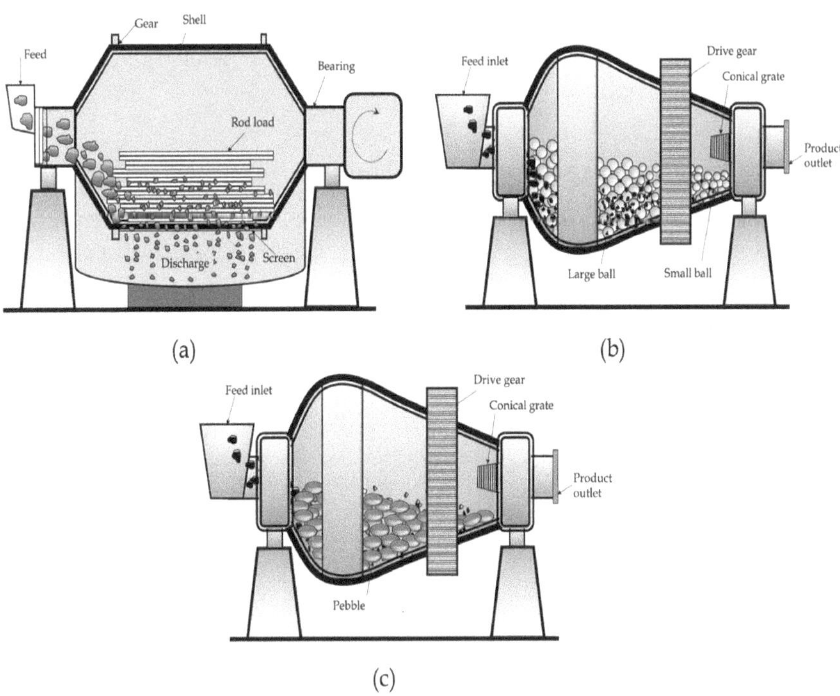

Figure 5.6 Schematics of different types of tumbling mills. (a) Rod mill, (b) Ball mill, (c) Pebble mill.

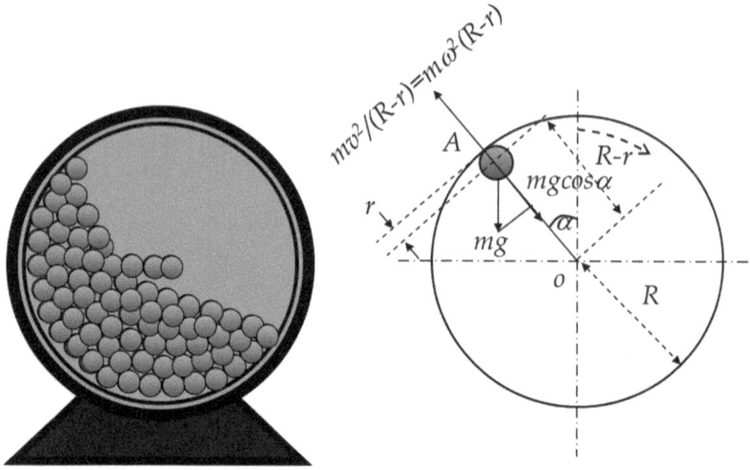

Figure 5.7 Schematic representation of force balance.

- The balls occupy 30%–50% of the mill volume, and their size depends on the feed and mill size.
- The large balls tend to break down the coarse feed materials, whereas the smaller balls help to form fine products by reducing void spaces between the balls.

5.3.5.2 *The critical speed of the ball mill*

- The critical speed of the ball mill can be obtained by balancing the respective components of gravity force in the opposite direction of centrifugal force (as shown in Figure 5.7):

$$mg\cos\alpha = m[4\pi^2 n^2 (R-r)]; \quad \text{where,} \quad \omega = 2\pi n, \tag{5.21}$$

  where m is the mass of the ball, n is the rotational speed of the mill, R is the radius of the mill, and r is the radius of the ball.
- At critical speed, $\alpha=0$, $\cos\alpha=1$, and n becomes the critical speed n_c as

$$n_c = \frac{1}{2\pi}\sqrt{\frac{g}{(R-r)}} \tag{5.22}$$

Example 5.6

A ball mill 8.53 m in diameter is being run at 0.2 Hz, and it is found that the mill is not working satisfactorily for a required size reduction operation because the actual rotational speed of the mill is much larger than the critical speed of the mill. The operating speed was 75% of the critical speed. What is the size of the particle that was reduced to its desired size?

Solution:

$$n = n_c \times 0.75 = \frac{1}{2\pi}\sqrt{\frac{g}{(R-r)}} \times 0.75 = 0.2 \text{ rps}$$

Implies

$$\frac{1}{2\pi}\sqrt{\frac{9.81}{(0.8-r)}} = 0.27, \text{ which implies, } r = 0.771 \text{ m} = 77.1 \text{ cm}$$

5.3.5.3 *Conical ball mill*

- Segregation of grinding units in a single chamber is done in a conical ball mill, as shown in Figure 5.6b.
- Feed enters from the left through a 60° cone into the primary grinding zone, where the diameter of the balls is large and leaves through the 30° cone to the right, where the diameter of the balls is less.
- Large balls become small as the mill is operated and migrates toward the discharge.
- New large balls are added periodically.
- Large particles are broken by large balls, and small particles by tiny balls.
- This design increases efficiency.

5.3.5.4 *Advantages of ball mill*

- Suitable for dry or wet material.
- Possible for grinding in a closed machine, with no dust flying.
- Operation is reliable, cheap, and accessible.
- It can be not only intermittent operation but also continuous operation.
- Instead of air, inert gas can smash the explosive materials.
- Milling toxic materials is possible as it is a completely enclosed form.
- It is also suitable for milling highly abrasive materials.

5.3.5.5 *Disadvantages of ball mill*

- Vibration and noise for bulky material grindings.
- Energy consumption is relatively high with low efficiency.
- Contamination of the product may occur due to wear and tear of the balls and the casing.
- Operation time for grinding is relatively long for a certain amount of material.
- Cleaning is somewhat tricky as it is a closed vessel.

5.3.5.6 *Applications of ball mill*

- Ball mills are used for grinding coal, pigments, and feldspar for pottery, minerals, glass, advanced ceramics, metal oxides, solar cell and semiconductor materials, nutraceuticals, and pharmaceuticals materials down to 1 μm or less.
- Ball mills are used for the size-reducing or milling of hard materials such as minerals, glass, advanced ceramics, metal oxides, and solar cells.

5.3.6 Fluidized Energy Mill

5.3.6.1 *Working principle*

- A fluidized energy mill (micronized or jet mill) consists of a hollow toroid of a 20–200 mm diameter with a loop height of 1.2–2.4 m, as shown in Figure 5.8.
- It operates by particle impaction and attrition.
- A fluid or milling gas, usually air or inert gas, is injected as a high-pressure jet through nozzles at the bottom of the loop.
- The powder particles in the mill are accelerated to high velocity for which the kinetic energy of the air or gas and the turbulence causes interparticle (particle-particle collision) and particle-wall collision, which results in particle size reduction at capacity up to 6,000 kg/h.

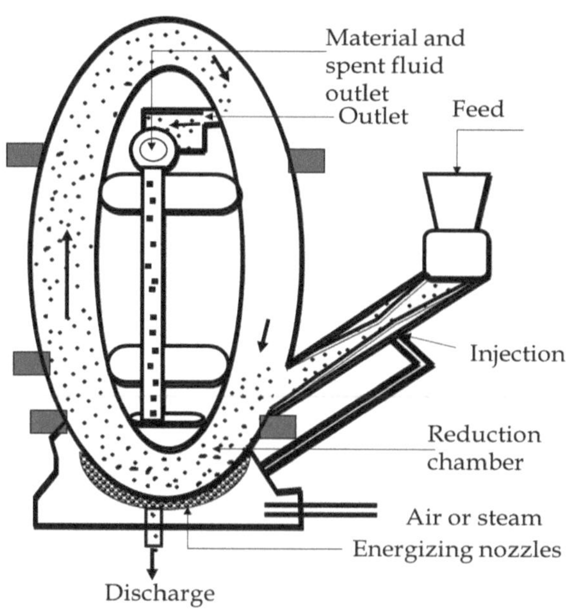

Figure 5.8 Schematic of fluidized energy mill.

- The suspending gas (6–9 kg of air/kg of product) is usually the compressed air or superheated steam admitted at a pressure of 7 atm through the energizing nozzles.

5.3.6.2 *Advantages of fluidized energy mill*

- The mill has no moving parts, and contamination due to the wear of parts is minimal.
- Optimizing energy due to interparticle-induced collision.
- Thermolabile materials can be milled with little degradation.

5.3.6.3 *Disadvantages of fluidized energy mill*

- Formation of aggregates or agglomerates after milling is undesirable.
- The generation of amorphous content due to high energy impact reduces the efficiency.

5.3.6.4 *Application*

- Fluidized energy is used in milling thermolabile materials, frits, kaolin, zircon, titanium, and calcium, alumina.
- It is chosen when a higher degree of drug purity is required.

5.3.7 Jet Mill

5.3.7.1 *Working principle*

- The air jet mill employs compressed air, gas, or high-pressure superheated steam to reduce size.
- The air jet mill works on fluid energy by creating a vortex inside the milling chamber.
- Material is fed into this vortex along an engineered tangent circle and accelerates at near supersonic speeds, as shown in Figure 5.9.
- High-speed rotation ensures the material-to-particle-on-particle impact to produce smaller particles.
- The centrifugal force drives large particles toward the perimeter, and fine particles move toward the center, where they exit through the vortex finder.

5.3.7.2 *Advantages*

- Suitable as it offers one-step grinding in a continuous or batch process with no dead zones to trap material.
- These mills have no moving parts, eliminating contamination from contact with external grinding media.

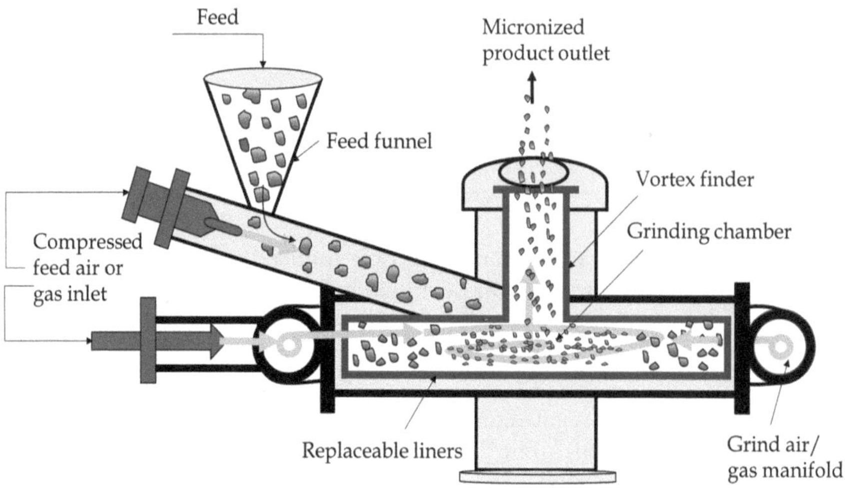

Figure 5.9 Schematic of jet mill.

- No additional heat is generated in grinding, making it the ideal process for grinding heat-sensitive and fragile materials.
- It is used in most bulk powder processing applications requiring sub-micron particle size.

5.3.7.3 *Applications*

- The air jet mills (fluid energy mills) are widely used in the manufacture of drugs and pharmaceuticals, organic and inorganic chemicals, agrochemicals, pigments, food color synthetic dyes, minerals, herbicides and pesticides, thermoplastics, and others like heat-sensitive, corrosive, and abrasive materials.

5.4 FACTORS AFFECTING THE CHOICE OF EQUIPMENT

- The choice of equipment depends on several factors, such as:
  - Stressing mechanism
  - Size of feed and product
  - Material properties
  - Carrier medium
  - Mode of operation
  - Capacity of equipment
  - Combination with other unit operations

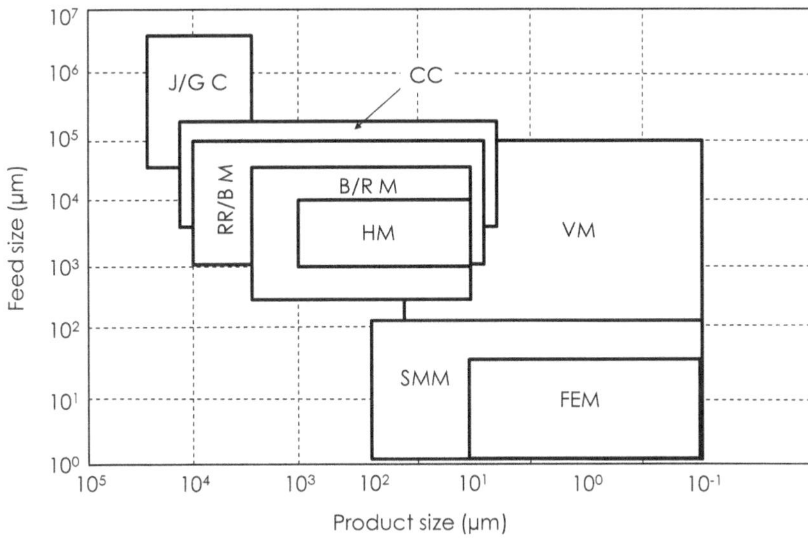

Figure 5.10 Equipment selection map based on product and feed sizes (Chauruka et al., 2015). J/G C, Jaw and gyratory crusher; CC, Cone crusher; RRBM, Ring roller and bowl mills; B/R M, Ball and rod mills; HM, Hammer mills; VM, Vibratory mills; SMM, Stirred media mills; FEM, Fluid energy mills.

- Based on the size of the product particle, the selection of equipment and the respective size-reducing mechanism are shown in Table 5.3.
- The suitability of equipment based on feed and product size can also be judged as per the map shown in Figure 5.10.

Table 5.3 Selection of equipment as per size of product particle

Size Range of Product	*Suitable Equipment*
Down to 3 mm	Crushers, Table mills, Edge runner mills
3 mm to 50 µm	Ball mills, Rod mills, Pin mills, Tube mills, Vibration mills, Hammer mills
<50 µm	Sand mills, Pearl mills, Colloid mills, Fluid energy mills
Size Range of Product	*Size-Reducing Mechanism*
1–0.1 m	Coarse crushing
0.1 m	Crushing
1 cm	Fine crushing, coarse grinding
1 mm	Intermediate grinding, milling
100 µm	Fine grinding
10 µm	Ultrafine grinding

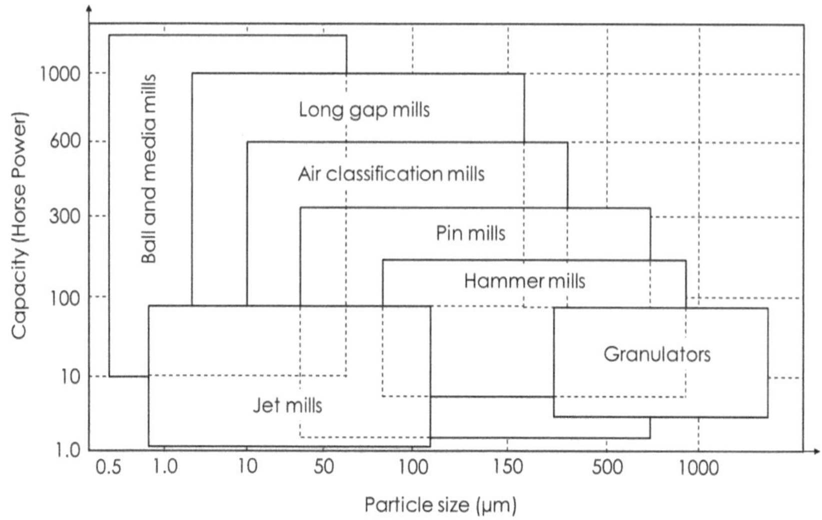

Figure 5.11 The choice of equipment as per their capacity based on product size (Neikov et al., 2009).

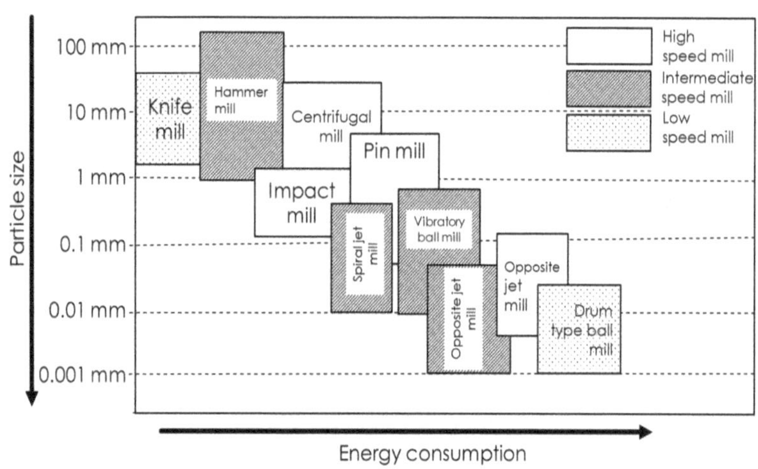

Figure 5.12 Milling equipment based on energy consumption and particle size (Mayer-Laigle et al., 2018).

- A vast range of mill-type equipment is available commercially, and the choice of equipment is based on various factors, as discussed earlier. The suitability of the equipment as per their capacity can be analyzed, as shown in Figure 5.11.
- The classification of different milling equipment based on energy consumption and particle size is shown in Figure 5.12.

IMPORTANT POINTS TO REMEMBER

- The choice of equipment for a given size reduction operation is influenced by the nature of the material, quantity, and size of the material to be handled.
- The crusher is distinguished as primary crushing (up to 100–300 mm), secondary crushing (25–100 mm), and pulverization (5–25 mm).
- The capacity of the gyratory cone crusher is independent of the compressive strength of the material crushed.
- This speed tumbling mill is usually 50%–75% of the critical speed.
- The critical speed of a ball mill is $n_c = \frac{1}{2\pi}\sqrt{\frac{g}{(R-r)}}$.
- Colloid mills work on the rotor-stator principle.
- The energy requirement for the unit mass of material is highest for fluid energy mills among jaw crushers, hammer mills, rod mills, ball mills, and fluid energy mills.
- Size reduction equipment in decreasing order of the average particle size produced by each is jaw crusher>ball mill>fluid energy mill.
- Size reduction of coarser, hard solids using a crusher is accomplished by compression.
- The power required for size reduction in crushing is proportional to the material's surface energy.
- To produce fine solid particles between 5 and 10 μm, the appropriate size-reducing equipment is a fluid energy mill.

SELF-TESTING QUESTIONS

I. Multiple-Choice Questions

Q1. Which one of the following is a fine crusher?
- (a) Ball mill
- (b) Pin mill
- (c) Disk crusher
- (d) All of the above

Q2. Which of the following is NOT a type of size reduction equipment?
- (a) Crushers
- (b) Grinders
- (c) Cutting Machines
- (d) Centrifuge

Q3. Which is the correct statement for universal crusher?
- (a) The swing jaw is fixed at an intermediate position
- (b) The swing jaw is fixed at the upper position
- (c) The swing jaw is fixed at the lower position
- (d) None of these

Q4. Which of the following are ultrafine grinders?
(a) Attrition mills
(b) Fluid energy mills
(c) Gyratory crusher
(d) Hammer mills

Q5. Which size reduction equipment is best for impact mechanism?
(a) Crushers
(b) Grinders
(c) Ultrafine Grinders
(d) Cutting machines

Q6. What mechanism do cutting machines use for size reduction?
(a) Shearing
(b) Impact
(c) Laser
(d) Both (a) and (b)

Q7. Where is the position of the pivoting of the swing jaw in a dodge crusher?
(a) The swing jaw is fixed at the upper position
(b) The swing jaw is fixed at the lower position
(c) The swing jaw is fixed at the intermediate position
(d) None of the above

Q8. The capacity of a roll crusher does not depend on ____________________.
(a) Size of the feed
(b) Size of the finished product
(c) Type of material
(d) Roll speed

Q9. Pick up the correct statement:
(a) In blake crusher, the swing jaw is fixed at the upper position
(b) In dodge crusher, the swing jaw is fixed at the lower position
(c) In universal crusher, the swing jaw is fixed at an intermediate position
(d) All of the above

Q10. Which of the following is a coarse crusher?
(a) Smooth-roll crusher
(b) Toothed-roll crusher
(c) Gyratory crusher
(d) Tube mill

Q11. Choose the correct statement among the following:

	Size Range of Product	*Term Used*
a.	1 cm	Coarse crushing.
b.	1 mm	Fine grinding.
c.	10 μm	Ultrafine grinding.
d.	100 μm	Intermediate grinding.

Q12. In which of the following size reduction equipment, mechanical pressure is achieved by the two jaws of the crusher of which one is fixed while the other reciprocates?
(a) Jaw crusher
(b) Ball mill
(c) Hammer mill
(d) None of the above

Q13. Which of the following types of size reduction equipment is correctly matched with its mechanism?
(a) Crushers – Compression
(b) Grinders – Shearing
(c) Ultrafine grinders – Impact
(d) Cutting machines – Impact

Q14. Which of the following is not a type of jaw crusher?
(a) Blake crusher
(b) Universal crusher
(c) Roll crusher
(d) Dodge crusher

Q15. The term used for particle size reduction that does not match with the size range of the product is________________.
(a) Coarse crushing: 1–0.1 m
(b) Fine crushing: 1 mm
(c) Crushing: 0.1 m
(d) Milling: 1 mm

Q16. At a speed below critical value, the capacity of the jaw or gyratory crusher varies with speed as ________________.
(a) The capacity varies directly with the speed of the machine
(b) The capacity varies inversely with the speed of the machine
(c) The capacity does not vary with the speed of the machine
(d) None of these

Q17. At a speed above a critical value, the capacity of the jaw or gyratory crusher varies with speed as ________________.
(a) The capacity varies directly with the speed of the machine
(b) The capacity varies inversely with the speed of the machine
(c) The capacity does not vary with the speed of the machine
(d) None of these

Q18. In a toothed-roll crusher, it is designed to reduce larger feed sizes to the desired product size while producing a considerably lower percentage of fines at a reduction ratio____________________.

(a) 1:2
(b) 5:2
(c) 6:1
(d) 1:6

Q19. The capacity of commercial hammer mills ranges from ____________________.

(a) 60 to 240 kg of solid per kilowatt hour of energy consumed.
(b) 5 to 50 kg of solid per kilowatt hour of energy consumed.
(c) 0 to 4 kg of solid per kilowatt hour of energy consumed.
(d) 250 to 300 kg of solid per kilowatt hour of energy consumed.

Q20. The operating speed of the tumbling mill is usually____________________.

(a) 50%–75% of the critical speed
(b) 75%–100% of the critical speed
(c) 25%–50% of the critical speed
(d) 0%–25% of the critical speed

Answers to multiple-choice questions.

Q1. Ans. (a)	Q8. Ans. (a)	Q15. Ans. (a)
Q2. Ans. (d)	Q9. Ans. (d)	Q16. Ans. (a)
Q3. Ans. (a)	Q10. Ans. (c)	Q17. Ans. (b)
Q4. Ans. (b) and (d)	Q11. Ans. (c)	Q18. Ans. (c)
Q5. Ans. (b)	Q12. Ans. (a)	Q19. Ans. (a)
Q6. Ans. (d)	Q13. Ans. (a)	Q20. Ans. (a)
Q7. Ans. (b)	Q14. Ans. (c)	

II. Objective Type Questions

1. What are the criteria to be considered for choosing equipment for size reduction?
2. Classify the size reduction equipment based on particle feed size.
3. What are the different types of crushers? Differentiate based on their principle of work.
4. What is the working principle of gyratory crusher?
5. What is the limiting size of particles that the rolls can nip depending on the coefficient of friction between the particle and the roll surface?
6. What is the working principle of a hammer mill?
7. What is a tumbling mill? How can the critical speed of the ball mill be calculated?
8. How does a fluidized energy mill work?
9. Explain the working principle of the colloid mill and its advantages.
10. How does a jet mill work?

III. Numerical Problems

Q1. A material of density $2.8 \times 10^3 kg/m^3$ and porosity 0.20 is crushed by a jaw crusher of jaw width of 0.16 m. The gap between the opening area of the jaws is 0.15 m. If the jaw crusher is operating at a speed of 300 rpm, what is the capacity of the jaw crusher? [Ans. 154.80 t/h]

Q2. A material of density $2.8 \times 10^3 kg/m^3$ and porosity 0.20 is crushed by a jaw crusher. The inner width of the crusher is 0.16 m with an open side setting of 0.11 m and jaw throw of 0.15 m. The angle of the nip is found to be 15°. The material constant is to be considered as 1.5. If the jaw crusher is operating at a speed of 300 rpm, what is the capacity of the jaw crusher in m^3/h as per Broman's (1984) correlation? [Ans. 266.02 m^3/h]

Q3. Find the capacity of the jaw crusher if the density of the material of particles is 3,600 kg/m^3, the porosity of the particles is 0.3, the area of swings is 0.021 m^2, the jaw width is 0.14 m, and it makes 250 swings per minute. [105.84 t/h]

Q4. If the gap between jaws in a jaw crusher is 6 mm, what would the jaw crusher's normal operating speed be? [280 rpm]

Q5. A crusher is designed in such a way that the width of the jaw (w) is 12 cm, the closed side setting (S) is 20 mm, and the gap (G) of the crusher jaws is 20 cm. The crusher is to be used for crushing a hard material, which is screened by a scalping screen with an opening size of 0.508 mm. Calculate the throw of the crusher using Sastri's (1984) concept. [Ans. 0.025 m]

Q6. A crusher is designed in such a way that the width of the jaw (w) is 12 cm, the closed side setting (S) is 20 mm, and the gap (G) of the crusher jaws is 20 cm. The crusher is to be used for crushing a hard material, which is screened by a scalping screen with an opening size of 0.508 mm. Calculate the critical speed of the crusher using Sastri's (1984) concept. [Ans. 231.38 rpm]

Q7. A crusher is designed in such a way that the width of the jaw (w) is 12 cm, the closed side setting (S) is 20 mm, and the gap (G) of the crusher jaws is 20 cm. The crusher is to be used for crushing a hard material, which is screened by a scalping screen with an opening size of 0.508 mm. Calculate the capacity of the crusher using Sastri's (1984) concept. The density of the material is 3.6 t/m^3. [Ans. 1.87 t/h]

Q8. A crusher is designed in such a way that it would be operating at a certain speed where its geometrical dimensions are suggested as the width of the jaw (w) is 10 mm, closed side setting (S) is 17 cm, and the gap (G) of the crusher jaws is 75 cm. The crusher is to be used for crushing a hard material of an average size and density of 2.50 t/m^3, which was screened by a scalping screen with an opening size of

0.084 mm. What is the ratio of critical speed to the operating speed of the crusher? [use Sastri's (1984) concept] [Ans. 1.18]

Q9. A crusher is designed in such a way that it would be operating at a certain speed where its geometrical dimensions are suggested as the width of the jaw (w) is 10 mm, closed side setting (S) is 17 cm, and the gap (G) of the crusher jaws is 75 cm. The crusher is to be used for crushing a hard material of an average size and density of 2.50 t/m^3, which was screened by a scalping screen with an opening size of 0.084 mm. Calculate the crusher's capacity (t/h) using Sastri's (1984) concept. [Ans. 22.07 t/h]

Q10. For a ball mill of 1,500 mm in diameter, 80 mm in diameter steel balls are used for grinding. Now, the mill is run at 25 rpm for the material to be ground. At what operating speed will the mill be run if 60 mm balls replace the 80 mm balls, with all the other conditions remaining the same? [Ans. 24.82 rpm]

Q11. If crushing rolls, 1 m in diameter, is set so that the crushing surfaces are 30 mm apart and the angle of nip is 30°, what is the maximum particle size that should be fed to the rolls? If the actual capacity of the machine is 20% of the theoretical, calculate the throughput in kg/s when running at 2.5 Hz if the working face of the rolls is 0.5 m long and the bulk density of the feed is 2,600 kg/m^3. [Ans. 19.5 kg/s]

Q12. In a smoothed-roll crusher, two rolls with a diameter of 60 cm are used for crushing a particle of a specific size where the particles pass through the crusher roll gap of 3 cm. If the coefficient of friction is 0.4, what would be the average radius of particles in the feed? [Ans. 0.039 m]

Q13. A crushing roller has rolls of 150 cm in diameter by 50 cm width of face. The crushing roller is set so that the crushing surface is 1.25 cm apart at the narrowest point. The manufacturer recommends 100 rpm as the roll speed. The crushing roller to crush a rock with a specific gravity of 2.35 and the angle of nip is 30°. What is the maximum permissible size (radius in cm) of the feed particle that was used for its size reduction? [Ans. 3.29 cm]

Q14. In a ball mill of 2,000 mm in diameter, 100 mm in diameter steel balls are used for grinding. The mill is run at 15 rpm for the material being ground. At what operating speed will the mill have to be run if 50 mm balls replace the 100 mm balls, with all the other conditions remaining the same? [Ans. 22.99 rpm]

Q15. In a smoothed-roll crusher, a feed (brass) equivalent to 4 cm in diameter spheres is fed through the roll gap of 2 cm to crush the feed into a desired size. The coefficient of friction between feed particles and the surface of the roll made of cast iron is 0.30. What is each roll's size (diameter in cm) used in the roll crusher to reduce the feed size? [Ans. 13.44 cm]

Q16. A roll crusher is to be designed to produce particles of an average diameter of 20 mm from a feed material of size approximately 80 mm. The gap of the roll in the roll crusher is to be fixed as equal to the product size. What would be the design value of the angle of the nip if the roll diameter of the crusher is considered as 25 cm? [Ans. 31.31°]

Q17. For the operation of particle size reduction, ceramic material was used for grinding in a ball mill with a diameter of 1,250 mm and a steel ball with a diameter of 25 mm. When the mill is run at 0.9555 Hz, it faces certain issues. However, it works properly when the mill is run at $0.975/x$ Hz. What would be the value of x? [Ans. 1.5]

Q18. A roll crusher reduces the material size at continuous and steady feeding conditions. The roll crusher is designed with a roller width and diameter of 250 and 400 mm, respectively. The crusher has a gap of 6 mm between rolls. If the rolls rotate at a speed of 400 rpm, find the capacity of the roll crusher. The density of the material is 3 tons/m^3. [Ans. 135.784 t/h]

Q19. What is the critical speed of a ball mill of 1.20 m diameter filled with an 8 cm ball? [Ans. 394.47 rpm]

Q20. A ball mill 1.60 m in diameter is being run at 0.8 Hz, and it is found that the mill is not working satisfactorily for a required size reduction operation because the actual rotational speed of the mill is much larger than the critical speed of the mill. The operating speed was 75% of the critical speed. What is the particle size that was reduced to its desired size? [Ans. Radius = 10.9 cm]

Q21. A roll crusher is used to reduce the size of the feed material to a size equivalent to spheres 10 cm in diameter. The crusher has a roll gap of 6 cm. If the coefficient of friction of the material with the roller solid is 0.2, what would be the roll crusher's radius of rolls considered for the design? [Ans. 0.981 m]

Q22. A roll crusher is to be used for crushing material of a maximum size where the roll crusher has a roll gap equal to 5 cm and a roll diameter of 50 cm. What would be the maximum particle size that the roll crusher can nip? [Ans. 3.5 cm]

Q23. A task is given to reduce the size of a charcoal of mean size 50 mm whose bulk density is 2,610 kg/m^3 to a product mean size of 20 mm by a hammer mill. The design specification of the hammer mill is given as follows. What is the mean residence time [h] and the maximum crushing capacity of the hammer mill? [Ans. 9.96×10^{-4} h and 14.532 t/h]

NOMENCLATURE

a	angle of nip [rad.] defined in Equation (5.2)
a	half of the roll's gap [m] defined in Equation (5.15)
a	parameter in the crusher per stroke [m] defined in Equation (5.4)
A	area of swing [m^2]
b	inner width of crusher [m]
C	compressive force [N]
D_e	sphere equivalent diameter of feed [mm]
d_r	diameter of roll [m]
d_v	sphere equivalent diameter of products defined in Equation (5.20) [mm]
h_v	vertical depth between jaws [m]
k	material constant [-]
L	length of the mill [m] defined in Equation (5.20)
L	roll's gap=$2a$ [m] defined in Equation (5.17)
l	width of a hammer in the rotational direction [m]
m	mass of the ball [kg]
n	distribution constant of feed [-] defined in Equation (5.20)
n	rotational speed of the ball mill [rps] defined in Equation (5.21)
N	speed of the crusher [rpm]
n_c	critical speed [rps]
N_j	number of swings per minute [min^{-1}]
N_{op}	operating speed of the crusher in [rpm or strokes/min]
N_r	number of hammers [-]
Q_m	theoretical capacity of a jaw crusher [kg/s]
Q_v	volumetric throughput (capacity) [m^3/s]
$Q_{S,m}$	the capacity of a jaw crusher as per Sastri (1994), [kg/h]
r	distance from the top of a hammer to the center of the rotational axis [m] defined in Equation (5.19)
r	radius of the ball [m] defined in Equation (5.21)
R	radius of the ball mill [m]
r_p	particle radius [m]
R_r	roll's radius [m]
S	closed side setting [m] as shown in Figure 5.7 defined in Equation (5.3)
S	jaw throw [m] defined in Equation (5.2)
S_c	opening of the scalping screen [m]
S_{os}	open side setting [m]
S_r	opening area ratio of the screen (%)
t	mean residence time defined in Equation (5.18) [h]
T	throw parameter [m]
w	width of jaws [m] defined in Equation (5.3)
w	width of the roll [m] defined in Equations (5.1 and 5.17)

GREEK LETTERS

θ angle of nip [degree] defined in Equations (5.13 and 5.15)
τ breakage characteristic time [h]
μ coefficient of friction [-]
ω rotational speed of the hammers [rad/s] defined in Equation (5.20)
ω speed of roll [rps] defined in Equation (5.17)
ε porosity of particles [-]
ρ_p density of particle or material [kg/m^3] defined in Equations (5.1 and 5.17)

SUBSCRIPTS

c scaling screen, critical
e equivalent
m mass
op operating
os open side
p particle
r roll, ratio
S Sastri
v sphere equivalent, vertical, volumetric

REFERENCES

Broman, J., Optimizing capacity and economy in jaw and gyratory crushers, *Eng. Min. J.*, 185, 69–71, 1984.

Chauruka, S. R, A. Hassanpour, R. Brydson, K. J. Roberts, M. Ghadiri, and H. Stitt, Effect of mill type on the size reduction and phase transformation of gamma alumina, *Chem. Eng. Sci.*, 134, 774–783, 2015.

Endoh, H., Estimation of maximum crushing capacity of hammer mills, *Adv. Powder Technol.*, 3 (4), 235–245, 1992.

McCabe, W. L., J. Smith, and P. Harriot, *Unit Operations of Chemical Engineering*, 7th Edition, McGraw-Hill, International Edition, 2017.

Neikov, O. D., S. Naboychenko, I. B. Mourachova, V. G. Gopienko, D. V. Frishberg, and D. V. Lotsko, *Handbook of Non-ferrous Metal Powders: Technologies and Applications,* Elsevier, 2009.

Richardson, J. F., J. H. Harker, and J. R. Backhurst, *Chemical Engineering*, Vol. 2, *Particle Technology and Separation Processes, A volume in Chemical Engineering Series*, 5th Edition, Elsevier, 2003.

Rose, H. E. and J. E. English, Theoretical analysis of the performance of jaw crushers, *IMM Trans.*, 76, C32, 1967.

Sastri, S. R. S., Capacities and performance characteristics of jaw crushers, *Min. Metall. Explor.*, 11, 80–86, 1994.

Wills, B. A. and J. A. Finch, *Wills' Mineral Processing Technology: An Introduction to the Practical Aspects of Ore Treatment and Mineral Recovery*, 8th Edition, Butterworth-Heinemann, Elsevier, Chapter 6, 2016. pp. 123–146.
Wills, B. A. and T. Napier-Munn, *Wills' Mineral Processing Technology*, Preface to 7th Edition, Butterworth-Heinemann, 2006.
Woodcock, C. R. and J. S. Mason, *Bulk Solids Handling: An Introduction to the Practice and Technology*, Springer, 1987.

Chapter 6

Energy for Size Reduction

Engineering stimulates the mind. Kids get bored easily. They have got to get out and get their hands dirty; make things, dismantle things, fix things. When the schools can offer that, you'll have an engineer for life.

—*Bruce Dickinson (airline pilot, musician)*

6.1 ENERGY EFFICIENCY OF SIZE REDUCTION

- The material is distorted and strained under applied stress.
- The energy required to strain is stored as mechanical energy until the ultimate strength is reached.
- When a material breaks up into pieces, the surface area increases by the energy stored, liberating excess energy as heat.
- The size reduction efficiency (η_c) is a function of surface energy creation by crushing or grinding, which can be expressed as

$$\eta_c = \frac{\text{Surface energy created by crushing}}{\text{Energy absorbed by the material}} = \frac{e_s(a_{s,p} - a_{s,f})}{E_s} \tag{6.1}$$

where e_s=surface energy per unit area (J/m^2), a_s=specific surface area (m^2/kg), and E_s=energy absorbed by unit mass of solid (J/kg). The subscripts "p" and "f" denote the product and feed, respectively.

- Energy absorbed by a solid cannot be 100% of shaft energy due to energy loss by conversion to heat energy. Hence, the mechanical efficiency of shaft energy used for absorption by material to achieve size reduction can be calculated as

Mechanical efficiency (η_m)

$$= \frac{\text{Energy absorbed or utilized by the material } (E_s)}{\text{Energy input to the machine } (E)} \tag{6.2}$$

DOI: 10.1201/9781003571025-8

Table 6.1 Standard range of efficiencies for size reduction equipment (Dhodapkar and Theuerkauf, 2011)

Equipment Type	*Typical Efficiency (%)*
Jaw and roll crushers	70–90
Impact crushers	30–40
Roller-ring mills	1–15
Ball mills	5–10
Impact mills	1–10

- Therefore, energy input to the size reduction machine is

$$E = \frac{E_s}{\eta_m} = \frac{e_s(a_{s,p} - a_{s,f})}{\eta_m \eta_c} \tag{6.3}$$

- The power requirement for the size reduction is

$$P = E\dot{m} = \frac{e_s(a_{s,p} - a_{s,f})\dot{m}}{\eta_m \eta_c} \tag{6.4}$$

- The standard range of efficiencies for size reduction equipment is shown in Table 6.1

Example 6.1

A machine is used to reduce the size of a particle in a sample of a solid mass of 100 kg. From a size reduction experiment, it is observed that a total of 0.5 Joule per surface area is required to reduce the size of the total amount of the solid. The difference in surface area production between the product and feed particles of the said sample is 96 m^2. For this reduction, the energy absorbed was found to be 60 J. Find the efficiency of the machine for the reduction of size.

Solution:

$$\eta_c = \frac{\text{Surface energy created by size reduction mechanism}}{\text{Energy absorbed by the material}} = \frac{e_s(a_{s,p} - a_{s,f})}{W_s}$$

$$e_s = 0.5 \text{ J/m}^2$$

$$a_{s,p} - a_{s,f} = 96/100 = 0.96 \text{ m}^2/\text{kg}$$

$$W_s = 60/100 = 0.60 \text{ J/kg}$$

Therefore,

$$\eta_c = \frac{e_s(a_{s,p} - a_{s,f})}{W_s} = \frac{0.5 \times 0.96}{0.60} = 0.80$$

6.2 LAWS OF ENERGY UTILIZATION TO REDUCE THE SIZE

- There are three laws based on which the energy consumption by the machine to reduce the size per unit mass is calculated (Rhodes, 2008). These are shown in Figure 6.1.

6.2.1 Rittinger's Law

- **Basic differential equation:** The energy (dE) required to effect a small change (dL) in the size of a unit mass of feed material is a simple power function of the size (Richardson et al., 2003).

$$\frac{dE}{dL} = -\alpha L^{\beta} [\text{-ve sign for size change from larger to smaller}] \quad (6.5)$$

- If β=–2, then integration gives

$$E = \alpha\left(\frac{1}{L_2} - \frac{1}{L_1}\right) \quad (6.6)$$

where $\alpha = K_R f_c$. The parameter f_c is the crushing strength of the material, and K_R=Ritinger's constant (Rittinger, 1867).

$$E = K_R f_c\left(\frac{1}{L_2} - \frac{1}{L_1}\right) \quad (6.7)$$

As per Equation (6.6), $E \propto \frac{1}{L}$. Again surface area $(S) \propto \frac{1}{L}$. Therefore, $E \propto S$.

- **Remember:** L is here a general term of size. It is the mean diameter of a particle.

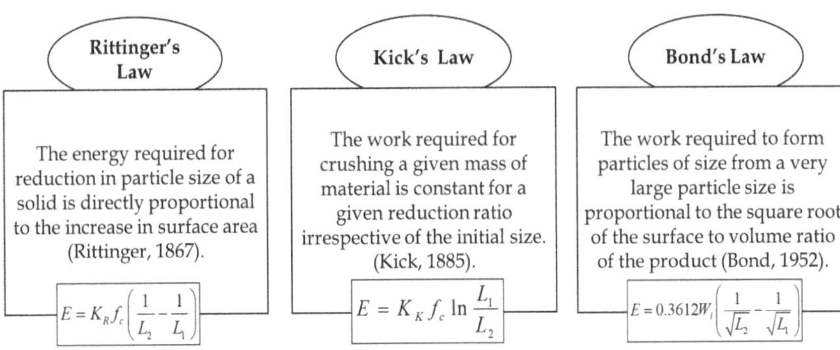

Figure 6.1 Laws of energy consumption for size reduction.

Example 6.2

A jaw crusher uses 50 kW/(kg/s) of energy for crushing material to reduce the particle size from an average feed size of 70×10^{-4}m to an average product size of 20×10^{-4}m. By assuming Rittinger's law, how much energy (in kW/kg) would be required to reduce the same material's average size of 45×10^{-4}m to 10×10^{-4}m?

Solution:

As per Rittinger's law

$$E = K_R f_c \left[\frac{1}{L_2} - \frac{1}{L_1}\right]$$

Therefore, as per problem,

$$50 = K_R f_c \left[\frac{1}{\left(20\times10^{-4}\right)} - \frac{1}{\left(70\times10^{-4}\right)}\right]$$

Therefore,

$$K_R f_c = 0.14 \text{ kW.m/kg}$$

Therefore, $E = 0.14\left[\frac{1}{\left(10\times10^{-4}\right)} - \frac{1}{\left(45\times10^{-4}\right)}\right] = 108.89 \text{ kW/kg}$

6.2.2 Kick's Law

- **Basic differential equation:** The energy (dE) required to effect a small change (dL) in the size of a unit mass of feed material is a simple power function of the size.

$$\frac{dE}{dL} = -\alpha L^{\beta} \qquad (6.5)$$

- From the basic Equation (6.5), if $\beta=-1$, then

$$E = \alpha \ln\frac{L_1}{L_2} \qquad (6.8)$$

- In this case, $\alpha=K_K f_c$ where f_c is the crushing strength of the material and K_K is the Kick's constant (Kick, 1885).

$$E = K_K f_c \ln\frac{L_1}{L_2} \qquad (6.9)$$

- As per Equation (6.9), the energy required is directly related to the reduction ratio L_1/L_2, which means that the energy needed to crush a given amount of material from a 6 cm mm to a 3 cm size is the same as that required to reduce the size from 2 to 1 cm.

Example 6.3

A material consisting originally of 15 cm particles is crushed to an average size of 7.5 cm and requires 25 kJ/kg. What should be the product size if the machine uses 45 kJ/kg energy to crush the same material from 10 cm? (Use Kick's law).

Solution:

Applying Kick's law, as per first condition

$$E = K_K f_c \ln \frac{L_1}{L_2} \Rightarrow 25 = K_K f_c \ln \frac{15}{7.5} \Rightarrow K_K f_c = 36.07 \text{ kJ/kg}$$

So, as per the second condition

$$45 = 36.07 \ln \frac{10}{L_2} \Rightarrow L_2 = 2.87 \text{ cm}$$

6.2.2.1 *Rittinger's vs. Kick's law*

- Rittinger's law is mainly suitable for the part of the process where a new surface is being created and is primarily used for fine grinding, where the increase in surface per unit mass of material is large.
- Kick's law more closely relates to the energy required to effect elastic deformation before a fracture occurs and is more accurate than Rittinger's law for coarse crushing (feed size greater than 50 mm), where the surface produced is considerably less.

6.2.3 Bond's Law

- **Basic differential equation:** The energy (dE) required to effect a small change (dL) in the size of a unit mass of feed material is a simple power function of the size.

$$\frac{dE}{dL} = -\alpha L^{\beta} \qquad (6.5)$$

- This is a law intermediate between Rittinger's and Kick's laws. Based on this, by putting β=−3/2 in basic Equation (6.5), one can get (Bond, 1952)

$$E = 2\alpha\left(\frac{1}{\sqrt{L_2}} - \frac{1}{\sqrt{L_1}}\right) = K_B\left(\frac{1}{\sqrt{L_2}} - \frac{1}{\sqrt{L_1}}\right) = K_B\sqrt{\left(\frac{1}{L_2}\right)\left(1 - \frac{1}{q^{1/2}}\right)} \qquad (6.10)$$

where $q = \frac{L_1}{L_2}$ is called the reduction ratio. $2\alpha = K_B$ is called the Bond's constant.

- **Work index:** The amount of energy required to reduce the unit mass of material from a large infinite particle size (L_1=infinity) to a size of 100 µm (L_2=100 µm=0.1 mm) is called work index (W_i). As per the definition, q=infinity. Then, from Equation (6.10), it can be written as

$$W_i = K_B\sqrt{\left(\frac{1}{0.1}\right)\left(1-\frac{1}{\propto}\right)} = \frac{K_B}{\sqrt{0.1}} = \frac{K_B}{0.3162} \tag{6.11}$$

Or

$$K_B = 0.3162 W_i \tag{6.12}$$

- The work index (W_i) depends on the type of material, as shown in Table 6.2.
- Substituting $K_B = 0.3162 W_i$, in Equation (6.10), one can write:

$$E = 0.3162\, W_i\left(\frac{1}{\sqrt{L_2}} - \frac{1}{\sqrt{L_1}}\right) = 0.3162\, W_i\sqrt{\left(\frac{1}{L_2}\right)\left(1-\frac{1}{q^{1/2}}\right)} \tag{6.13}$$

- The size of the material is considered the size of the square hole (mesh) through which 80% of the material will pass.
- The specific energy requirement for size reduction as per different laws with the particle size is shown in Figure 6.2.

Table 6.2 Work index of some materials

Material	*Specific Gravity*	*Work Index (for Wet Grindings)*[a]	*Material*	*Specific Gravity*	*Work Index (for Wet Grindings)*[a]
Bauxite	2.20	8.78	Cement clinker	3.15	13.45
Cement raw material	2.67	10.51	Clay	2.51	6.30
Coal	1.4	13.00	Granite	2.66	15.13
Coke	1.31	15.13	Gravel	2.66	16.05
Gypsum rock	2.69	6.73	Iron ore (hematite)	3.53	12.84
Limestone	2.66	12.74	Phosphate rock	2.74	9.92
Quartz	2.63	15.87	Slate	2.57	14.30
Trap rock	2.87	19.32			

Source: Allis Chalmers Corp., Solid Processing Equipment Div., Appleton, Wisconsin.

[a] For dry grinding, you need to multiply by 4/5.

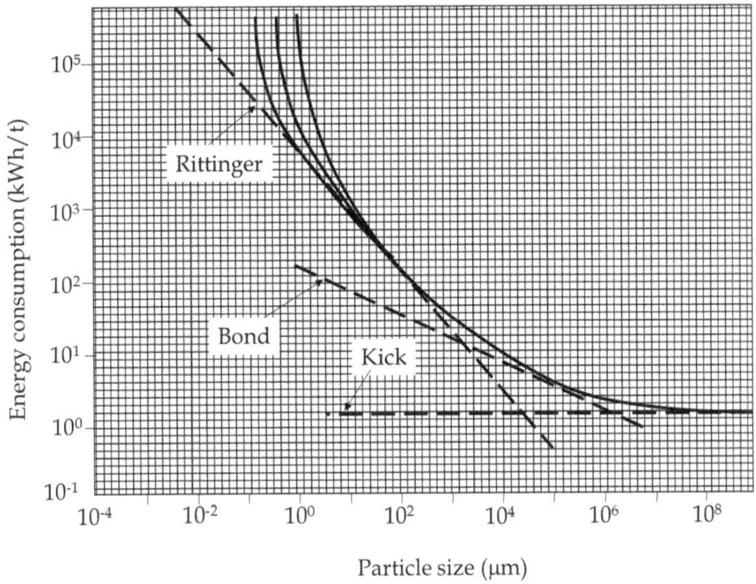

Figure 6.2 Relationship between specific energy requirement and particle size in comminution (Hukki, 1962).

Example 6.4

What is the energy required to crush quartz at a rate of 100 tons/h of size from 20 to 5 mm in diameter? The work index of bauxite is 15.87 kWh/t.

Solution:

Using Bond's law

$$E/m = 0.3612W_i\left[\left(1/\sqrt{L_2}\right)-\left(1/\sqrt{L_1}\right)\right]$$

$$= 0.3612 \times 15.87[(1/\sqrt{5})-(1/\sqrt{20})]$$

Implies, $E/m = 1.28$ kWh/t
Therefore, $E = 1.28$ kWh/t $\times$ 150 tons $= 128.18$ kWh

6.3 APPLICABILITY OF DIFFERENT LAWS

The applicability of the three basic laws of size reduction is given as follows:

Rittinger's law

- It is applied where surface area created is significant-fine grinding.
- Energy input is not very high.
- Applicable to brittle materials undergoing fine milling.
- The theory ignore deformation before fracture.
- It applies to sizes < 100 μm.

Kick's law

- Applied if plastic deformation within the elastic limit.
- Suitable for course grindings in which there is a relatively small increase in surface area per unit mass.
- For compression of large particles Kick's theory is useful.
- It applies to course sizes > 10 mm

Bond's law

- Generally used for hard materials.
- It is useful for rough mill sizing.
- The work index is useful for comparing efficiency of milling operation
- It applies to sizes 100 μm to 10 mm

IMPORTANT POINTS TO REMEMBER

- When a material breaks up into pieces, the surface area increases by the energy stored, liberating excess energy as heat.
- The size reduction efficiency (η_c) is a function of surface energy creation by crushing or grinding.
- Rittinger's law is suitable mainly for the part of the process where a new surface is being created and is primarily used for fine grinding, where the increase in surface per unit mass of material is large.
- The energy required for size reduction is directly proportional to the increase in surface area as per Rittinger's law.
- The work required for size reduction of a given mass of solid is constant for a given reduction ratio as per Kick's law.
- The reduction ratio is the ratio of initial size to final size, not final to initial size.
- The work required to reduce size from a very large particle is proportional to the square root of the surface-to-volume ratio of the product as per Bond's law.
- Kick's law more closely relates to the energy required to effect elastic deformation before a fracture occurs and is more accurate than Rittinger's law for coarse crushing (feed size greater than 50 mm), where the surface produced is considerably less.

- The amount of energy required to reduce the unit mass of material from a large infinite particle size (L_1=infinity) to a size of 100 μm (L_2=100 μm=0.1 mm) is called work index (W_i).
- The work index (W_i) depends on the type of material.
- Rittinger's law deals with measurements of surface areas.
- Kick's law deals with volumes of product particles.
- Bond's theory deals with the lengths of cracks formed.
- Kick's law applies to coarse sizes (>10 mm).
- Bond's theory applies down to 100 μm.
- Rittinger's law applies to sizes <100 μm.

SELF-TESTING QUESTIONS

I. Multiple Choice Questions

Q1. The power required for size reduction in crushing is proportional to ____________________.
   (a) 1/Surface energy of the material
   (b) (1/Surface energy of the material)$^{1/2}$
   (c) Surface energy of the material
   (d) Independent of surface energy of the material

Q2. The work required to form particles of definite size from a very large particle is proportional to the square root of the surface-to-volume ratio of the product. This law is called ____________________.
   (a) Bond's law
   (b) Rittinger's law
   (c) Kick's law
   (d) All of the above

Q3. For assessing the energy requirement for a size reduction, the size of the material is taken as the size of square holes through which ____________________.
   (a) 30% of it will pass
   (b) 50% of it will pass
   (c) 80% of it will pass
   (d) More than 95% of it will pass

Q4. What is the relation between Bond's constant and work index if a material of average size is reduced to 0.1 mm?
   (a) $K_B=0.3162W_i$
   (b) $K_B=0.3162/W_i$
   (c) $W_i=0.3162\ K_B$
   (d) $K_B=\sqrt{0}.3162\ W_i$

Q5. Which of the following statement(s) is false?
(a) Rittinger's law is applicable for elastic deformation before the fracture occurred
(b) Reduction ratio is the ratio of the final particle size to the initial particle size
(c) Both (a) and (b)
(d) None of these.

Q6. Bond's law states that the energy required for size reduction is proportional to the
(a) The square root of the product
(b) Cube of the product
(c) The inverse of the product
(d) The logarithm of the product

Q7. Rittinger's number is a measure of the ____________.
(a) Temperature change during a chemical reaction
(b) Pressure drop in a fluid flow
(c) The energy required for particle size reduction
(d) Heat transfer coefficient of a process

Q8. The constants (K_R, K_K, and K_B) employed in the laws of crushing depend upon the ____________.
(a) Feed material
(b) Type of crushing machine
(c) Both (a) and (b)
(d) Neither (a) nor (b)

Q9. The work index of the material gypsum rock with a specific gravity of 2.69 is ____________
(a) 6.73
(b) 2.20
(c) 2.67
(d) 1.31

Q10. The unit of Kick's constant is ____________.
(a) kWh/kg
(b) kWh/s.kg
(c) kg/s
(d) kW.s/kg

Q11. Which of the following laws is most applicable for coarse crushing?
(a) Rittinger's law
(b) Kick's law
(c) Bond's law
(d) None of the above

Q12. Which of the following laws is most applicable for fine grinding?
(a) Rittinger's law
(b) Kick's law

(c) Bond's law
(d) None of the above

Q13. Crushing efficiency is the ratio of ____________________?
(a) Surface energy is created by crushing to the energy absorbed by the solid
(b) Energy absorbed by the solid to the energy input to the machine
(c) Surface area of the product to the surface area of the feed
(d) None of the above

Q14. What is the value of p for Bond's law $\frac{dE}{dL} = -CL^p$?
(a) −1
(b) −3/2
(c) −2
(d) None of the above

Q15. What is the value of p for Kick's law in $\frac{dE}{dL} = -CL^p$?
(a) −1
(b) −3/2
(c) −2
(d) None of the above

Q16. Define energy consumption by Rittinger's law.
(a) The energy required for the reduction in particle size of a solid is directly proportional to the increase in surface area.
(b) The work required for crushing a given mass of material is constant for a given reduction ratio, irrespective of the initial size.
(c) The work required to form particles of size from a very large particle size is proportional to the square root of the surface-to-volume ratio of the product.
(d) None of these.

Q17. What is true about Kick's law?
(a) The law is based on stress analysis if plastic deformation is within the elastic limit.
(b) It is more accurate than Rittinger's law for coarse crushing.
(c) It is applicable for feed sizes greater than 50 mm.
(d) All of the above.

Q18. Among the following, what is true about the energy consumption by Kick's law?
(a) The energy required for a solid's particle size reduction is directly proportional to the increase in surface area.
(b) The work required for crushing a given mass of material is constant for a given reduction ratio, irrespective of the initial size.

(c) The work required to form particles of size from a very large particle size is proportional to the square root of the surface-to-volume ratio of the product.
(d) None of the above.

Q19. According to Kick's law, the energy required to crush a given amount of material from a 10 mm to a 5 mm size is the ____________ as that needed to reduce the size from 100 to 50 mm.
(a) Half
(b) Same
(c) Double
(d) None of the above

Q20. Which of the following statements is true?
(a) Rittinger's law applies mainly to that part of the process where new surfaces are being created.
(b) Kick's law relates to the energy required to effect elastic deformation before a fracture occurs.
(c) Kick's law is more accurate than Rittinger's law for coarse crushing, where the amount of surface produced is considerably less.
(d) All of these.

Answers to multiple-choice questions

Q1. Ans. (c)	Q8. Ans. (c)	Q15. Ans. (a)
Q2. Ans. (a)	Q9. Ans. (a)	Q16. Ans. (a)
Q3. Ans. (c)	Q10. Ans. (d)	Q17. Ans. (d)
Q4. Ans. (a)	Q11. Ans. (b)	Q18. Ans. (b)
Q5. Ans. (c)	Q12. Ans. (a)	Q19. Ans. (b)
Q6. Ans. (a)	Q13. Ans. (a)	Q20. Ans. (d)
Q7. Ans. (c)	Q14. Ans. (b)	

II. Objective Type Questions

1. Define the size reduction efficiency of the equipment.
2. How is the size reduction efficiency related to specific surface energy?
3. State the three laws based on which the energy consumption by the machine to reduce the size per unit mass is calculated.
4. Write the basic differential equation by which the energy required to effect a small change in the size of a unit mass of feed material.
5. Differentiate the limitations of Rittinger's law, Kick's law, and Bond's law.
6. What is the work index? What is its physical significance?
7. Explain the applicability of different laws for the energy consumption of size reduction.

8. How is energy consumption related to the feed size of the material as per Rittinger's law?
9. How is energy consumption related to the product size of the material as per Bond's law?
10. What is the reduction ratio?

III. Numerical Problems

Q1. A machine is used to reduce the size of a particle in a sample of solid mass of 500 kg. The size reduction experiment shows that a total of 0.8 Joule per surface area is required to reduce the size of the total amount of the solid. The difference in surface area production between the product and feed particles is 100 m^2. For this reduction, the energy absorbed was found to be 90 J. Find the machine's efficiency for the size reduction. [Ans. 89%]

Q2. What would be the energy input to the machine to reduce the size of a material if the change in specific surface area is 0.8 m^2/kg? The surface energy of the material per unit area is 20 J/m^2. Mechanical and crushing efficiencies are 0.6 and 0.9, respectively. [Ans. 29.63 J/kg]

Q3. The power required to crush 100 tons/h of limestone from 100 to 6.25 mm is 170 kW. Using Kick's law, what power is required to crush 100 tons/h of limestone from 50 to 3.125 mm? [Ans. 170 kW]

Q4. A material is crushed in a jaw crusher, and the average size of the particle is reduced from 5 to 1 cm, with an energy consumption of 1.32×10^4 J/kg. What will energy consumption be used to crush the material of an average size of 7.5–1.5 cm, assuming Rittinger's law? [Ans. 8.8 kJ/kg]

Q5. A Blake jaw crusher is used for crushing material, reducing the average particle size from an average feed size of 50×10^{-4} m to an average product size of 10×10^{-4} m while using 20 kW/(kg) of energy. Assuming Rittinger's law, how much energy (in kW/kg) would be required to reduce the same material's average size of 65×10^{-4} to 15×10^{-4} m? [Ans. 12.8205 kW/kg]

Q6. One hundred and fifty tons of material per hour must be crushed using 270 kW of power. Calculate the material's work index if 80% of the feed passes through a 50-mm screen and 80% of the final product passes through a 3 mm screen. What power is required (in kW) for the same feed at 150 tons/h to be crushed to a product such that 80% is to pass through a 1.5-mm screen? [Ans. 418.16 kW]

Q7. A Blake jaw crusher is used for crushing a sample of materials, reducing the average particle size from 50 to 10 mm while using only 13 kW/(kg/s) of power. What would the Kick's constant be in terms of kWh/ton? [Ans. 2.243 kWh/t]

Q8. The energy required per unit mass to grind limestone particles of very large size to 100 μm is 12.7 kWh/ton. What is the energy needed to reduce the particles from an extremely large size to 50 μm (use Bond's law)? [Ans. 17.96 kWh/t]

Q9. It is supposed to crush bauxite in a jaw crusher such that the average particle size is reduced from 60 to 10 mm with an energy consumption of 25.0 kW/kg. What would be the consumption of energy (kW/kg) needed to crush the same material of an average size of 75 mm to an average size of 20 mm? The work index for the bauxite is 8.78 units.

(a) Assuming Rittinger's law applies? (b) Assuming Kick's law applies? (c) Assuming Bond's law? [Ans. (a) 11 kW/kg; (b) 18.44 kW/kg; (c) 10.84 kW/kg]

Q10. A material consisting originally of 100 mm particles is crushed to an average size of 50 mm and requires 10 kJ/kg for this size reduction. Determine the energy required to crush the particles from 10 to 5 mm using Rittinger's law? [Ans. 100 kJ/kg]

Q11. A material consisting originally of 100 mm particles is crushed to an average size of 50 mm and requires 10 kJ/kg for this size reduction. Determine the energy required to crush the particles from 10 to 5 mm using Kick's law. [Ans. 10 kJ/kg]

Q12. A material consisting originally of 100 mm particles is crushed to an average size of 50 mm and requires 10 kJ/kg for this size reduction. Determine the energy required to crush the particles from 10 to 5 mm using Bond's law. [Ans. 31.62 kJ/kg]

Q13. A material consisting originally of 200 mm particles is crushed to an average size of 100 mm and requires 20 kJ/kg for this size reduction. 40 kJ/kg of energy is required to crush the same material from 200 mm to L_2 mm. Determine L_2 (as per Kick's law). [Ans. 50 mm]

Q14. A material consisting originally of 200 mm concrete particles is crushed to an average size of 100 mm and requires 40 kJ/kg for this size reduction. Determine Kick's constant if the crushing strength of the material is 2×10^3 kN/m^3. [Ans. 2.89×10^{-6} kJ/kg/Pa]

Q15. A jaw crusher crushes a material such that the average size of the particle is reduced from 10 to 5 mm with the consumption of 10 kW/(kg/s) of energy. Calculate the energy consumption (in kW/(kg/s) to crush the same material of average size 20–10 mm, assuming Rittinger's law applies. [Ans. 5 kW/kg/s]

Q16. The energy consumption needed for some amount of quartz particles which are 2 mm and dry ground in a ball mill to its size reduced to 1 mm is 11.75 [kW·s/kg] (mm)$^{-0.5}$. What is the work index of the quartz material in (kWh/t)? [Ans. 16.70 kWh/t]

Q17. In the food industry, sugar crystals are typically ground from an average Sauter mean diameter of 400 μm to its powder with an average Sauter mean diameter of 80 μm. This size reduction process's net energy consumption rate was 0.55 kWh/ton. What would be the net energy consumption for reducing sugar crystals to 40 μm size by Rittinger's law? [Ans. 1.24 kWh/t]

Q18. A material is crushed in a jaw crusher such that the average size of the particle is reduced from 12 to 6 mm with the consumption of 15 kWh/t of energy. Calculate the energy consumption (in kWh/t) to crush the same material of average size 30–15 mm, assuming Rittinger's law applies. [Ans. 6.0 kWh/t]

Q19. A material is crushed in a jaw crusher such that the average size of the particle is reduced from 12 to 6 mm with the consumption of 5 kW/(kg/s) of energy. Select the correct option regarding energy consumption (in kW/(kg/s)) to crush the same material of average size 30–15 mm, assuming Kick's law applies.

[Ans. For both cases, the reduction ratio is identical. Therefore, as per Kick's law, the energy consumption rate to crush the same material of the same reduction ratio will be the same.]

Q20. A certain crusher takes rock whose average particle diameter is 0.025 m and crushes it into a product whose average particle diameter is 0.018 m at 20000 kg/h. The mill takes 60.28 kW of energy to run at this rate. For the same capacity, if the average particle diameter in the product of size 0.008 m is obtained by consuming 32.79 kW. Which law of energy utilization to reduce the size is applicable? [Ans. 32.79 kWh/t]

Q21. If $P = E\dot{m} = \dfrac{e_s(a_{s,p} - a_{s,f})\dot{m}}{\eta_m \eta_c}$ so, what energy consumption law can be derived for size reduction? where P=power, E=energy per unit mass flow rate, e_s=surface energy per unit area, a_s=specific surface area (area/mass), η=efficiency. [Ans. Rittinger's law]

Q22. It is seen from a size reduction experiment that to get the product of particles, the average ratio of surface to volume is 1.0 (1/mm) from an infinite size. The energy per unit mass was required to be 1 kW/kg for the operation as per Bond's law. Assume 100% efficiency of the machine and the particle is spherical. What should be the Bond's constant (kW·√mm)/kg and product particle size (mm), respectively? [Ans. 1.0]

Q23. A magnetite ore particle is crushed in a jaw crusher, thereby reducing the average particle size from 40 to 10 mm with an energy consumption of 15 kW.s/kg. How much energy is required to consume the same material of the average particle size reduced from 65 to 35 mm? Apply Rittinger's law. [Ans. 2.64 kW.s/kg]

Q24. Using Bond's law, calculate the energy required to crush bauxite at a rate of 120 tons/h from a of size 30 to 10 mm diameter. The work index of bauxite is 8.78 kWh/t. [Ans. 50.86 kWh]

NOMENCLATURE

a_s specific surface area [m²/kg]
E energy input to the size reduction machine [J]
E_s energy absorbed by unit mass of solid [J/kg]
e_s surface energy per unit area [J/m²]
f_c crushing strength of the material
K_K Kick's constant
K_R Rittinger's constant
K_B Bond's constant
L size of the material [m]
$\dot{m}$ mass flowrate [kg/s]
η_m mechanical efficiency [-]
η_c size reduction efficiency [-]
P power requirement for the size reduction [W]
q reduction ratio [-]
W_i work index

GREEK LETTERS

α coefficient used in Equation (6.5)
β coefficient used in Equation (6.5)
η size reduction efficiency [%]

SUBSCRIPTS

B Bond
c crushing
f feed
i index
K Kick
p product
R Rittinger
s surface, solid, shaft

REFERENCES

Bond, F. C., The third theory of comminution, *AIME Trans.*, 193, 484–494, 1952.
Dhodapkar, S. and J. Theuerkauf, Maximizing performance in size reduction: Expect more in selection and scaleup, prediction of energy consumption and reliability. *Chem. Eng.*, 118 (6), 45–48, 2011, Gale Academic OneFile.
Hukki, R. T., Proposal for a solomnic settlement between the theories of von Rittinger, *Kick and Bond. Trans. AIME*, 223, 403–408, 1962.

Kick, F., *Das Gesetz der Proportionalen Widerstande und Sein Enwendung*, Artuz Felix, 1885.

Rhodes, M., *Introduction to Particle Technology*, John Wiley & Sons Ltd., 2008.

Richardson, J. F., J. H. Harker and J. R. Backhurst, *Chemial Engineering, Vol. 2, Particle Technology and Separation Processes, A Volume in Chemical Engineering Series*, 5th Edition, Elsevier, 2003.

Rittinger, P. R., *Lehrbauch der Aufbereitungskunde*, Ernst and Korn, 1867.

Section III

Size Enlargement

Chapter 7

Introduction to Size Enlargement

> At its heart engineering is about using science to find creative practical solutions. It's a noble profession.
>
> —*Queen Elizabeth II*

7.1 WHAT IS SIZE ENLARGEMENT?

- The definition of size enlargement is given in Figure 7.1.

7.2 WHY IS SIZE ENLARGEMENT SIGNIFICANT?

- The significance of the size enlargement is pointed out as follows:
  - to improve flow properties
  - to reduce caking and lump formation

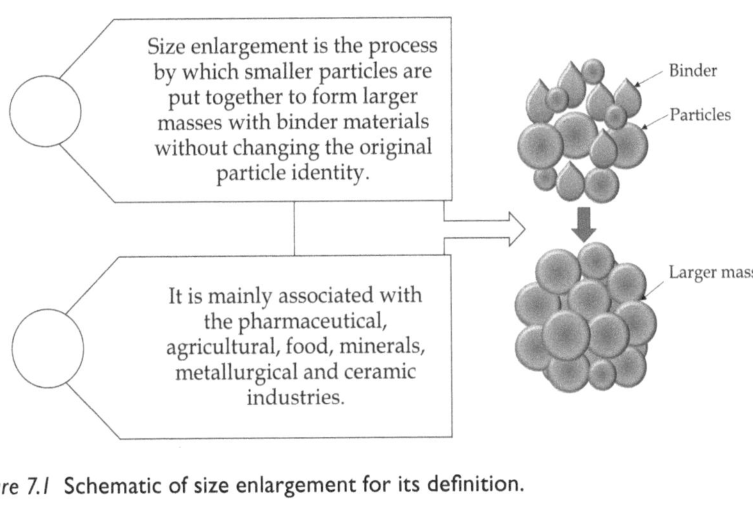

Figure 7.1 Schematic of size enlargement for its definition.

DOI: 10.1201/9781003571025-10

- to increase bulk density for storage
- to create non-segregating mixtures of ingredients
- to provide a defined metered quantity of active ingredient
- to control the surface-to-volume ratio in catalyst supports
- to produce a useful structural form
- to improve product appearance
- to control solubility, porosity, and hardness

7.3 INTERFACIAL BONDING FOR SIZE ENLARGEMENT

- The particles are adhered to by sufficiently strong interfacial bonds (due to interparticle forces) to form granules.
- The strong bond prevents the breakdown of the final dried granules.
- The magnitude of interparticle forces depends on the particles' size, the granule's structure, the moisture content, and the surface tension of the liquid.
- There are five mechanisms responsible for interparticle forces operating during and after agglomeration (Rumpf, 1962). These include:

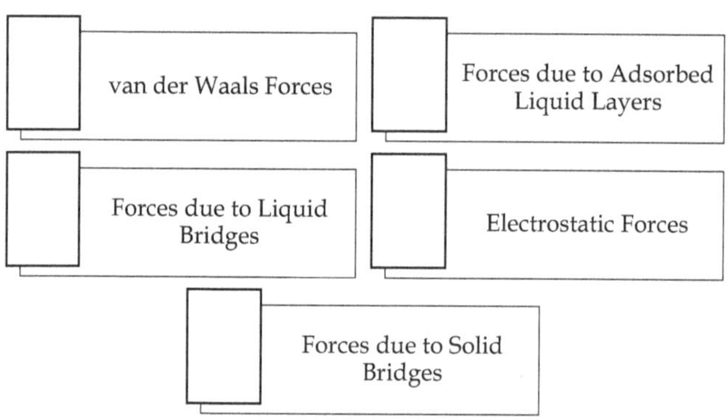

7.3.1 van der Waals Force

- The interparticle attractive force between solids is the van der Waals force.
- The energy of the force is of the order of 0.1 eV with the sixth power of the distance between molecules.
- The van der Waals force range is more prominent than chemical bonds.
- The attractive force as a van der Waals force, as per Hamaker (1937), is usually presented in the form:

$$F_{VW} = \frac{K_H R_1 R_2}{(R_1 + R_2)6r^2} \tag{7.1}$$

where F_{VW}=van der Waals force and K_H=the Hamaker constant. Most of the solids have $K_H = (0.4-4.0)\times 10^{-19}$ J. The constant is a function of liquid wettability, adhesion, friction, adsorption, colloidal stability, polymer flow, and deformation. R_1 and R_2 are the radius of the spheres 1 and 2, respectively, and r is the gap between the spheres.

Example 7.1

Find the van der Waals attractive force (F_{vw} in N) between two spherical particles whose radii are 10 and 50 μm, respectively. The gap between the spheres is 1 μm, and Hamaker's constant (K_H) value is 0.8×10^{-19} J.

Solution:

The attractive force as a van der Waals force as per Hamaker (1937),

$$F_{VW} = \frac{K_H R_1 R_2}{(R_1+R_2)6r^2} = \frac{0.8\times 10^{-19}(10\times 10^{-6})(50\times 10^{-6})}{(10\times 10^{-6}+50\times 10^{-6})6(1.0\times 10^{-6})^2} = 1.11\times 10^{-13}\text{ N}$$

7.3.2 Forces Due to an Adsorbed Liquid Layer

- The bond is also formed between the particle surfaces covered by a layer of adsorbed vapor on their surfaces in the presence of condensable vapor.
- The strength of the bond depends on the area of contact and the tensile strength of the adsorbed layers.
- The thickness and strength of the layers increase with the partial pressure of the vapor in the surrounding atmosphere.
- The adsorbed layer bonding gives way to liquid bridge bonding at a certain critical pressure (Coelho and Harnby, 1978).

7.3.3 Forces Due to Liquid Bridges

- The presence of liquid on the surface of particles affects the interparticle forces:
  - by the smoothing effect on the surface
  - by increasing particle–particle contact
  - by its effect of reducing the interparticle distance.
- These forces form an interparticle liquid bridge, as shown in Figure 7.2.
- According to Newitt and Conway-Jones (1958), there are generally four types of liquid states for liquid bridge formation depending on the proportion of fluid present between groups of particles, which are known as pendular, funicular, capillary, and droplets, as shown in Figure 7.3.
- Increasing the proportion of liquid to particles, the attractive force between particles decreases. This type of state is called funicular.

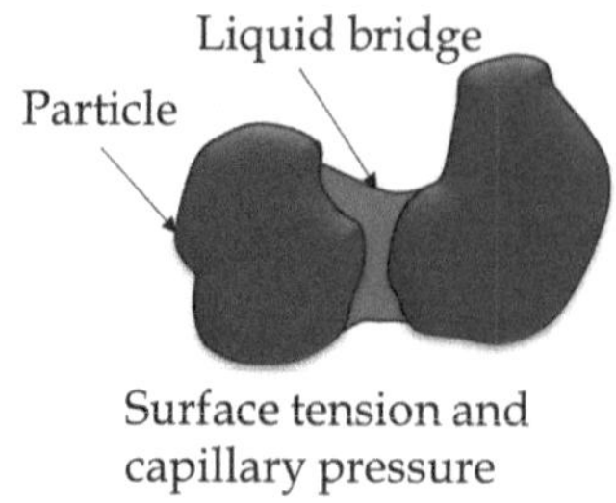

Figure 7.2 Schematic of interparticle liquid bridges under surface tension and capillary pressure.

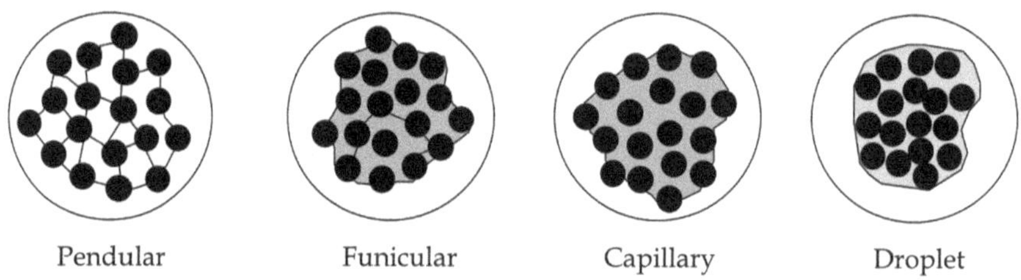

Figure 7.3 Different states for liquid bridge formation (adapted from Newitt and Conway-Jones, 1958).

- When liquid completely fills the interstitial pores between the particles, the granule strength falls due to less curved liquid surfaces and fewer boundaries for surface tension forces to act on. This type of state is called a capillary state.
- If the particles are completely dispersed in the liquid, the strength of the structure is reduced. This phenomenon forms the state of the droplet.
- In the case of a pendular bridge state, strong granules are formed in which the quantity of liquid is not critical but should be less than that required to change into the funicular and capillary states.

7.3.4 Electrostatic Force

- Friction due to interparticle collisions and frequent rubbing of particles against equipment surfaces result in the electrostatic charging of particles.
- The transfer of electrons between the bodies causes the charging.
- As per Coulomb's law, the electrostatic force (Coulomb) between two charged spheres is proportional to the product of their charges q_1 and q_2 at a distance r. The relation is expressed as (Huray, 2010).

$$F_c = \frac{kq_1q_2}{r^2} \tag{7.2}$$

- The proportionality constant (called Coulomb's constant) depends on the permittivity of free space, which is related as

$$k = \frac{1}{4\pi\varepsilon_0}\,[\mathrm{Nm^2/C^2}] \tag{7.3}$$

where ε_0 is the permittivity of the free space $= 8.854 \times 10^{-12}$ Farade/m. In SI units, the constant k has the value $k = 8.99 \times 10^9\ \mathrm{Nm^2/C^2}$ for air.

Example 7.2

What is the electrostatic force between two objects of −7 and −13 μC that are 0.05 m apart in air?

Solution:

$$q_1 = -7 \times 10^{-6}\ \mathrm{C}$$
$$q_2 = -13 \times 10^{-6}\ \mathrm{C}$$
$$k = 8.99 \times 10^9\ \mathrm{Nm^2/C^2}$$
$$r = 0.05\ \mathrm{m}$$

Therefore,

$$F = \frac{kq_1q_2}{r^2} = \frac{8.99 \times 10^9 \times (7 \times 10^{-6}) \times (13 \times 10^{-6})}{0.05^2} = 327.24\ \mathrm{N}$$

7.3.5 Solid Bridges

- Granules formed by liquid bridges are usually not the end product of a granulation process. More permanent bonding within the granule is created by solid bridges formed as the liquid is removed from the original granule, as shown in Figure 7.4.
- Three main solid bridges between particles are (Rhodes, 2008):
  - crystalline bridges

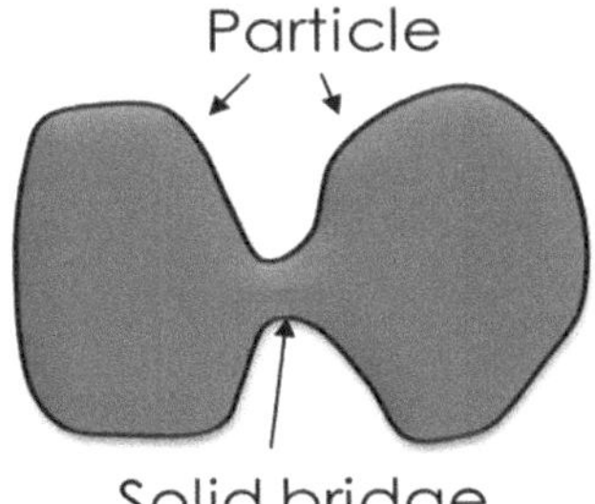

Figure 7.4 Schematic of solid bridges.

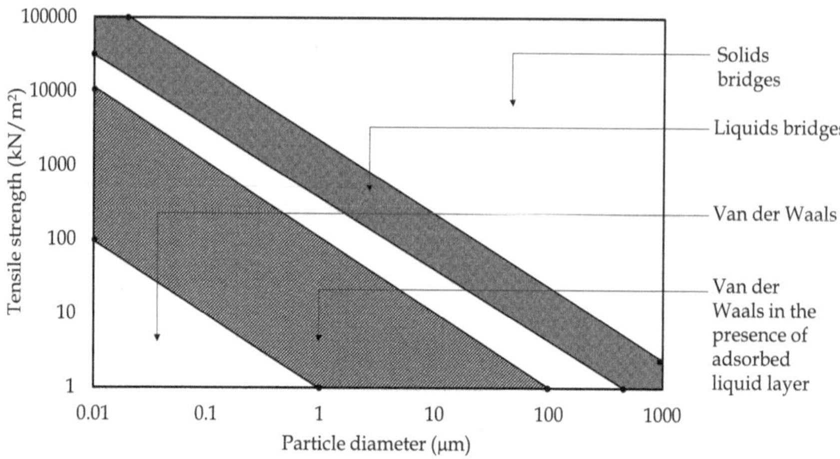

Figure 7.5 Map of the different bonds based on particle size and tensile strength (Rhodes, 2008).

- liquid binder bridges
- solid binder bridges

- If the particles are soluble in the liquid added to create granules, crystalline bridges may be formed when the liquid evaporates.
- The relative magnitude of the different bonds as a function of particle size is shown in Figure 7.5.
  - The van der Waals forces become essential only for particles below 1 μm in size.
  - The adsorbed vapor forces are relevant below 80 μm.
  - The liquid bridge forces are active below about 500 μm.

7.4 EXCIPIENTS FOR SIZE ENLARGEMENT

The particle design and the properties of the particles acquired after granulation depend on the particle size of the drug and excipients.

- Excipients for granulation can be primarily divided into the following categories:
  - Bulking agents
  - Functional additives
  - Other (colorants, coating aids, stabilizers, pH modifiers, and release rate of modifiers).

- Bulking agents or fillers serve to form the core or structure of a dosage form.
- Functional additives include binders, disintegrants, lubricants, colorants, and stabilizing agents.
- The selection of excipients depends on several factors, such as the drug used, the process involved, the formulator, and the cost of the excipient.

7.4.1 Bulking Agents

- Very common filler/bulking agents are as follows:

  - Sugar
  - Lactose
  - Pregelatinized starch
  - Micro-crystalline cellulose
  - Starch
  - Dicalcium phosphate

7.4.2 Functional Additives

- Very common functional additives are as follows:

  - Binders
  - Disintigrants
  - Lubricants
  - Stabilizing agents
  - Colorants

7.4.3 Binders

- Binders used in granulation consist of various sugars and polymers (natural, semisynthetic, and synthetic), as shown as follows:

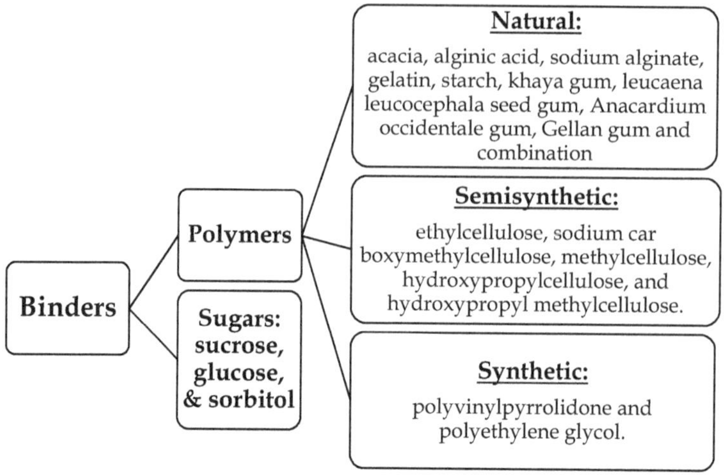

- The binders are used to provide the cohesiveness required for the bonding of the solid particles under compaction to form the required particle size.
- Binders promote size enlargement to produce granules and improve the flowability of the blend.
- The cohesive properties of binders reduce the enlarged particle's friability, durability, and elegance.

7.4.4 Disintegrants

- Disintegrants increase the hydrostatic pressure in the formulation when it comes in contact with water.
- It ensures the breakup of the tablet matrix upon ingestion.
- Some important disintegrants are shown as follows:

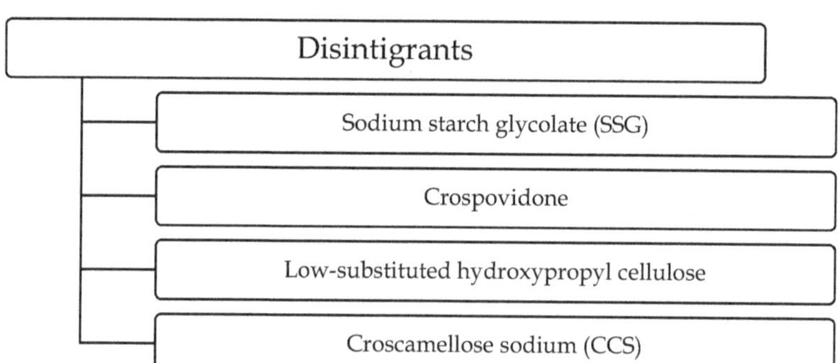

7.4.5 Lubricants

- Lubricants reduce the friction between the tablet and the die metal surface.
- They reduce the ejection force and help ensure the granules are cleanly ejected without cracking or breakage.
- The lubricants that are generally used for size enlargement are as follows:

Lubricants

- Calcium stearate
- Magnesiu-m stearate (most common)
- Wax
- Stearic acid
- Hydroge-nated vegetable oil
- Talc
- Starch

7.5 SIZE ENLARGEMENT RATE PROCESSES

- There are several names of size enlargement process adapted typically in different types of industries, which is named as follows:

Compaction	• The process involving large deformations, strain, nonlinear material behavior and friction.
Granulation	• It transforms a powder into larger entities to end up with an aggregate larger in size and with a porous structure.
Encapsulation	• The process of coating of microscopic particles with another material.
Pelletizing	• Process of compressing or molding a material into the shape of a pellet.
Agglomeration	• Formation of assemblages in a suspension and particle stick to each other and form irregular clusters, flocs or aggregates.

7.6 APPLICATIONS OF SIZE ENLARGEMENT

Various applications of the size enlargement process in different industries are briefed as follows:

7.6.1 Applications in Pharmaceutical Industries

- Size enlargement is used to manufacture tablets. Specifically, granulation is one of the size enlargement processes utilized to improve flow and compression characteristics.
- Agglomeration is used in pharmaceutical products to contain low amounts of dust to provide increased safety when handling and processing toxic materials.

7.6.2 Application in Chemical Industries

- Agglomeration enhances a product by providing desired size distributions or product geometry for better consumer perception.
- It protects the end user from dust and possibly toxic or hazardous effects. An example is pigments, which tend to be dusty and exhibit unfavorable flow and bad metering characteristics.
- They are often microagglomerated. Thus, they became dust-free, smoothly flowing, and withstand handling and shipping without degradation.

7.6.3 Application in Food Industries

- Size enlargement through agglomeration is used to manufacture instant food products.
- This process is used since it decreases the density, where the size of the agglomerates usually ranges from 0.2 to 2.0 mm.
- Thus, instant foods reduce drudgery and time preparation and induce improved functional properties like quick rehydration.

7.6.4 Application in the Animal Feed Industry

- It is used in the animal feed industry to improve flow, storage, transportation, metering, and feeding behavior. It is also used to prevent losses and dust annoyance.
- The size enlargement by agglomeration is used in animal feeding in veterinary medicine to treat sick animals.
- A milk replacer is a typical animal feed processed by agglomeration. This replacer replaces whole milk (e.g., butter) with a cheaper animal or vegetable fat.

7.6.5 Application in Mining and Metallurgical Industries

- Iron ore is preferred as pellets rather than particles since the iron ore as the powder used in a blast furnace is more tightly packed and restricts the airflow.
- Size enlargement through wet agglomeration and post-treatment combine metal or metal-bearing powders by dry mixing to yield formulated alloying components for a dust-free and easily handleable product.

7.6.6 Application in Ceramic Industries

- In the ceramic industry, size enlargement is executed to:
  - improve the storage and handling properties of ceramic raw materials and additives
  - enhance fine powder's flow properties and packing efficiency.

7.6.7 Application in Agricultural Industries

- Size enlargement of fertilizers through granulation is utilized.
- Fertilizer granulation will ensure:
  - friability
  - freedom from caking
  - improved agrotechnical properties of the fertilizer.

7.6.8 Application in Construction Industries

- Particulate solids such as sand and naturally occurring binders such as clay are mixed and formed into rectangular shapes, dried, and later fired to obtain readily usable building material.

7.6.9 Application in Cosmetic Industries

- Size enlargement is used in the cosmetic industry to enhance the product's appearance.
- It is essential to make the particle size to create pearly and shimmering looks.
- The smaller particle yields the less lustrous powder. Thus, larger particles are preferred for larger masses since they give a glitterier luster and are more transparent.

IMPORTANT POINTS TO REMEMBER

- Size enlargement is the process by which smaller particles are put together to form larger masses without changing the original particle identity.
- The particles are adhered to by sufficiently strong interfacial bonds (due to interparticle forces) to form granules.
- The van der Waals force range is more prominent than chemical bonds.
- The van der Waals forces become essential only for particles below 1 μm in size.
- The strength of the bond is dependent on the area of contact and the tensile strength of the adsorbed layers.
- There are generally four types of liquid states for liquid bridge formation depending on the proportion of fluid present between groups of particles, which are known as pendular, funicular, capillary, and droplets.
- The liquid bridge forces are active below about 500 μm.
- Binders used in granulation consist of various sugars and polymers (natural, semisynthetic, and synthetic).
- The cohesive properties of binders reduce the enlarged particle's friability, durability, and elegance.
- Lubricants reduce the ejection force and help ensure the granules are cleanly ejected without cracking or breakage.

SELF-TESTING QUESTIONS

I. Multiple Choice Questions

Q1. ____________________ is the process whereby small particles are agglomerated or otherwise brought together into larger, relatively permanent masses in which original particles can still be distinguished.

(a) Size reduction
(b) Size enlargement
(c) Filtration
(d) Crushing

Q2. Fertilizer granulation will ensure____________________.

(a) Improved agrotechnical properties of the fertilizer
(b) Friability
(c) Both (a) and (b)
(d) Liquid bridges

Q3. Fertilizer granulation will ensure ____________________.

(a) Improved agrotechnical properties of the fertilizer
(b) Friability

(c) Both (a) and (b)
(d) Liquid bridges

Q4. What ingredients are necessary for tablet manufacturing by wet granulation?
(a) Binder
(b) Granulating fluid
(c) Diluent
(d) All of them

Q5. Which of the following can be used as a binder for a granulation process?
(a) Sodium chloride
(b) Calcium carbonate
(c) Toluene
(d) Polyethylene glycol

Q6. Which mechanisms are NOT responsible for interparticle forces after agglomeration?
(a) Electrostatic force
(b) Van der Waals force
(c) Force due to adsorbed liquid layers
(d) Pressure force

Q7. Which of the following is NOT TRUE regarding size enlargement?
(a) It improves flow properties
(b) It decreases the bulk density for storage
(c) It controls the surface-to-volume ratio
(d) It reduces caking lump formation

Q8. Which mechanism is NOT responsible for the interparticle force during agglomeration?
(a) Van der Waals force
(b) Electrostatic force
(c) Gravitational force
(d) Force due to adsorbed liquid layers

Q9. Which of the following is an application of size enlargement in the food industry?
(a) Shrink wrapping
(b) Freeze drying
(c) Puffing
(d) Fermentation

Q10. Which of the following is a disadvantage of using high levels of disintegrants in the granulation process?
(a) Reduced tablet weight variation
(b) Increased tablet hardness

(c) Reduced tablet friability
(d) Increased tablet disintegration time

Q11. Which of the following factors can affect the deformation of granules?
(a) Moisture content
(b) Particle size distribution
(c) Both (a) and (b)
(d) Neither (a) nor (b)

Q12. Lubrication is essential in granulation to ________________.
(a) Improve powder flowability during compression
(b) Reduce tablet disintegration time
(c) Increase drug absorption in the gastrointestinal tract
(d) Enhance granule size distribution

Q13. ______________ is the process by which small particles are put together to form large masses without changing the original particles' identity.
(a) Flotation
(b) Size enlargement
(c) Filtration
(d) Crushing

Q14. Fertilizer granulation will ensure ________________.
(a) Improved agrotechnical properties of the fertilizer
(b) Friability
(c) Both (a) and (b)
(d) Liquid bridges

Q15. In ________________ industries, size enlargement enhances fine powder flow properties and packing efficiency.
(a) Metallurgical
(b) Ceramic
(c) Agricultural
(d) Cosmetic

Q16. Which of the following attractive forces exist between the different layers of graphite?
(a) Van der Waals forces
(b) Columbic forces
(c) Electromagnetic forces
(d) Gravitational forces

Q17. The following are the temperatures of milk in Celsius (°C). Which of the following do you think intermolecular forces are more predominant than thermal energy?
(a) 35
(b) 82
(c) 50
(d) 9

Q18. Typical values of the Hamaker constant (K_H) lie in the range of ________________J.

(a) 10^{-20} to 10^{-18}
(b) 10^{-15} to 10^{-12}
(c) 10^{-20} to 10^{-19}
(d) 10^{-18} to 10^{-16}

Q19. Which of the following is not accurate regarding size enlargement?

(a) It improves flow properties
(b) It decreases the bulk density for storage
(c) It controls the surface-to-volume ratio
(d) It reduces caking lump formation

Q20. Which of the following size enlargement process involves large deformation and strain?

(a) Compaction
(b) Granulation
(c) Encapsulation
(d) Agglomeration

Answers to multiple choice questions

Q1. Ans. (b)	Q8. Ans. (c)	Q15. Ans. (b)
Q2. Ans. (c)	Q9. Ans. (c)	Q16. Ans. (a)
Q3. Ans. (c)	Q10. Ans. (d)	Q17. Ans. (d)
Q4. Ans. (d)	Q11. Ans. (c)	Q18. Ans. (c)
Q5. Ans. (d)	Q12. Ans. (a)	Q19. Ans. (b)
Q6. Ans. (d)	Q13. Ans. (b)	Q20. Ans. (a)
Q7. Ans. (b)	Q14. Ans. (c)	

II. Objective Type Questions

1. Define the size enlargement process.
2. Why is the size enlargement process important?
3. What are the different mechanisms responsible for the interparticle forces operating during and after agglomeration?
4. The intermolecular bond is formed between the particle surfaces covered by a layer of adsorbed vapor on their surfaces in the presence of condensable vapor. How are the thickness and strength of the layers related to the partial pressure?
5. How does the presence of liquid on the surface of particles affect the interparticle forces?
6. According to Newitt and Conway-Jones (1958), explain the different types of liquid states for liquid bridge formation.

7. What are the main solid bridges between particles?
8. Explain the different types of excipients required for size enlargement with examples.
9. Write a short note on different types of binders to use the size enlargement.
10. What are the different rate processes of size enlargement?

III. Numerical Problems

Q1. Find the attractive force (F_{vw} in N) due to van der Waals forces between two spherical particles whose radii are 6 and 8 μm, respectively. The gap between the spheres is 0.1 μm, and Hamaker's constant (K_H) value is 0.8×10^{-19} J. The attractive force, $F_{vw,}$ between two spherical particles as a result of van der Waals forces was derived by Hamaker as: $F_{VW} = \dfrac{K_H R_1 R_2}{(R_1 + R_2) 6r^2}$. [Ans. 4.57×10^{-12} N]

Q2. Consider spheres of radius 1 and 2 μm, respectively, and a plane surface at a distance of 10 nm. The Hamaker constant for the materials in contact is 10^{-19} J. What would be the van der Waals force (in N) between the sphere and the plane surface? [Ans. 1.11×10^{-10} N]

Q3. The force between the two charges is 200 units. What will be the force if the distance between the charges is doubled? [Ans. 50]

NOMENCLATURE

F_{VW} van der Waals force [N]
k Coulomb's constant [Nm^2/C^2]
K_H Hamaker constant [for most solid, $K_H = (0.4 - 4.0) \times 10^{-19}$ J]
r gap between the spheres for van der Waals force or charge particles
R radius of the sphere

GREEK LETTERS

ε_0 permittivity of the free space [F/m]

SUBSCRIPTS

0 free space
vw van der Waals

REFERENCES

Coelho, M. C. and N. Harnby, The effect of humidity on the form of water retention in a powder. *Powder Technol.*, 20 (2), 197–200, 1978.

Hamaker, H. C., The London—van der Waals attraction between spherical particles, *Physica*, 4 (10), 1058–1072, 1937.

Huray, P. G., *Maxwell's Equations*, Wiley, pp. 8, 57, 2010.

Newitt, D. M. and J. M. Conway-Jones, A contribution to the theory and practice of granulation, *Trans. Inst. Chem. Eng.*, 36, 422–441, 1958.

Rhodes, M., *Introduction to Particle Technology*, John Wiley & Sons Ltd., 2008.

Rumpf, H., The strength of granules and agglomerates, in: W.A. Knepper, (Ed.), *Agglomeration, Proceedings of the 1st International Symposium on Agglomeration*, Philadelphia, PA, John Wiley & Sons, 1962. pp. 379–418.

Chapter 8

Mechanism of Size Enlargement

Science is about knowing, engineering is about doing.

—*Henry Petroski*

8.1 SIZE ENLARGEMENT PROCESS

- Agglomeration is the formation of agglomerates or aggregates by sticking together smaller particles, and granulation is agglomeration by agitation.
- Different size enlargement processes are shown as follows.

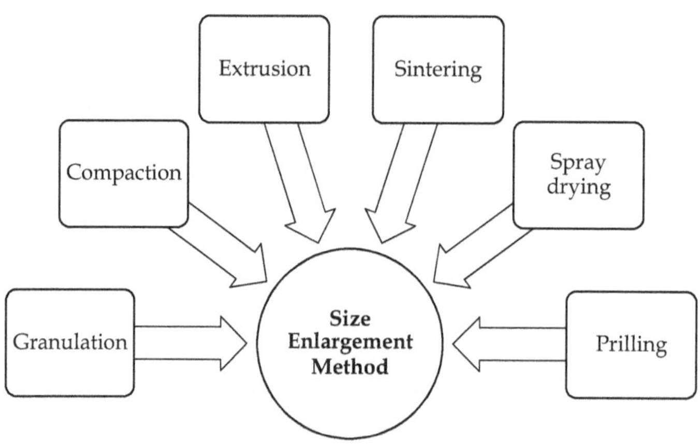

- The process selection requires thorough knowledge of the physico-chemical properties of the drug, excipients, required flow, release properties, etc.

DOI: 10.1201/9781003571025-11

8.2 GRANULATION

- Granulation is a size enlargement process for designing particles of the required size (agglomerated granules) from powder particles depending on particle size, the concentration of excipients, granulation time, granulator type, drying rate, etc.
- The primary methods for the granulation process include solid bridges, sintering, chemical reaction, crystallization, and deposition of colloidal particles.
- The mechanisms by which granules are formed from the powder particles are wetting and nucleation, coalescence or growth, consolidation, and attrition or breakage.

8.3 TYPES OF GRANULATION TECHNIQUES

- Based on the method used to facilitate the agglomeration of powder particles, the granulation technique is widely categorized into two basic types: wet granulation and dry granulation. The wet granulation technique is widely used.
- A schematic diagram of conventional granulation is shown in Figure 8.1.

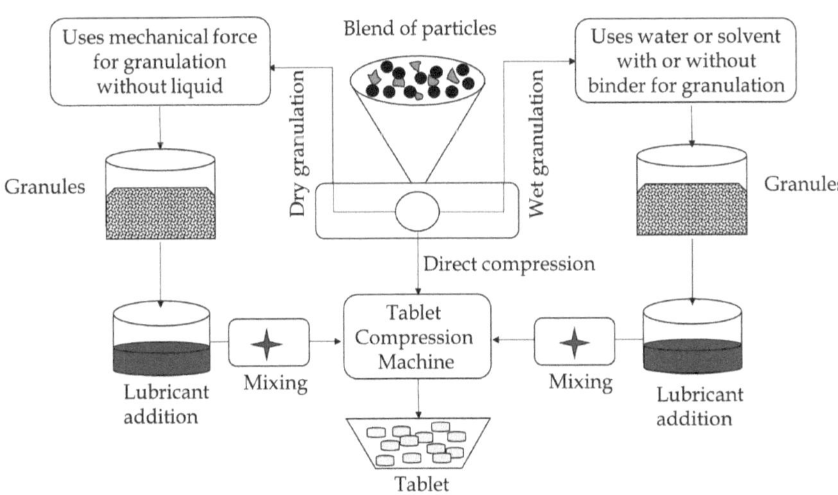

Figure 8.1 Schematic diagram of conventional granulation.

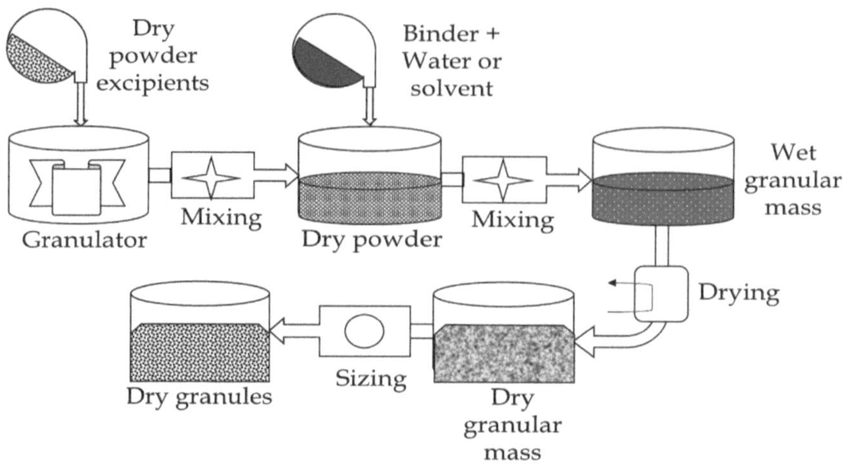

Figure 8.2 Schematic diagram of conventional wet granulation.

8.4 WET GRANULATION

- It uses granulation liquid (with or without binder/solvent) to facilitate agglomeration by forming wet mass by adhesion.
- It involves multiple unit processes such as wet massing, drying, and screening, which are complex, time-consuming, and expensive, requiring large spaces and numerous pieces of equipment as shown in Figure 8.2.
- Various wet granulation techniques are steam granulation, moisture-activated dry granulation or moist granulation, thermal adhesion granulation, melt granulation, freeze granulation, foamed binder or foam granulation, and reverse wet granulation (Shanmugam, 2015).

8.5 MECHANISM OF WET GRANULATION

- The successive mechanisms of the wet granulation technique by which granules formed from the powder particles are wetting and nucleation, coalescence and growth, consolidation, and attrition or breakage (Parikh, 2021), as shown in Figure 8.3.

8.5.1 Wetting and Nucleation Stage

- In this process, the void between particles is replaced by liquid. As shown in Figure 8.4, the liquid is added as a liquid drop.
- The wetting depends on the liquid's surface tension and the particles' contact angle.

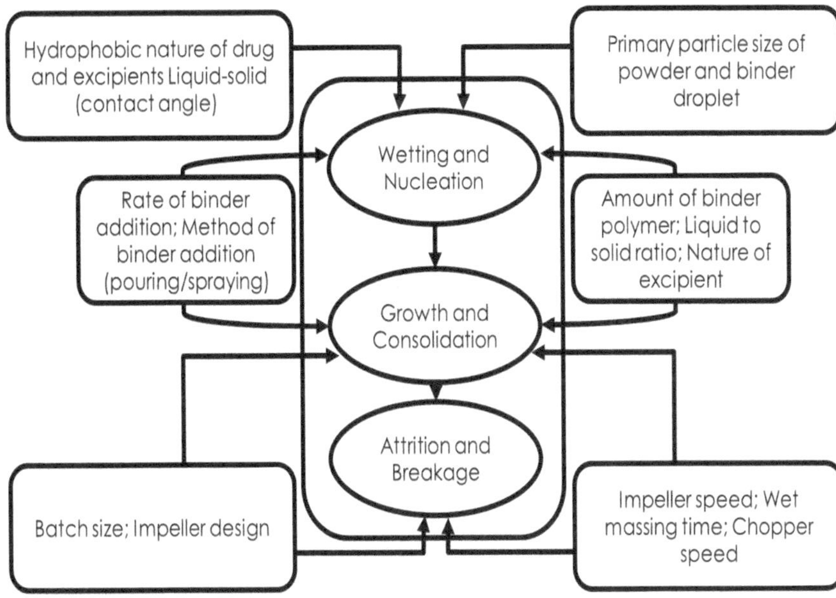

Figure 8.3 The outline of the mechanism of the wet granulation process.

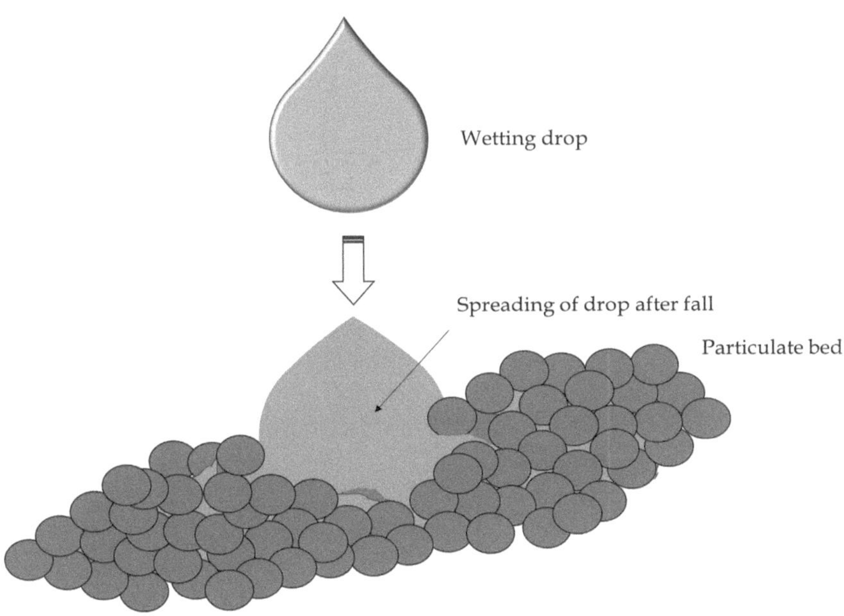

Figure 8.4 Schematic of wetting and nucleation.

- The initial coalescence or growth of primary particles happens in the close region of the wetting drop. It is called nucleation.
- The rate at which wetting occurs, known as the Washburn equation (Washburn, 1921), is expressed as

$$\frac{dz}{dt} = \frac{R_p \gamma \cos\theta}{4\mu z} \tag{8.1}$$

or

$$z = \sqrt{\frac{R_p \gamma \cos\theta}{2\mu} t} \tag{8.2}$$

- where z is the distance to which the liquid is penetrated at a time t, and θ is the dynamic contact angle of the liquid with the powder. The average pore radius (R_p) is a function of packing density and the size distribution of the powder, γ is the surface tension of the liquid, and μ is the viscosity of the liquid.
- The stages of wetting (Hapgood et al., 2002, 2007) are shown schematically in Figure 8.5.
- The time (t_p) taken for the drop of a known volume V_d to sink into the powder bed of porosity (ε_b) completely can be calculated by a correlation suggested by Hapgood et al. (2003):

$$t_p = 1.35 \frac{V_d^{2/3} \mu}{\varepsilon_b^2 R_p \gamma \mathrm{Cos}\theta} \tag{8.3}$$

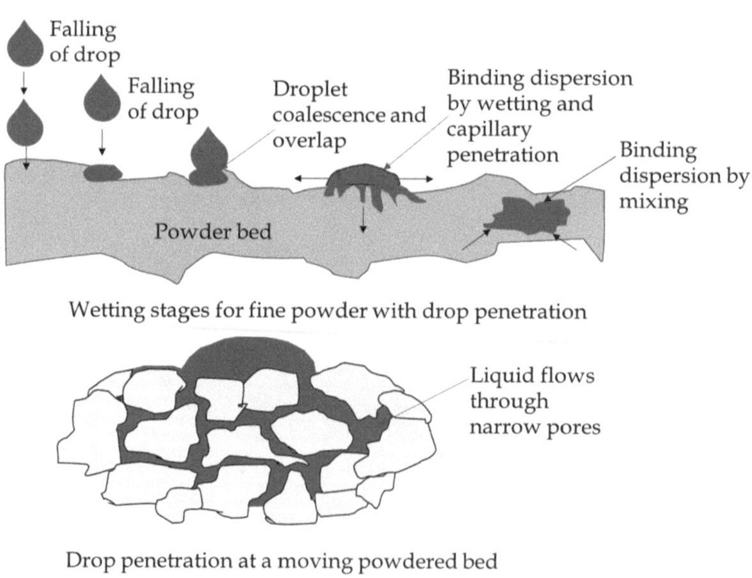

Figure 8.5 The schematic of the stages of wetting.

$$R_p = \frac{\varphi d_{32}}{3}\left(\frac{\varepsilon_b}{1-\varepsilon_b}\right) \tag{8.4}$$

where ε_b is the porosity or void fraction, d_{32} is the Sauter mean particle diameter, and φ is the sphericity of the particle.

- An increase in wetting results in narrower granule size distribution and better product quality.
- The liquid with less viscosity and higher surface tension increases the wetting rate in the powder of the particles of lower contact angle and with increasing size of pores.
- Smaller particles ensure a high liquid penetration rate but yield a lower extent of wetting.

Example 8.1

In a nucleation process of size enlargement, initial coalescence or growth of primary particles happens near the larger wetting liquid drop of size 10 mm and dynamic contact angle of 35° in the wet granulation process in a powder bed of void fraction 0.02 and average pore diameter is 75 µm. The surface tension and viscosity of the liquid are 0.072 N.m and 0.01 Poise, respectively. Find the distance the liquid penetrated at the sinking time of drop.

Solution:

The volume of the drop=$V_d = \frac{1}{6}\pi d_{\text{drop}}^3 = \frac{1}{6}\pi(0.020)^3 = 5.24\times10^{-7}\ \text{m}^3$

The viscosity of fluid=$\mu = 0.01$ Poise

The porosity of the powder bed=$\varepsilon_b = 0.02$

The average pore radius=$R_p = 75/2\times\text{m} = 37.5\times10^{-6}$ m

The surface tension of the liquid=$\gamma = 0.072$ N/m

The dynamic contact angle=$\theta = 35°$

Time it takes for the drop to sink into the powder bed completely:

$$t_p = 1.35\frac{V_d^{2/3}\mu}{\varepsilon_b^2 R_p \gamma\cos\theta} = 1.35\times\frac{(5.24\times10^{-7})^{2/3}(0.01)}{(0.02)^2(37.5\times10^{-6})(0.072)\cos(35°)}$$

$$= 734.31\ \text{seconds} = 12.24\ \text{minutes}$$

The distance to which the liquid is penetrated at this time is as per Washburn equation (1921)

$$z = \sqrt{\frac{R_p\gamma\cos\theta}{2\mu}t} = \sqrt{\frac{37.5\times10^{-6}\times0.072\cos(35°)}{2\times0.01}734.31} = 0.285\ \text{m}$$

8.5.2 Drop-Controlled Nucleation

- The nucleation is said to be "drop-controlled" if each droplet lands on the powder without touching other droplets and sinks quickly into the powder to form a new nucleus granule.
- This drop-controlled nucleation regime occurs at low penetration time and low dimensionless spray flux (ψ_a), which is given by (Litster and Ennis, 2004)

$$\psi_a = \frac{3Q}{2v_s w_s d_d} \tag{8.5}$$

where Q is the solution flow rate, v_s is the powder velocity in the spray zone, d_d is the average drop diameter, and w_s is the width of the spray.
- At a very low spray flux $\psi_a << 1$, drop footprints will not overlap and each drop will form a separate nucleus granule, whereas, at a very high spray flux, $\psi_a >> 1$, there will be a significant overlap of drops hitting the powder bed.
- The fraction (f_{wet}) of the powder surface that is wetted by spray drops depends on the spray flux, which is given by (Litster and Ennis, 2004)

$$f_{\text{wet}} = 1 - \exp(-\psi_a) \tag{8.6}$$

- The fraction of (f_n) of the powder surface formed depends on the number of drops (n) that fall to wet the powder, which can be calculated using (Litster and Ennis, 2004)

$$f_n = \exp(-4\psi_a)\left[\frac{(4\psi_a)^{n-1}}{(n-1)!}\right] \tag{8.7}$$

- The different regimes for nucleation, such as shear–controlled (mechanical dispersion), drop-controlled, and intermediate, are shown in Figure 8.6.
- Drop-controlled nucleation occurs when one drop forms one nucleus and should occur at low spray flux and penetration time.
- Other than these conditions, powder mixing and shear characteristics will dominate as a mechanical dispersion regime where the liquid binder is dispersed by powder shear and agitation (Figure 8.6).

Example 8.2

Find the spray flux value if a given fraction of powder surface (f_{wet}) of 0.25 is wetted by the spray drops to execute the granulation process as it passes beneath the spray zone. At this spray flux, if in the spray zone of width 50 cm, the solution is moving at a flowrate of $1\times10^{-4}\,\text{m}^3/\text{min}$ and the powder velocity in the spray zone is 0.01 m/s, find the average diameter of the drop penetrating in the spray zone.

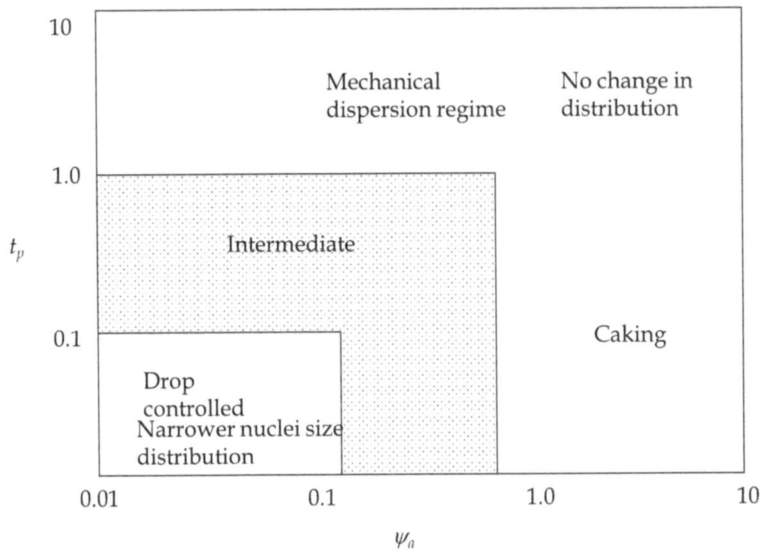

Figure 8.6 Nucleation regime map (Hapgood et al., 2003).

Solution:

The fraction of the powder surface that is wetted by spray drops (f_{wet}) is given by

$$f_{\text{wet}} = 1 - \exp(-\psi_a)$$

Accordingly

$$0.25 = 1 - \exp(-\psi_a)$$

$$\Rightarrow \exp(-\psi_a) = 0.75$$

$$\Rightarrow \psi_a = 0.288$$

$Q = 0.001\ \text{m}^3/\text{min} = 0.001/60\ \text{m}^3/\text{s}$
$v_s = 0.01\ \text{m/s}$
$w_s = 50\ \text{cm} = 0.50\ \text{m}$
Therefore, as per

$$\psi_a = \frac{3Q}{2v_s w_s d_d}$$

$$d_d = \frac{3Q}{2v_s w_s \psi_a} = \frac{3(0.0001/60)}{2(0.01)(0.50)(0.288)} = 0.0017\ \text{m} = 1.7\ \text{mm}$$

8.5.3 Granule Consolidation Stage

- In this stage, the granule density is increased because of the closer packing of primary particles.
- In this stage, the granule is saturated at a certain porosity of the bed (ε_b) and the liquid level (w, wt.%). The following equation can quantify the granule saturation (S):

$$S = \frac{w\rho_s(1-\varepsilon_b)}{\rho_l \varepsilon_b} \tag{8.8}$$

where ρ_s is the solid density, andρ_l is the liquid density.

- The saturation increases as the porosity decreases, and once the saturation exceeds 100%, further consolidation pushes liquid to the granule surface, making the surface wet and changing granule growth rates.

Example 8.3

In the granule consolidation stage of a wet granulation process, the granule is saturated at a specific porosity of the bed and the liquid level. What would be the liquid level in wt.% if the granule is 100% saturated with a liquid of density 1000 kg/m^3 and solid density 2540 kg/m^3? The pore volume fraction of the granule is 25%.

Solution:

The following equation can quantify the granule saturation (S):

$$S = \frac{w\rho_s(1-\varepsilon_b)}{\rho_l \varepsilon_b}$$

As per this equation and with the given condition, one can write

$$1 = \frac{w(1540)(1-0.25)}{(1000)(0.25)}$$

$$\Rightarrow w = \frac{(1000)(0.25)}{(1540)(1-0.25)} = 0.2164 = 21.64\%$$

8.5.4 Growth Stage

- Due to agitation and dissipation of kinetic energy, granules coalesce by collision and change surface area for bonding.
- A critical maximum granule size beyond which coalescence is impossible during a collision.
- The parameter which determines whether coalescence will occur is a Stokes number St:

$$\text{St} = \frac{\rho_{\text{gr}} V d_{\text{gr}}}{16\mu} = \frac{\text{Collisional kinetic energy}}{\text{Energy dissipaion through viscous dissipation}} \tag{8.9}$$

where ρ_{gr} is the granules' density, V is the approach velocity, d_{gr} is the diameter of granules, and μ is the viscosity of liquid.

- The coalescence must occur if the Stokes number is less than a critical value St*, given by (Litster and Ennis, 2004)

$$St^* = \left(1+\frac{1}{e}\right)\ln\left(\frac{h}{h_a}\right) \tag{8.10}$$

where h is the thickness of the coated layer of each granule, h_a is a measure of surface roughness of the granule, and e is the coefficient of restitution for the collision.

- The ratio of relative speed after collision to the relative speed before collision between two granules in a particular direction is known as the coefficient of restitution in that direction. The restitution coefficient is a unit less quantity, and its values range between 0 and 1.
- If $e=0$, then it is a perfectly inelastic collision. If $0<e<1$, then it is a real-world inelastic collision, in which some kinetic energy is dissipated, whereas if $e=1$, then it is a perfectly elastic collision in which no kinetic energy is dissipated, and the granules rebound from one another with the same relative speed with which they approached.

Example 8.4

A wet granulation process yields granules of mean size 1 cm, which are moving at a velocity of 0.3 m/s and colliding with each other in a medium of kinematic viscosity 10.2 cSt. Also, from the analysis, it is seen that the granules are coated with a layer of thickness 10 µm with 0.1 µm surface roughness. The coefficient of restitution for the collision can be considered as 0.8. Assess whether the coalescence of the granules will happen or not?

Solution:

$$\text{Kinematic viscosity} = \frac{\mu}{\rho} = 10.2 \text{ cSt} = 0.102 \text{ St}$$

$$St = \frac{Vd_{gr}}{16\mu/\rho_{gr}} = \frac{(0.3)(0.010)}{16(0.102)} = 0.0018$$

$$St^* = \left(1+\frac{1}{e}\right)\ln\left(\frac{h}{h_a}\right) = \left(1+\frac{1}{0.8}\right)\ln\left(\frac{10\times10^{-6}}{0.1\times10^{-6}}\right) = 10.36$$

The coalescence must occur since the Stokes number is less than a critical value Stokes number as calculated above.

8.5.5 Granule Deformation

- One cannot ignore the granule deformation in the case of high agitation intensity systems.

- Granule structure growth results are based on the degree of deformability.
- The growth behavior of granule structure depends on the following behaviors.
  - **Steady growth behavior:** It occurs if the granule size increase is roughly proportional to granulation time.
  - **Induction behavior:** It occurs when there is a long period, during which no increase in size occurs.

8.5.6 Growth Regime

- Various granule growths can be described using the saturation and deformation Stokes number (St_{def}) and the granule growth regime map (Hapgood et al., 2007), as shown in Figure 8.7.
- The deformation Stoke's number (St_{def}) is defined as

$$St_{def} = \rho_{gr} U_c^2 / (2Y_d) \tag{8.11}$$

where U_c is the representative collision velocity in the granulator, ρ_{gr} is the average granule density, Y and d represent the dynamic yield stress of the granule.

- Maximum saturation is defined as

$$S_{max} = w\rho_s(1 - \varepsilon_{min}) / (\rho_l \varepsilon_{min}) \tag{8.12}$$

8.5.7 Granule Breakage Stage

- Breakage of granules is basically the fracture of a granule to form two or more pieces.
- Breakage can be controlled by altering the granule properties such as:
  - by increasing fracture toughness

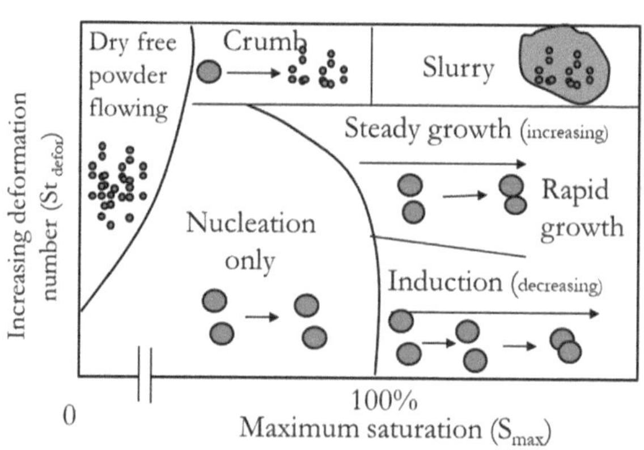

Figure 8.7 Growth regime map of granules (Hapgood et al., 2007).

- by increasing resistance to attrition
- by reducing agitation intensity.

8.6 DIFFERENT TYPES OF WET GRANULATION

Research and development on granulation techniques reported the different types of wet granulation recently (Shanmugam, 2015) are briefed as follows.

8.6.1 Reverse Wet Granulation

- It is a granulation process by which the granules are formed by the immersion of the dry powder into the binder liquid, followed by controlled breakage.
- The binder solution is prepared initially with a hydrophilic polymer, and the dry powder is added to the binder solution or slurry mixed in a granulator.
- Granules are then formed by immersing them into the granulator with other dry excipients.
- The resulting wet granules are then milled and sized after drying. The schematic diagram of this process is presented in Figure 8.2.
- This technique improves the dissolution characteristics of poorly water-soluble drugs due to uniform contact between the drug and a hydrophilic polymer, allowing uniform binder distribution.
- This method produces small and spherical-shaped granules with improved flow properties, uniform wetting, and erosion of the granules.

8.6.2 Steam Granulation

- In this technique, water steam is used as a binder instead of liquid.
- Steam is used to enhance the diffusion of binder into the powder and favorable thermal balance during the drying step.
- The schematic diagram of steam granulation is shown in Figure 8.8.
- This process provides the higher ability of the steam to distribute uniformly and diffuse into the powder particles, production of spherical granules with larger surface area, and shorter processing time.
- There is no involvement of organic solvents in this process.

8.6.3 Moisture-Activated Granulation or Moist Granulation

- It uses very little water, less than 5% (1%–4% preferably), to activate the mixture of drug, binder, and other excipients and facilitate agglomeration.

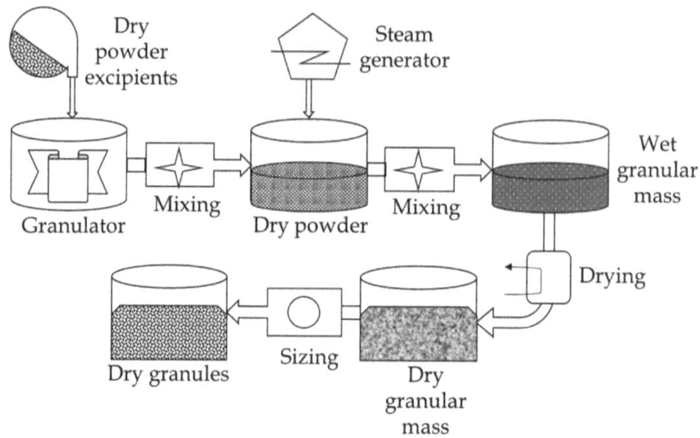

Figure 8.8 Schematic diagram of reverse wet granulation.

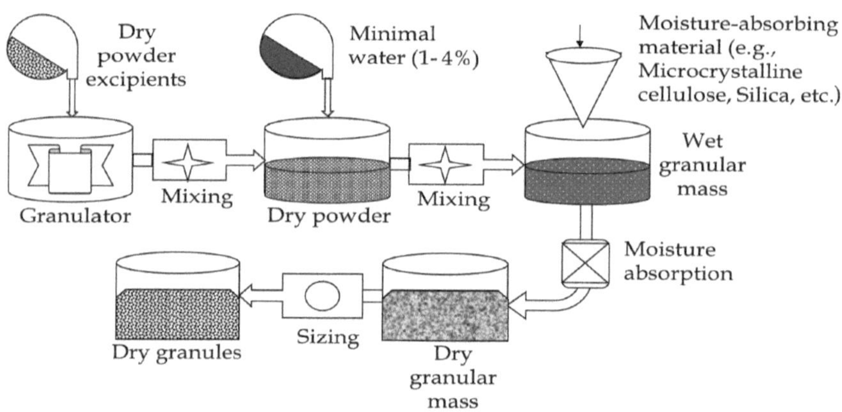

Figure 8.9 Schematic diagram of moisture-activated granulation.

- This technique involves two steps, which are shown in Figure 8.9. The steps are as follows:
  - wet agglomeration of the powder particles
  - moisture absorption or distribution.
- After the agglomeration, moisture-absorbing material, such as microcrystalline cellulose, silicon dioxide, etc., is added to facilitate the absorption of excess moisture.
- The absorption of moisture results in moisture redistribution within the powder mixture, leading to a relatively dry granule mixture.
- The process does not lead to larger lumps formation since the water used is very small compared to usual wet granulation. Also, it increases particle size, better flow, and compressibility.

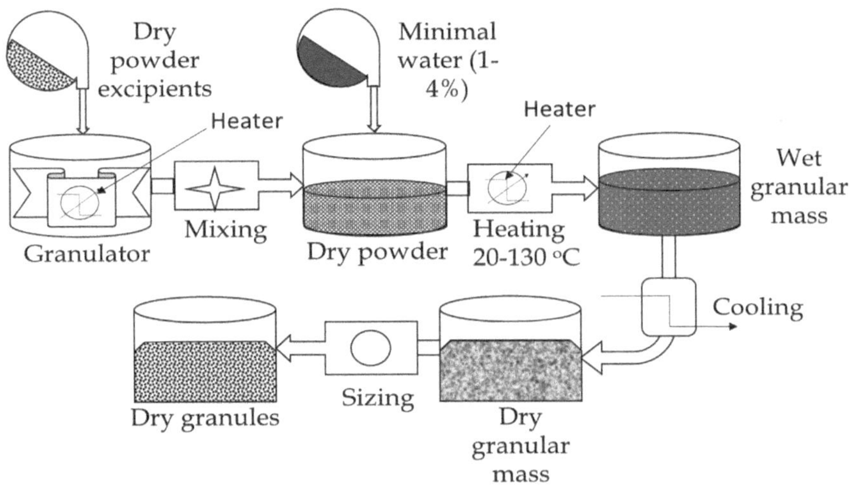

Figure 8.10 Schematic diagram of thermal adhesion granulation.

8.6.4 Thermal Adhesion Granulation

- It is executed by adding a small amount of water and solvent as granulation liquid and heat for agglomeration. The schematic diagram of the process is shown in Figure 8.10.
- The drug and excipient mixture is heated to a temperature range of 30°C–130°C in a closed system under tumble rotation to facilitate the agglomeration of the powder particles.
- The process eliminates the drying process due to the addition of a low amount of granulation liquid.
- The process is simple and convenient, with low moisture and binder contents in a closed system for preparing highly compressible materials.

8.6.5 Melt Granulation

- The technique facilitates the agglomeration of powder particles using binders (either hydrophilic or hydrophobic material), which melt or soften at relatively low temperatures (50°C–90°C). The schematic diagram of melt granulation is shown in Figure 8.11.
- The subsequent cooling of the agglomerated powder and the solidification of the molten or softened binder complete the granulation process.
- The method can be applied to enhance the stability of the moisture-sensitive drug and further improve the poor physical properties of the drug substance.

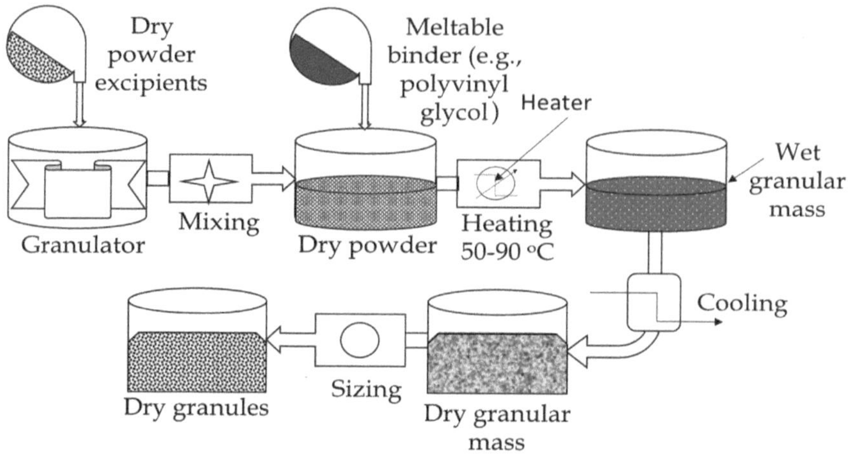

Figure 8.11 Schematic diagram of melt granulation.

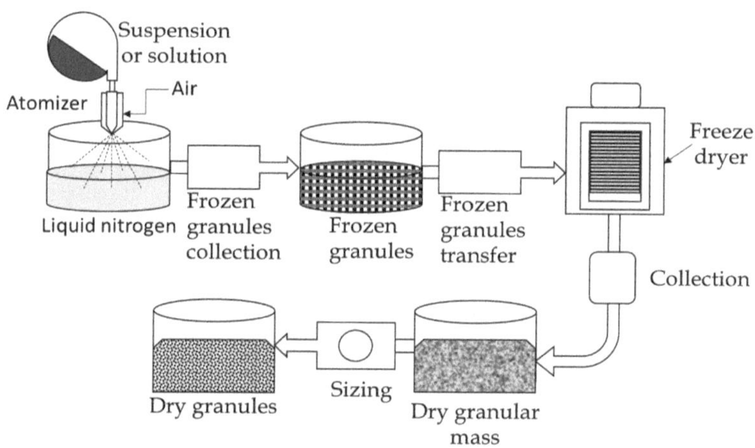

Figure 8.12 Schematic diagram of freeze granulation.

8.6.6 Freeze Granulation

- This granulation is executed by spraying a liquid slurry or suspension droplets into liquid nitrogen.
- After spraying, the drops are instantly frozen into granules, followed by freeze-drying of the frozen droplets.
- In the subsequent freeze-drying process, the granules are dried by sublimation of ice without any segregation, as shown in Figure 8.12.
- The process's advantages are the formation of free-flowing spherical granules with no cavities and a high degree of granule homogeneity by using both water-based and solvent-based slurries.

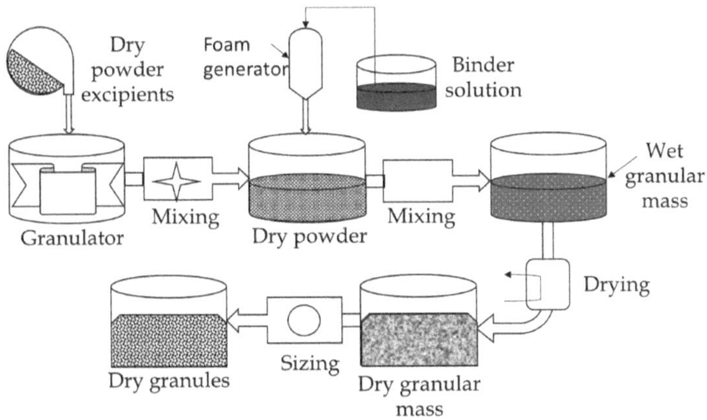

Figure 8.13 Schematic diagram of foam granulation.

8.6.7 Foam Granulation

- In this process, the liquid/aqueous binder is added as foam instead of spraying or pouring liquid onto the powder particles. The schematic diagram of this process is shown in Figure 8.13.
- The addition of the binder solution as foam reduces the problems of binder distribution that can affect tablet hardness and drug release.
- In this process, the binders are used at a high spread-to-soak ratio, and because of this, the binders are coated onto the particles rather than soaked.
- The high spread-to-soak ratio results in consistent binder distribution, improving reproducibility and shortening the processing time.

8.7 DRY GRANULATION

- Dry granulation is executed either by roller compaction or by slugging. The two different types of dry granulations (Shanmugam, 2015) are illustrated in the schematic diagram in Figure 8.14.
- Progress in the dry granulation technique compared to wet granulation is still scarce. However, one important innovation is known as pneumatic dry granulation technology developed by Atacama LabsOy, Finland (Politi and Heilakka, 2008) which was reported by Shanmugam (2015).

8.7.1 Pneumatic Dry Granulation

- This method produces granules from powder particles by applying compaction to create a compacted mass comprising a mixture of fine particles and granules.

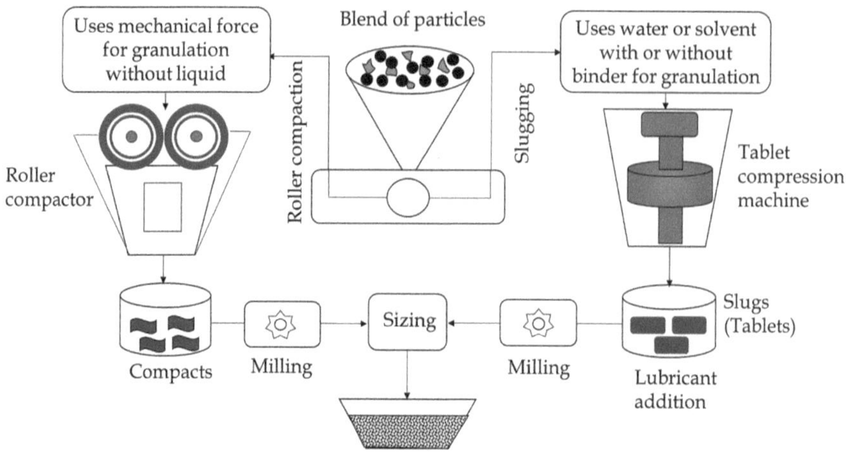

Figure 8.14 Schematic diagram of dry granulation.

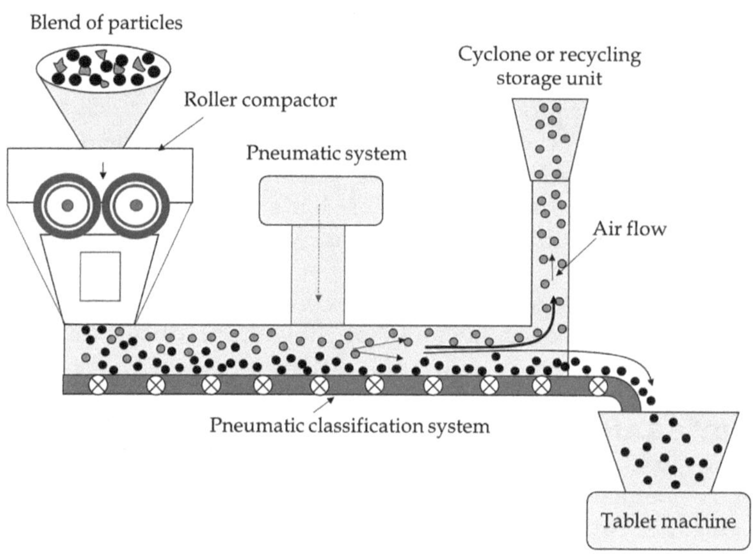

Figure 8.15 Schematic diagram of pneumatic dry granulation.

- Subsequently, the fine particles and/or smaller granules are separated from the intended-size granules in a fractioning chamber by entraining them in a gas stream (pneumatic system).
- The intended-size granules pass through the fractioning chamber and are then compressed into tablets. In contrast, the entrained fine particles are then separated by a cyclone separator and are returned to the roller compactor for immediate re-processing. The schematic diagram of this process is represented in Figure 8.15.

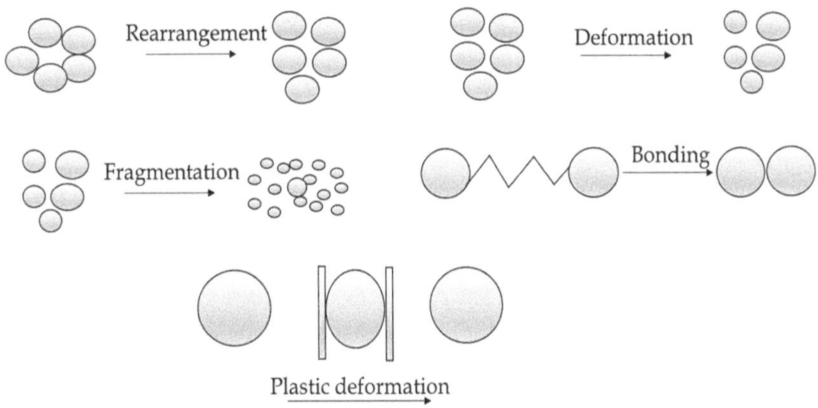

Figure 8.16 Mechanism of the dry granulation process.

- This process produces suitable flowing granules for compaction formulations with a tensile strength of ~0.5 MPa.
- It is also reported that this technology enables high drug loads of up to 70%–100% (Shanmugam, 2015).

8.8 MECHANISM OF DRY GRANULATION

- The dry granulation process is based on interparticle bond formation, as shown in Figure 8.16. The bond formation is characterized by different stages (Parikh, 2021), which usually occur in the following order:
  - **Particle rearrangement:** Particle rearrangement occurs by filling void spaces.
  - **Particle deformation:** It occurs as compression forces are increased.
  - **Particle fragmentation:** The particle fracturing creates multiple new surface sites, additional contact points, and potential bonding sites.
  - **Particle bonding:** The bonding takes place at the molecular level due to the effect of van der Waals forces.
  - **Simultaneous plastic and elastic deformation:** The granules attempt to return to their original shape or form, referred to as elastic deformation, whereas a deformation that does not recover after the stress is released is plastic deformation.

IMPORTANT POINTS TO REMEMBER

- Agglomeration is the formation of agglomerates or aggregates by sticking together smaller particles and granulation is agglomeration by agitation.

- The primary methods for the granulation process include solid bridges, sintering, chemical reaction, crystallization, and deposition of colloidal particles.
- The successive mechanisms of the wet granulation are wetting and nucleation, coalescence and growth, consolidation, and attrition or breakage.
- Nucleation is the initial coalescence or growth of primary particles that happens in the close region of the wetting drop.
- Smaller particles ensure a high liquid penetration rate but yield a lower extent of wetting.
- The fraction of the powder surface that is wetted by spray drops depends on the spray flux, which is given by $f_{\text{wet}} = 1 - \exp(-\psi_a)$.
- Drop-controlled nucleation occurs when one drop forms one nucleus and should occur at low spray flux and low penetration time.
- The granulation saturation increases as the porosity decreases, and once the saturation exceeds 100%, further consolidation pushes liquid to the granule surface, making the surface wet and changing granule growth rates.
- Null coefficient of restitution reflects the perfectly inelastic collision.
- Unit coefficient of restitution is a perfectly elastic collision in which no kinetic energy is dissipated, and the granules rebound from one another with the same relative speed with which they approached.
- Granule structure growth results are based on the degree of deformability.
- Reverse wet granulation is a granulation process by which the granules are formed by the immersion of the dry powder into the binder liquid, followed by controlled breakage.
- In steam granulation, steam is used to enhance the diffusion of binder into the powder and favorable thermal balance during the drying step.
- Thermal adhesion granulation is executed by adding a small amount of water and solvent as granulation liquid and heat for agglomeration.
- Melt granulation can be applied to enhance the stability of the moisture-sensitive drug and further improve the poor physical properties of the drug substance.
- The addition of the binder solution as foam reduces the problems of binder distribution that can affect tablet hardness and drug release.
- The high spread-to-soak ratio results in consistent binder distribution, improving reproducibility and shortening the processing time.
- The dry granulation process is based on interparticle bond formation, which is characterized by particle rearrangement, particle deformation, particle fragmentation, particle bonding, and simultaneous plastic and elastic deformation.

SELF-TESTING QUESTIONS

I. Multiple Choice Questions

Q1. The desired condition for drop-controlled nucleation is____________________.
- (a) When one drop forms one nucleus
- (b) The spray density is high, and relatively few drops overlap
- (c) The drop must not penetrate completely into the powder bed before it touches other drops
- (d) Both (a) & (b)

Q2. In a dry granulation process, the granule bond formation is executed by different stages. Which of the following is the correct order of the dry granulation process?
- (a) Rearrangement → Deformation → Fragmentation → Bonding → Plastic deformation
- (b) Deformation → Bonding → Rearrangement → Plastic deformation → Fragmentation
- (c) Rearrangement → Deformation → Fragmentation → Plastic deformation → Bonding
- (d) Rearrangement → Deformation → Bonding → Fragmentation → Plastic deformation

Q3. The term ____________________is typically applied to the initial coalescence or growth of primary particles near the larger wetting drop.
- (a) Nucleation
- (b) Wetting
- (c) Granulation
- (d) Extrusion

Q4. At low spray flux $\psi_a \ll 1$, drop footprint will not overlap, and each drop will form a separate nucleus granule. The statement is true or false?
- (a) True
- (b) False

Q5. In the granule consolidation stage, which statement is appropriate?
- (a) The granule viscosity is increased by closer packing of primary particles
- (b) The granule porosity is increased by closer packing of primary particles
- (c) The granule density is increased by closer packing of primary particles
- (d) The thickness of each layer of granule is increased by closer packing of primary particles

Q6. Which of the following factors can affect the deformation of granules?
- (a) Moisture content
- (b) Particle size distribution

(c) Both (a) & (b)
(d) Neither (a) nor (b)

Q7. Which of the following size enlargement process involves large deformation and strain?
(a) Compaction
(b) Granulation
(c) Encapsulation
(d) Agglomeration

Q8. For coalescence to occur, the Stokes number must be ___________________.
(a) less than the critical value
(b) more than the critical value
(c) equal to the critical value
(d) none of the above

Q9. The ___________________ is a measure of the ratio of collisional kinetic energy to energy dissipated through viscous dissipation.
(a) Stokes number
(b) Low deformation
(c) Induction behavior
(d) Particle deformation

Q10. Which one of the following is the correct statement for a wet granulation process?
(a) The liquid with less viscosity and higher surface tension increases the wetting rate in the powder of the particles of lower contact angle and with increasing size of pores.
(b) The liquid with less viscosity and higher surface tension decreases the wetting rate in the powder of the particles of lower contact angle and with increasing size of pores.
(c) The liquid with high viscosity and lower surface tension increases the wetting rate in the powder of the particles of lower contact angle and with increasing size of pores.
(d) The liquid with less viscosity and higher surface tension increases the wetting rate in the powder of the particles of higher contact angle and with increasing size of pores.

Q11. Smaller particles ensure a ___________________ liquid penetration rate but yield a ___________________extent of wetting.
(a) Low, higher
(b) High, lower
(c) High, higher
(d) Low, lower

Q12. An increase in wetting results in ___________________and better product quality.
(a) no change in granule size distribution
(b) wider granule size distribution

(c) narrower granule size distribution
(d) none of these

Q13. Which one of the following relations between spray flux and the fraction of the powder surface that is wetted by spray drops is correct?
(a) $\psi_a = -\ln(1 - f_{\text{wet}})$
(b) $\psi_a = -\ln(1 + f_{\text{wet}})$
(c) $\psi_a = 1 - \ln(1 - f_{\text{wet}})$
(d) $\psi_a = \ln(1 - f_{\text{wet}})$

Q14. At a given spray flux value, the fraction of the powder surface that is wetted by spray drops as it passes beneath the spray zone (f_{wet}) is given by________________.
(a) $f_{\text{wet}} = -1/\exp(-\psi_a)$
(b) $f_{\text{wet}} = 1 - \exp(\psi_a)$
(c) $f_{\text{wet}} = 1 - \exp(-\psi_a)$
(d) $f_{\text{wet}} = 1 + \exp(-\psi_a)$

Q15. At a very high spray flux, $\psi_a >> 1$, there will be significant ________________ hitting the powder bed.
(a) breakage of drops
(b) coalescence of drops
(c) distortion of drops
(d) overlap of drops

Q16. The fraction of the powder surface formed in a wet granulation process depends on the ________________.
(a) number of drops that fall to wet the powder
(b) sphericity of drops that fall to wet the powder
(c) type of liquid for making drops
(d) the temperature of the atmosphere

Q17. Drop-controlled nucleation occurs when one drop forms one nucleus and should occur at ________________.
(a) high spray flux and low penetration time
(b) low spray flux and low penetration time
(c) low spray flux and high penetration time
(d) high spray flux and high penetration time

Q18. The saturation increases as the porosity decreases, and once the saturation exceeds 100%, what is the correct option to be matched?
(a) further consolidation hinders liquid to the granule surface
(b) further consolidation does not happen to the granule surface
(c) further consolidation pushes liquid to the granule surface
(d) none of these

Q19. A real-world inelastic collision between granules, in which some kinetic energy is dissipated in a condition of restitution coefficient as________________(where e is the restitution coefficient).
(a) $0<e<1$
(b) $e=0$

(c) $e=1$
(d) none of these

Q20. The condition for a perfectly elastic collision of granules in which no kinetic energy is dissipated, and the granules rebound from one another with the same relative speed with which they approached is_________________ (where e is the restitution coefficient).

(a) $0<e<1$
(b) $e=0$
(c) $e=1$
(d) none of these

Answers to the multiple choice questions

Q1. Ans. (a)	Q8. Ans. (a)	Q15. Ans. (d)
Q2. Ans. (a)	Q9. Ans. (a)	Q16. Ans. (a)
Q3. Ans. (a)	Q10. Ans. (a)	Q17. Ans. (b)
Q4. Ans. (a)	Q11. Ans. (b)	Q18. Ans. (c)
Q5. Ans. (c)	Q12. Ans. (c)	Q19. Ans. (a)
Q6. Ans. (c)	Q13. Ans. (a)	
Q7. Ans. (a)	Q14. Ans. (c)	

II. Objective Type Questions

1. What is agglomeration?
2. What are the subsequent mechanisms of granulation?
3. What are the different types of wet granulation processes?
4. Explain the schematic process of conventional wet granulation.
5. Explain each of the mechanisms of wet granulation.
6. Explain the nucleation regime map of granulation.
7. What is the critical Stokes number? How is it significant in the growth stage of granulation?
8. Explain the growth regime map of the granulation process.
9. How does the reverse wet granulation process work?
10. Describe the thermal adhesion granulation process.
11. Explain the working principle of the melt granulation process with a schematic diagram.
12. What is freeze granulation?
13. Why is the high spread-to-soak ratio maintained in the foam granulation process?
14. Explain the working principles of the pneumatic dry granulation process.
15. Explain with an outline of the mechanism of the dry granulation process.

III. Numerical Problems

Q1. A given fraction of powder surface (f_{wet}) of 0.282 is wetted by the spray drops to execute the granulation process. What would the spray flux value be as it passes beneath the spray zone? [Ans. 0.331]

Q2. Consider a particle with a diameter of 10 μm and a density of 1000 kg/m^3, suspended in water at a temperature of 20°C. The fluid velocity is 0.1 m/s. The dynamic viscosity of water at 20°C is 1.002×10^{-3} Pa.s. What is the Stokes number of the particle? [Ans. 0.0623]

Q3. Nucleation is typically applied to the initial coalescence or growth of primary particles near the larger wetting liquid drop of size 20 mm and dynamic contact angle of 60° in the wet granulation process in a powder bed of void fraction 0.01 and average pore diameter is 100 μm. The surface tension and viscosity of the liquid are 0.072 N.m. and 0.01 Poise, respectively. Find the time it takes for the drop to sink entirely into the powder bed (in minutes). [Ans. 32.48 minutes]

Q4. Nucleation is typically applied to the initial coalescence or growth of primary particles near the larger wetting liquid drop of size 15 mm and dynamic contact angle of 89° in the wet granulation process in a powder bed of void fraction 0.02 and average pore diameter is 100 μm. The surface tension and viscosity of the liquid are 0.072 N.m. and 0.01 Poise, respectively. What is the distance the liquid penetrated at the sinking time of drop? [Ans. 49.66 cm]

Q5. In a wet granulation process, it is observed that the penetration depth (in m) of binder liquid drop changes with the time of penetration in the porous bed of powder as per the following equation:

$$z = 0.011\sqrt{t}$$

The dynamic contact angle of the liquid droplet is 75°. The surface tension of the liquid and viscosity are 0.072 N/m and 0.001 Pa.s, respectively. What is the pore diameter of the bed? [Ans. 25.96 μm]

Q6. In a wet granulation process, it is observed that the penetration depth (in m) of binder liquid drop (60% w/w sucrose solution) changes with the time of penetration in the porous bed of powder as per the following equation:

$$z = 0.017\sqrt{t}$$

The pore diameter of the bed is 105 μm. The surface tension of the liquid and viscosity are 0.072 N/m and 0.001 Pa.s, respectively. What is the contact angle of the particle? [Ans. 85.61°]

Q7. In a wet granulation process, it is observed that the penetration depth (in m) of binder liquid drop (60% w/w sucrose solution) changes with the time of penetration in the porous bed of powder as per the following equation:

$$z = 0.017\sqrt{t}$$

The dynamic contact angle of the liquid droplet is 35°. The surface tension and viscosity of the liquid solution are 66 mN/m and 48 mPa.s, respectively. The Sauter mean particle diameter of the bed is 100 μm of sphericity 0.89. What is the pore volume fraction of the bed at a penetration depth of 2 cm? [Ans. 0.1215]

Q8. In the granule consolidation stage of a wet granulation process, the granule is saturated at a specific bed porosity and a liquid level. What would be the liquid level in wt.% if the granule is 100% saturated with a liquid of one-third solid density and a pore volume fraction of 20%? [Ans. 8.33%]

Q9. A wet granulation process yields granules with a mean size of 1.5 mm, which are moving at a velocity of 0.2 m/s and colliding with each other in a medium of kinematic viscosity 15.7 cSt. Also, the analysis shows that the granules are coated with a layer of thickness 1 μm with 0.1 μm surface roughness. The coefficient of restitution for the collision can be considered as 0.8. Assess whether the coalescence of the granules will happen or not. [Ans. $St = 0.001 < St^* = 5.18$. The coalescence must occur since the Stokes number is less than a critical value 5.18]

NOMENCLATURE

d_d	average drop diameter [m]
f_n	fraction of the powder surface formed depends on the number of drops (n) that fall to wet the powder [-]
f_{wet}	fraction of the powder surface that is wetted by spray drops [-]
d_{32}	Sauter mean particle diameter [m]
d_{gr}	diameter of granules [m]
e	coefficient of restitution for the collision [-]
h	thickness of the coated layer of each granule [m]
h_a	measure of surface roughness of the granule [m]
n	number of drops [-]
Q	solution flow rate [m³/s]
R_p	average pore radius [m]
S	granule saturation [%]
S_{max}	maximum saturation [%]
St	Stoke's number [-]
St*	critical Stoke's number [-]
St_{def}	deformation Stoke's number [-]
t	time [s]
t_p	total sink time [s]
U_c	representative collision velocity in the granulator [m/s]
V	approach velocity [m/s]

v_s	powder velocity in the spray zone [m/s]
w	liquid level [wt.%]
w_s	width of the spray [m]
Y_d	dynamic yield stress of the granule [N/m^2]
z	distance to which the liquid is penetrated [m]

GREEK LETTERS

ρ_l	liquid density [kg/m^3]
ε_b	porosity or void fraction of bed [-]
ρ_s	solid density [kg/m^3]
φ	sphericity of the particle [-]
γ	surface tension of the liquid [N/m]
θ	dynamic contact angle of the liquid with the powder [rad]
ψ_a	spray flux [-]
ρ_{gr}	granules density [kg/m^3]
ε_{min}	minimum porosity [m]
μ	viscosity of the liquid [Pa.s]

SUBSCRIPTS

32	volume to surface
b	bed
c	collision
d	dynamic, drops
def	deformation
gr	granules
l	liquid
max	maximum
min	minimum
n	number
p	sink, pore
s	solid, spray
wet	wet

REFERENCES

Hapgood, K., S. Iveson, J. Litster and L. Liu, Granulation rates processes, in: A. Salman, M. Hounslow, & J. P. K. Seville (Eds.), *Handbook of Powder Technology*, Elsevier Publishing, Vol. 11, 2007. pp. 897–977.

Hapgood, K. P., J. D. Litster and R. Smith, Nucleation regime map for liquid bound granules, *AIChE J.*, 49 (2), 350–361, 2003.

Hapgood, K. P., J. D. Litster, S. R. Biggs and T. Howes, Drop penetration into porous powder beds, *J. Colloid Interface Sci.*, 253 (2), 353–366, 2002.
Litster, J. and B. J. Ennis, *The Science & Engineering of Granulation Processes*, Kluwer Academic, 2004.
Parikh, D. M., *Handbook of Pharmaceutical Granulation Technology*, 4th Edition, CRC Press, 2021.
Politi, G. and E. Heilakka, *Granules, tablets and granulation*, Google Patents, 2008.
Shanmugam, S., Granulation techniques and technologies: Recent progresses, *Bioimpacts*, 5 (1), 55–63, 2015.
Washburn, E. W., The dynamics of capillary flow, *Phys. Rev.*, 17 (3), 273, 1921.

Chapter 9

Equipment for Size Enlargement

> Strive for perfection in everything you do. Take the best that exists and make it better. When it does not exist, design it.
>
> —*Sir Henry Royce, English engineer and car designer*

9.1 COMMON CATEGORIES OF GRANULATION EQUIPMENT

- Common categories of granulators that are mainly used:
  - Tumbling granulators (the most common types are disk and drum granulators)
  - Mixer granulators (continuous high shear, e.g., Shugi mixer; high batch shear, e.g., paddle mixer)
  - Fluidized granulators (bubbling beds, spouted beds, Wurster coaters)
  - Centrifugal granulators (centrifugal fluidizing granulators)
  - Spray granulators
  - Pressure compaction (extrusion, roll press, tablet press, melding press, pellet mill)
- Different types of granulators, their features, and their applications (as per Ennis and Litster, 1997) are summarized in Table 9.1.

9.2 TUMBLING GRANULATOR

- The particles are passed through an inclined cylinder (drum) or pan (disk), set in motion by a tumbling action caused by the balance between gravity and centrifugal force, as shown in Figure 9.1.
- The tumbling action gives rise to a natural classification of the contents according to size.
- The equipment consists of a rotating, tilted disk or pan with a rim. Solids and binder solutions (to promote growth) are fed continuously

DOI: 10.1201/9781003571025-12

Table 9.1 Different types of granulators, their features, and applications (as per Ennis and Litster, 1997)

Method	*Product Granule Size (mm)*	*Granule Density*	*Scale Throughput*	*Comments*	*Typical Applications*
Tumbling (disk, drum)	0.5–20	Moderate	0.5–800 t/h	Very spherical	Fertilizers, iron ore, agricultural chemicals
Mixer (continuous and batch high shear)	0.1–2.0	Low < 200 kg/batch	<50 t/h	Gently handling	Chemicals, detergents, pharmaceuticals, ceramics
Fluidized bed (bubbling, spouting)	0.1–2.0	High	<600 kg/batch	Suitable for coating and easy to scale up	Continuous (fertilizers, detergents) batch (pharmaceuticals, agricultural chemicals)
Centrifugal granulators	0.3–3.0	Moderate to high	<200 kg/batch	Layering and coating of powder	Used in the pharmaceutical, agricultural, and chemical industry
Spray granulator	0.05–0.5	Low		The morphology of the product varies widely	Used for processing foods, dyes, detergents, ceramics, pharmaceuticals
Prilling	0.7–2.0	Moderate			Used in urea, ammonium nitrate process
Pressure compactors (e.g., extrusion, roll press, tablet press, molding press, pellet mill).	>0.5 >1 10	High to very high	<5 ton/h	Narrow size distributions of products, sensitive to powder flow and material properties	Suitable for pharmaceuticals, catalysts, inorganic and organic chemicals, plastic preforms, metal parts, ceramics, clays, minerals, and animal feeds

to the granulator. A coating of the feed material builds up on the disk, and this layer's thickness is controlled by scrapers, which may oscillate mechanically. The pan's surface may also be lined with expanded metal or an abrasive coating to promote proper lifting and cascading of the particulate bed.

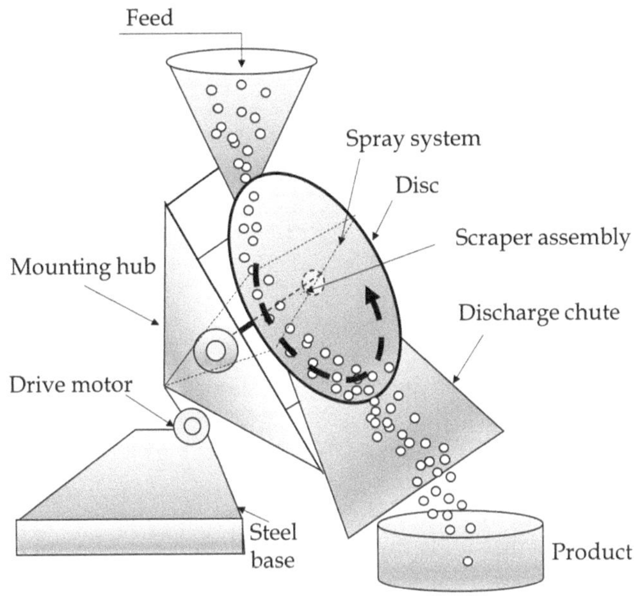

Figure 9.1 Tumbling granulator.

- The pan can be of different shapes (Capes, 1980) such as:
  - an outer reroll ring that allows granules to be simultaneously coated or densified without further growth
  - multi-stepped sidewalls
  - a pan as a truncated cone.
- The typical operating range for disks is 50%–75% of critical speed, with angles of 45°–55°. The critical speed can be calculated as given in an earlier lecture for its calculation in the case of a ball mill.
- The advantage of this process is that the product can be with a narrow size distribution.

Typical example: Iron ore granulation in drum granulators (You et al., 2022)

- The primary raw materials composing iron ore for its granulation, such as mixed iron ore, dolomite, limestone, coke breeze, and return fines, can be used with a typical composition (67.96, 2.86, 2.67, 3.48, 3.38, 19.65 wt.% respectively). The chemical compositions of raw materials are given in Table 9.2.
- The mixed raw materials of typically 12.5 kg can be fed into the granulator (typically, a cylindrical container of 400 mm in diameter and 500 mm in length) for granulation experiments. In the granulation process, after the raw materials are fed into the granulator, the drum starts to rotate to mix the raw materials.

Table 9.2 Chemical compositions of raw materials (wt.%)

Raw Materials	*TFe*	*CaO*	*SiO_2*	*Al_2O_3*	*MgO*	*LOI*
Mixed iron ore	59.88	0.78	4.22	1.73	0.24	8.92
Dolomite	0.36	30.53	1.17	1.03	29.43	45.93
Limestone	0.35	52.25	2.08	1.03	3.21	42.32
Quicklime	0.28	74.63	1.62	1.47	0.67	12.57
Coke breeze	0.45	1.04	3.96	4.60	0.13	86.62
Return fines	57.44	9.74	5.34	1.99	1.33	2.15

LOI, loss on ignition.

- The rotation speed of the drum granulator can be fixed at 18 rpm to achieve the best mixing and granulation performance. There should be a small hole of diameter 10 mm at the 1 o'clock direction of the granulators to add water.
- The atomized water is to be sprayed to the surface from the hole after the raw materials are mixed for 1 minute. The spray rate can be approximately 10 g/s. The water content should be about 8.8% in weight to achieve optimal granulation performance.
- The granulation time can be set to 10 minutes. At this granulation time of 10 minutes, the granules can be formed within the size range of 3 and 8 mm, which are the preferred granule sizes for the sintering process. Increasing the granulation time may result in coarser granules (>12 mm), which can decrease the sintering performance. After granulation, the granules were removed from the granulators for further analysis.
- Mixed iron ore and return fines have high iron content and low CaO, SiO_2, Al_2O_3, and MgO impurities. Dolomite and limestone have high MgO or CaO contents, which control the basicity. Quicklime reacts with water to form calcium hydroxide colloidal particles with high dispersion and high specific surface area. These colloidal particles fill the voids between iron ore fine particles to increase cohesion between particles and facilitate granules' coalescence and growth. Coke breeze is a typical solid fuel providing heat during the sintering process.

9.3 MIXER GRANULATOR

- In mixer granulators, the motion of the particles is governed by some form of agitator rotating at speed on a vertical or horizontal axis, as shown in Figure 9.2.
- It comprises a mixing chamber, impeller, grinder, outlet, filter unit, air, and solution purifier. The mixing granulator has three types: cone, dome, and round to eliminate dead corners and wall sticks.

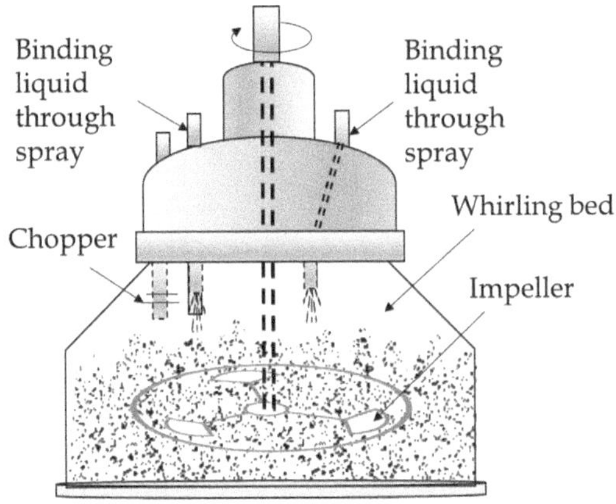

Figure 9.2 Schematic of mixer granulator.

Table 9.3 Ratios of different compounds used for the granulation

Component	*IND*	*PCS*	*Xylitol*	*Ethanol*
(mg/tablet)	30	60	120	180
Component	Distilled water	Silicified microcrystalline cellulose	Crospovidone	Magnesium stearate
(mg/tablet)	60	73.5	15	1.5

- The granulator utilizes the impeller's mixing (compression) force on the powder material to provide the composition, density, and aggregation degree.
- Operating the horizontal axis granulation at 50 rpm is suitable for fertilizer granulation.
- Vertical Schugi high-shear continuous granulator is generally used for detergents and agricultural chemicals, operating at around 3000 rpm.
- The pharmaceutical industry uses vertical axis mixers operating at speeds ranging from 500 to 1500 rpm for mixers less than 30 cm in diameter and 50–200 rpm for mixers larger than 1 m in diameter.

Typical example: Production of porous calcium silicate (PCS) granules-based tablets with poorly water-soluble indomethacin (IND) by wet granulation method in a high-speed mixer granulator (Fujimoto et al., 2016)

- The PCS granules can be prepared using IND, PCS, and xylitol (as a binder) along with other components in the ratio, as shown in Table 9.3.

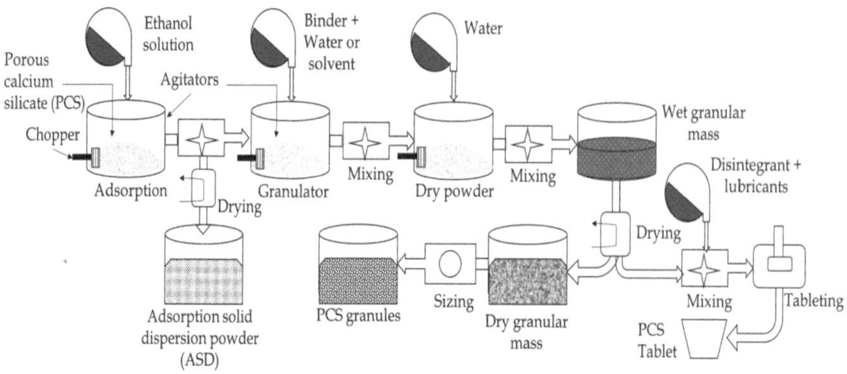

Figure 9.3 Preparation of porous calcium silicate (PCS) granules-based tablets with poorly water-soluble indomethacin (IND) by wet granulation method.

- To prepare PCS granules, IND will be completely dissolved in ethanol by heating at 60°C. The solution is then added to PCS and mixed for 5 minutes using a high-speed mixer granulator at 250 rpm (typically) with an agitator and then at 2500 rpm with a chopper.
- The binder (Xylitol) is then dissolved in distilled water and added to the PCS/IND mixture. If required, distilled water will be added to equalize the weight ratio of liquid (ethanol and water) to PCS. After the granulation endpoint, PCS granules will be dried at 70°C for 12 hours and pulverized in a speed mill.
- PCS tablets are prepared using a mixture of PCS granules, silicified microcrystalline cellulose, crospovidone, and magnesium stearate in ratios specified in Table 9.3. The total weight of the mixture can typically be taken as 100 g. PCS granules are mixed with silicified microcrystalline cellulose and crospovidone using a blender for 5 minutes, further mixed with magnesium stearate for 2 minutes, and then compressed using a rotary tableting machine at a rotation speed of 20 rpm with a compression load of 5 kN for each batch of tablet making. The schematic diagram of the whole process is shown in Figure 9.3.

9.4 FLUIDIZED BED GRANULATOR

- The fluidized bed granulation process involves the suspension of particles. It is set in motion by fluidizing air to improve the powder properties for preparing pharmaceutical dosage form with enhanced functional properties, i.e., solubility and dissolution, as shown in Figure 9.4.
- The air is passed through an air distributor (such as a nozzle, perforated plate, porous plate, tuyer, or grid type)

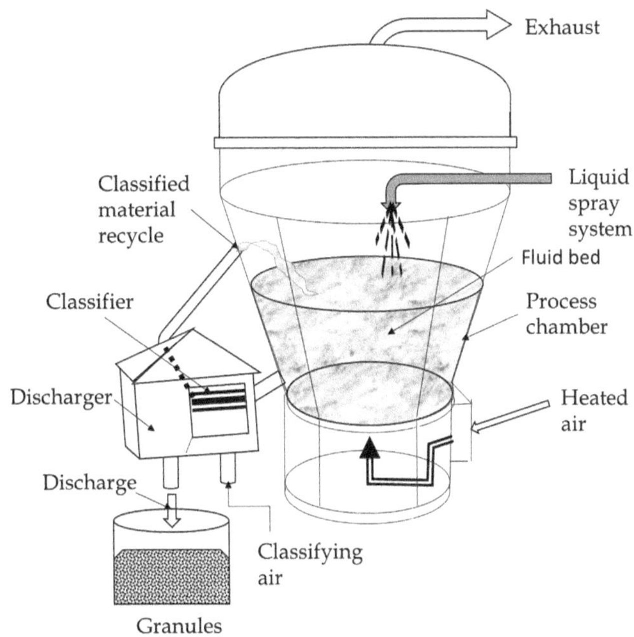

Figure 9.4 Fluidized bed granulator.

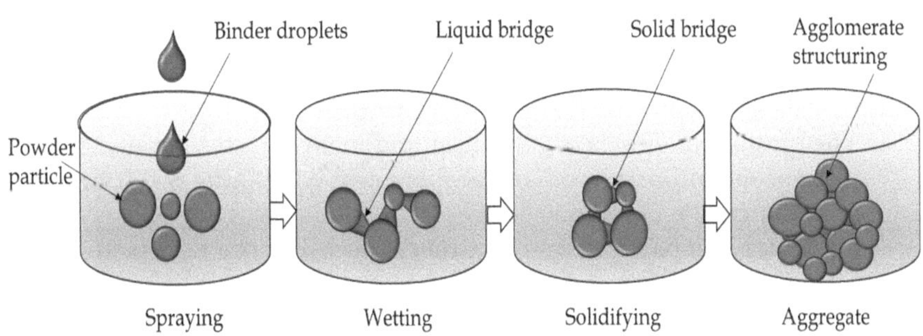

Figure 9.5 Steps in fluidized bed granulation.

- The liquid binder and wetting agents (hydrophilic granulating agent is used for hydrophobic starting material for better contact) are sprayed in the atomized form above or within the bed, which wet the particles.
- At fluidized conditions, the particles collide with each other, and then the formation of the liquid bridges leads to the acquisition of granules.
- The liquid bridge adheres the particles to each other in two mechanisms: (a) by surface tension at the interface and (b) by hydrostatic suction.
- The solidification of liquid bridges is dried, then a cluster of solid particles, such as granules or agglomerates, are formed. The steps of granulation are shown in Figure 9.5.

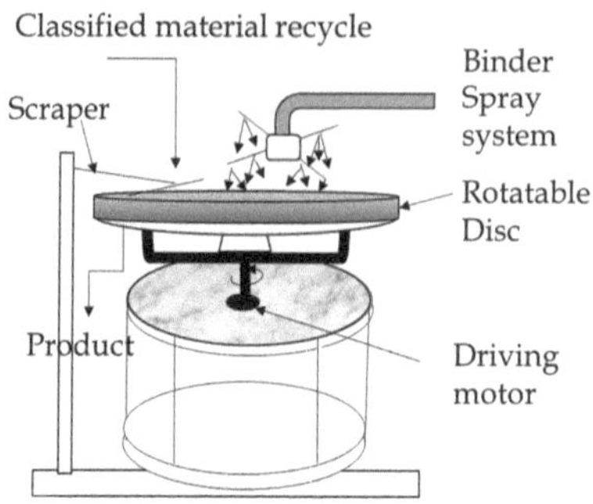

Figure 9.6 Rotating-disk fluid bed granulator.

- So, fluidized bed granulation involves continuously blending dry excipients and wetting and drying particles.
- Two types of fluidized bed granulators are generally used, as follows.
  - **Top spray fluid bed granulator:** At the top of the equipment, the spray nozzle is located, and at the base, the powder material is loaded into the product container with a fine mesh retention screen to allow small particles.
  - **Rotating-disk fluid bed granulator:** The product moves under centrifugal force to the outer position when the disk is rotated at variable speed and lifted into the expansion chamber by fluidizing the air stream. The spraying of liquid is applied to the particles in a tangential position. The schematic of the rotating-disk fluidized bed granulator is shown in Figure 9.6.
- The granulator provides good heat and mass transfer because it provides proper mixing, mechanical simplicity, combining the drying and granulation stages, sufficient fines to fill void spaces between granules for better compaction, and continues the bed more or less at a uniform temperature.

Typical example: Granulation of gellan gum (microbial hydrocolloid) in fluidized bed granulator (Dev et al., 2023)

- Clerigel powder along with binder Maltodextrin (MD) or D-Glucose or anhydrous lactose form can be used for granulation in a fluidized bed granulator, as shown in Figure 9.4.
- The granulator is charged up to 70% (75 g gellan gum) of its capacity. The bed is fluidized with an upward-flowing compressed air stream at 2 mL/min air velocity at a 1.5 bar pressure drop.
- The bed is heated for 10 minutes to ensure a material temperature of 50°C. The binder solution (100 mL) is then sprayed through the nozzle in the fluidized bed.

- The solution can be fed in pulse-feeding mode using a peristaltic pump operated at a 20 mL/min flow rate. The air stream temperature at the inlet and outlet can be maintained at 75°C and 50°C, respectively.
- After the complete addition of the binding agent, the material is to be dried for the next 10 minutes at 50°C.
- The binder solutions should be prepared by dissolving the high-purity binder in distilled water at 2.5%–10% in w/v. After that, the granule's flowability and cohesiveness are to be evaluated.

9.5 CENTRIFUGAL GRANULATOR

- In the centrifugal granulation unit, a granulator, a connecting shaft, a frequency conversion motor, and a granulation chamber are employed to actualize the centrifugal granulation process.
- The granulator is a rotary cup, as shown in Figure 9.7, made of stainless steel (with a typical diameter of 126 mm, a depth of 40 mm, and an inner leaning angle of 125°).
- The granulator is connected to the frequency conversion motor with an adjustable rotating speed range of 0–3000 rpm via the connecting shaft.
- The granulation chamber with an octagonal section is specially designed, and the maximum diameter of the cross-section where the granulator is located is 1.5 m. A high-speed camera can be set at the observation window to shoot the granulation process.

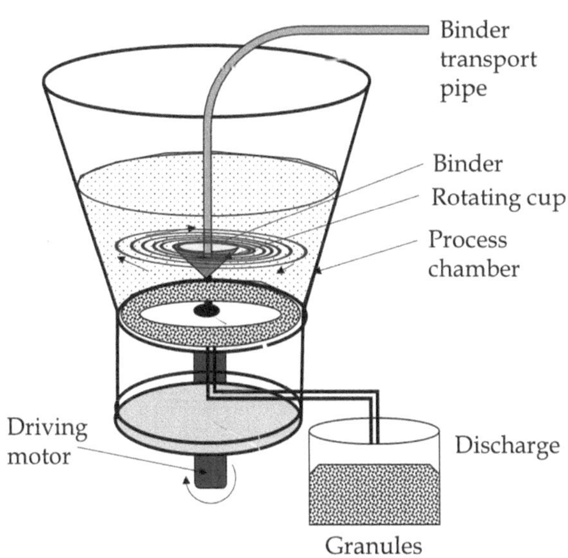

Figure 9.7 Centrifugal granulator.

- The binder solution flows down, is sprayed on the rolling cores from the upper nozzle or discharging pipe, and spreads on the action of the centrifugal force generated by the rotating granulator at a specific flow rate.
- Hot air is blown upward between the turntable and the granulation area. The air causes the cores to roll.
- Liquid filaments are then formed at a specific flow rate and separated into tiny droplets. After adequate cooling, the droplets become a solid bridge with the particles and make granules.
- When the flow rate is very low, the film breaks into rivulets. As the flow rate increases, a smooth, uniform film is formed in the form of circumferential rings. The process is also called solid layering.
- The centrifugal granulation technique is quite promising due to its advantages of simple and compact structure, effortless control, and low energy consumption.

Typical example: Centrifugal granulation of molten blast furnace (BF) slag (Tan et al., 2018)

- Molten BF slag is almost the last high-temperature liquid waste heat resource that has not been recycled in the iron and steel-making industry. Dry granulation techniques are generally proposed to aim at heat recovery of BF slag. The experimental system for the centrifugal granulation of molten slag is shown in Figure 9.8.

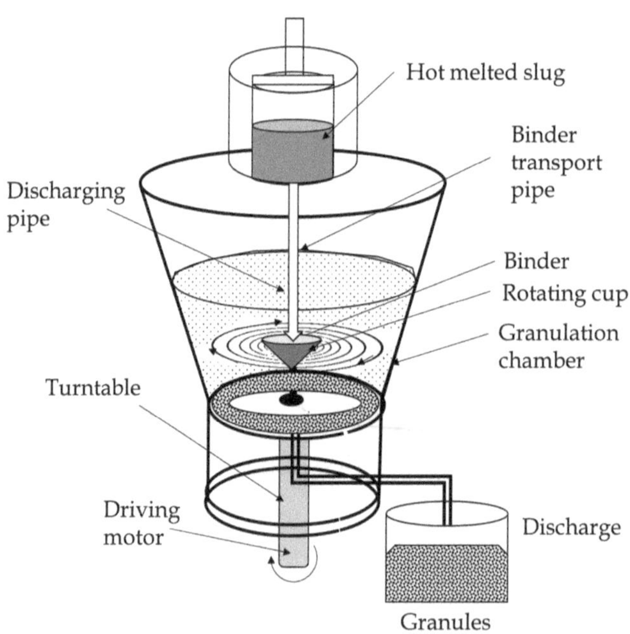

Figure 9.8 Centrifugal granulation of molten blast furnace (BF) slag.

- The working medium is industrial water-quenched BF slag.
- The typical centrifugal granulation unit consists of a granulator, a connecting shaft, a frequency conversion motor, and a granulation chamber to actualize the centrifugal granulation process. As shown in Figure 9.8, the granulator is a rotary cup manufactured of stainless steel with a diameter of 126 mm, a depth of 40 mm, and an inner leaning angle of 125°.
- The granulator is connected to the frequency conversion motor with an adjustable rotating speed range of 0–3000 rpm via the connecting shaft. The granulation chamber with an octagonal section is specially designed, and the maximum diameter of the cross-section where the granulator is located is 1.5 m. A high-speed camera (typically phantom V 5.1) is set at the observation window to shoot the granulation process of molten slag at 3100 fps.
- Before experiments, the BF slag is remelted in the high-temperature furnace at 1500°C and kept for 1 hour to ensure temperature uniformity. After that, the experiment is triggered by pulling the stopper to release the molten BF slag through the discharging pipe toward the rotating cup.
- A high-speed camera is used to shoot the granulation evolutions. The temperature distributions of the molten slag during granulation are acquired using the image gray-level processing method. The products are collected and sorted. After sieving and weighting, the average diameter of the obtained slag particles is calculated and analyzed further.

9.6 PRILLING EQUIPMENT

- Prilling is a method to produce reasonably uniform spherical granules from molten solids, robust solutions, or slurries (e.g., nitrogen fertilizers/urea prilling).
- It consists of two operations: producing liquid drops and solidifying them individually by cooling counter. The cooling is done by falling them through a rising ambient air stream in a prilling tower. During cooling, they will crystallize into spherical solids. A typical example of urea prilling in a prilling tower is shown schematically in Figure 9.9.
- Concentrated urea solution comes from the evaporator and is sprayed downward from the top of the prilling tower. The inlet air draft rising inside the prilling tower cools the falling, sprayed solution, forming solid "prills." The prills continue to fall to the bottom of the tower for collection.
- At the top of the tower, around the sprayer heads, the urea micro prills (~100 μm) and average 1-μm diameter mist are entrained by the air going on up to the roof fans and thus out, which is further filtered in the ground.

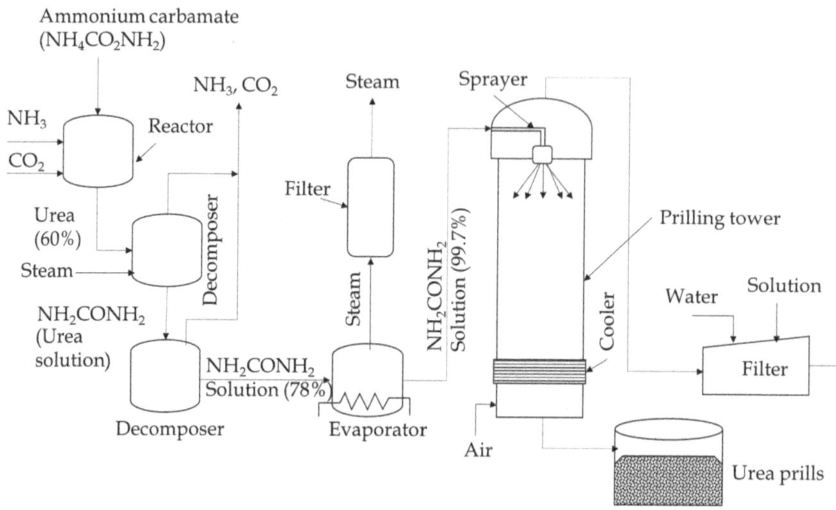

Figure 9.9 Typical prilling in a prilling tower.

Typical example: Formation of iron-encapsulated microstructured emulsion particles by the prilling process (Dubey and Windhab, 2013)

- Encapsulation of functional components is commonly used to protect them from the environment or to control their release. Iron is a model nutrient encapsulated in a water-in-oil emulsion.
- Before prilling, the emulsion (approx. 40% (w/w) water-in-oil, W/O) is prepared with palm stearin (melting point 50°C–55°C), which contains 1.9% (w/w) Span-20 as an emulsifier and 2.86% (w/w) PGPR (poly-glycerol polyricinoleate) as a stabilizer and an emulsifier.
- The O/W emulsion is prepared by rotor–stator device at a very high rotational speed (about 7000 rpm) for 30 minutes at a constant temperature (60°C). This long-term acting high rotational speed leads to the dispersed phase's equilibrium mean droplet size.
- In general, the droplet size is about 1–4 μm depending on the viscosity of the dispersed phase at the given rotational speed. Due to the low stability of fat-based emulsions at high temperatures, the emulsions are used in the prilling experiment in the prilling tower (Figure 9.9) immediately after preparation.
- The solid particles from emulsions are produced in a prilling tower (typical height of about 4.5 m and diameter 1 m). Two types of twin-fluid nozzle geometries (internal and external mixing) can be used for the experimental investigations with different GLR gas–liquid mass ratios of typically 0.282–1.377 at constant liquid flow (typical 138 mL/min).

- The corresponding applied air pressures can be varied from 1.22 to 3.7 bar. The average temperature inside the tower is maintained at about –10°C. After the prilling operation, particles are collected from the bottom of the tower and stored at room temperature in a closed container for further analysis.

9.7 EXTRUSION MACHINE

- Extrusion is a compressive metal-forming process. In this process, a piston or a plunger is used to apply a compressive force to the metal piece. First, a metalwork piece of standard size is produced. Then it is heated in a hot extrusion chamber and placed into an extrusion press, as shown in Figure 9.10.
- The extrusion press is like a piston-cylinder device in which a metal piece is placed in the cylinder and pushed by the piston.
- The upper portion of the cylinder is fitted with a die. After that, a compressive force is applied to this part by a plunger fitted into the press, which pushes the metal piece toward the die.
- The die is a small opening of the required cross-section. This high compressive force allows the work metal to flow through the die and convert into the desired shape.
- There are generally three types of extrusion machines commercially available which are briefed as follows:

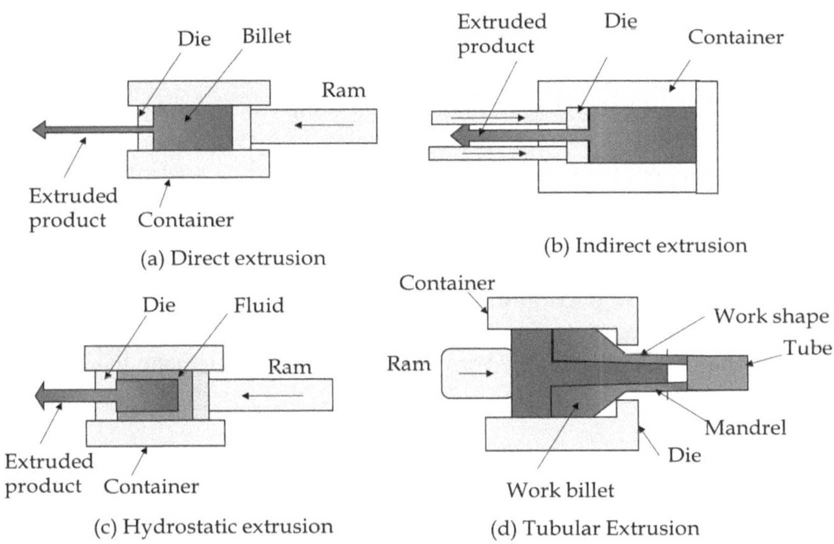

Figure 9.10 Different configurations of extrusion machine.

Direct Extrusion:	• In this case, metal is forced to flow in the direction of feed of punch. • The punch moves toward die during extrusion. • This process required higher force due to higher friction (Figure 9.10a).
Indirect Extrusion:	• In this process, metal is allowed to flow through annular space between punch and container toward opposite direction of plunger movement. • The die is fitted at opposite side of punch movement (Figure 9.10b).
Hydrostatic Extrusion:	• This process uses fluid to apply pressure on metal. • There is a fluid between the metal piece and plunger without friction. • The plunger applies force on fluid which further applied on workpiece (Figure 9.10c).
Tube Extrusion:	• A tube is formed by pressing the extruded polymer through a circular die, cooling it in a water bath, and then cutting it or winding it up. This is mainly used in the production of railings, tubes, shafts. (Figure 9.10d)

- The machine is widely used in the production of especially plastic objects like tubes and hollow pipes, structures, frames, doors, windows, etc.

9.8 ROLLER COMPACTION MACHINE

- In this machine, the fine powders are forced between two counter-rotating rolls, and the raw materials are pressed into a solid compact (e.g., flakes, sheets, strips), as shown in Figure 9.11. Roll compactors are also called dry granulators.
- A roller compactor generally consists of three major types, as follows.
  - **Type 1:** A feeding system transports the powder to the compaction area between the rolls.
  - **Type 2:** The powder is compacted by force between two counter-rotating rolls to a ribbon.
  - **Type 3:** A size reduction unit for milling the desired particle size.
- Roller compactors may also be divided into two categories: one is equipped with a fixed gap, and the other with a floating gap. Both consist of the three major units as explained above but differ in how the smallest distance (gap) between the rolls is realized.
- The roller compactor is used suitably to prepare directly compressible excipients, compaction drugs and drug formulations, granulation of dry herbal extracts, immediate-release formulations, and enhanced dissolution and controlled-release formulations (Kleinebudde, 2004).

9.9 TABLET PRESS

- A tablet press machine is used to make the powder into different types of tablets or pellets of a particular size, shape, and weight by compression. The machine is called a tablet punching machine. All tablet compressing machines employ a similar basic principle of compression.

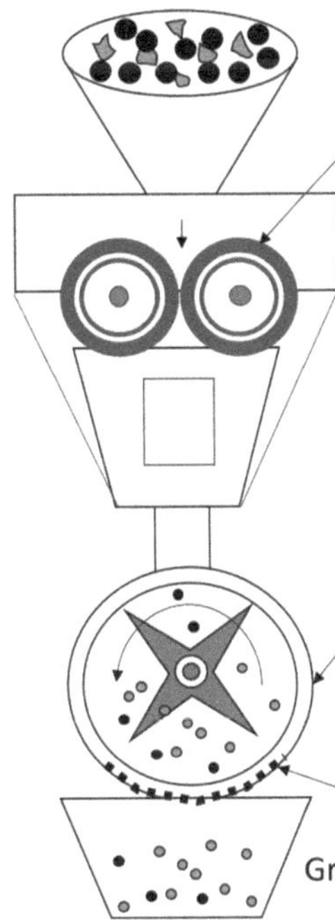

Figure 9.11 Schematic of roller compaction machine.

- There exist two types of tablet presses, namely:
- **Single-punch tablet press**
  - Usually, this equipment, as shown in Figure 9.12, entails a single tooling station having a die and pairs of bottom and top punches. During the process of compression, the bottom punch stays stationary. The working principle is the same as that of the hammering movement.
- **Rotary tablet press**
  - It comes in the form of a rotating turret that carries any number of dies and punches. This type of press is designed to enhance the output of tablets.
  - They spin about the turret. The punches come into contact with cams, which modulate the vertical posture of each punch. In a rotary tablet press, both the upper and the lower punches exert the force of compaction on the fill material leaving the powder granules you want to compress in the middle. The shape of the tablets depends on the punch design.
  - Contrary to the single-punch tablet compressing machine, this type of press has various tooling stations. Usually, these stations rotate to compress powder or granule mixture into tablets of uniform size, shape, and weight.

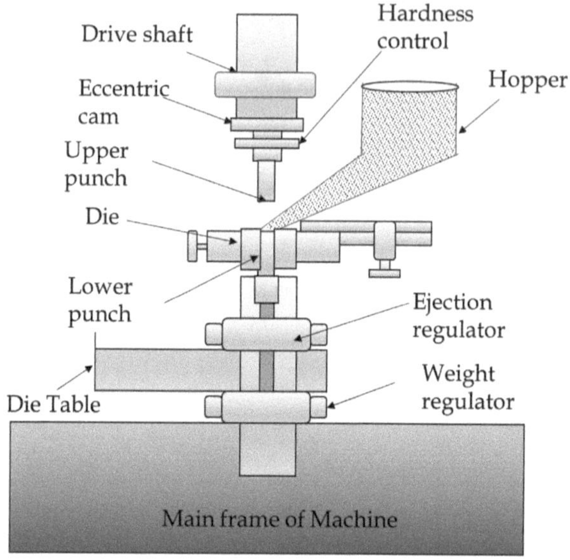

Figure 9.12 Schematic diagram of single-punch tablet press.

- The number of stations on the machine and the speed of rotation of the turrets determines the capacity of a rotary tablet press. A schematic diagram of the rotary tablet press is shown in Figure 9.13.
- There are two main types of rotary tablet presses.
  - **Single rotary:** In a single rotary tablet press, the tablet dies are mounted on a central rotating turret. It can only be used to produce round tablets. Single rotary presses also have a higher maximum tablet output than double rotary presses.
  - **Double rotary:** In a double rotary press, two sets of dies are mounted on two separate turrets. It can be used to produce a variety of shapes and sizes. Double rotary presses offer greater precision and accuracy and the ability to make complex shapes. Double rotary tablet presses can produce two tablets simultaneously, increasing production efficiency. Double rotary tablet presses typically have larger die cavities than single rotary tablet presses, which means they can accommodate more active pharmaceutical ingredients (API) and produce larger tablets.

9.10 MOLDING PRESS

- The molding press produces thermoplastic resin, thermoset polymers, etc.

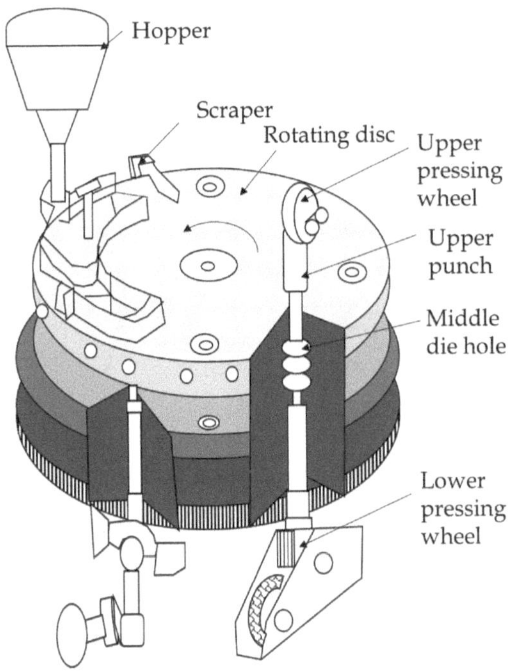

Figure 9.13 Schematic diagram of the rotary tablet press.

- The plastic material is heated above its melting point, converting the solid polymer to a molten material. It is then forced into a closed mold that defines the shape of the material to be produced.
- There are two basic types of plastic molding process (Parikh, 2021).
    - **Injection blow molding:** During the injection molding process, a blow molding machine (also called a plastic injection molding machine) produces molded plastic parts by converting plastic pellets into molten material and injecting that material into a mold. After cooling, the plastics become solid. The main components of the injection molding machine are shown in Figure 9.14.
    - **Extrusion blow molding:** During the extrusion molding, extrusion equipment shapes plastics by forcing molting material through a shaped die. After the hot material exits the die, it is cooled and cut to length.
- The injection molding machine consists of a barrel (cylindrical pipe). A hopper is located at one end of the barrel. A hydraulic ram or rotating screw runs by an electric motor that provides force inside the barrel. A heating element (heater) is attached to the barrel and circumstantially used to melt the molding material from the hopper.

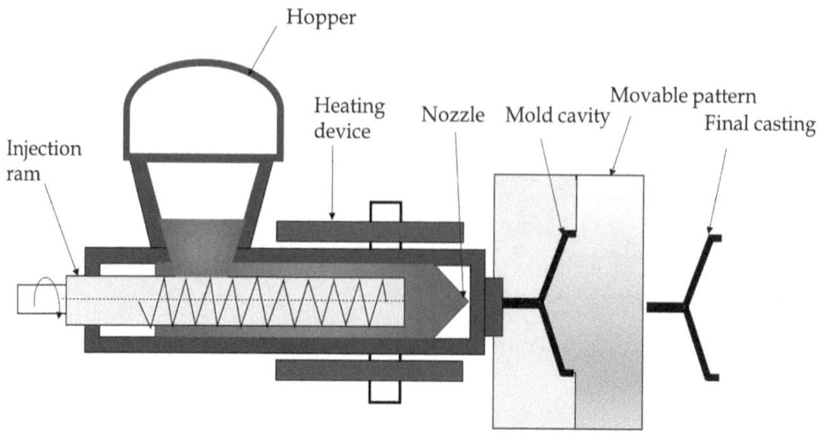

Figure 9.14 The schematic of the injection molding machine.

- The molding material/raw material is poured into the hopper by a feeding device. After that, molding material goes down under gravity into the cylinder (barrel), as shown in Figure 9.14. A circumferential heater located on the barrel is used to melt the material. When the powder form of molding material goes down into the barrel from the hopper, it starts melting, and a hydraulic ram or rotating screw pushes the material forward into the mold by applying some pressure from the hydraulic system. Molten plastic material is injected into a closed mold attached to the other side of the barrel. Injection pressure is generally 100–150 MPa. After injection, pressure is applied for some time or locked at the same position with some force. After the whole process is done, the parts manufactured are cooled sufficiently. Then the mold is opened, and some ejectors are used to properly remove the part without damage. After removing the part, the mold is closed again.

9.11 PELLET MILL OR PRESS

- The machine is used to create pellets from powdered material. The mills can be generally grouped into large-scale and small-scale types (Ghebre-Sellassie, 1989).

9.11.1 Large-Scale Type

- Large-scale pellet mills are suitable for producing animal feed, wood, and fuel pellets for a pellet stove.

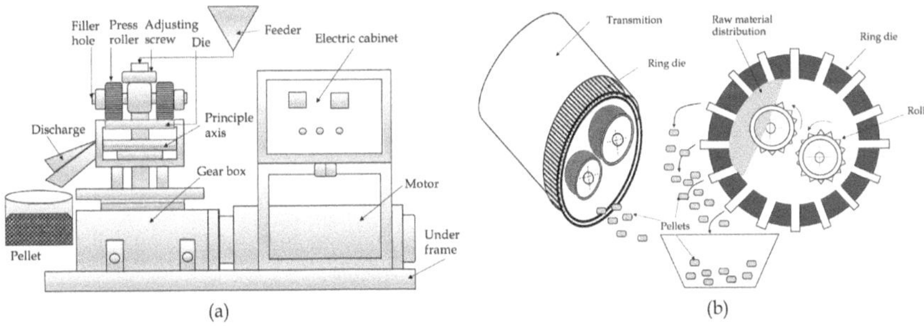

Figure 9.15 Schematic of die pellet machine. (a) Flat die mills and (b) ring die mill.

- Two common types of large-scale pellet mills are flat die mills and ring die mills.
    - **Flat die mill:** In the flat die mill, a flat die is used with slots, as shown in Figure 9.15a. The powder is added to the top of the die. The feeding relies on the material's weight, and the feedstock enters the pelleting room directly and evenly. The mill has a circular perforated disk on which two or more rollers rotate and force the material through the holes to form pellets. As the die rotates, a roller presses the powder through the holes in the die. A cutter on the other side of the die cuts the exposed pellet free from the die.
    - **Ring die mill:** The ring die press features a rotating perforated ring on which rollers (normally two or three) press on the material to the inner perimeter to form pellets (Figure 9.15b). The powder is fed into the die, and the powder is evenly distributed by the spreaders. Then the rollers compress the powder through the die holes. The cutters cut the pellets free from the outside of the die. The feeding is forced into the pelleting chamber and distributed by the high-speed rotary scrapers.
- The output capability of the flat die pellet mill ranges from around 80 to 1000 kg/h at most, while the ring dies pellet mill can produce around 500–8000 kg/h at most.

9.11.2 Small-Scale Type

- Small-scale mills are categorized as screw and hydraulic presses, as shown in Figure 9.16. The basic process is the same for both cases.
- A die holds the uncompressed powder in a shaped pocket. As per the pocket shape, the final pellet shape is formed.

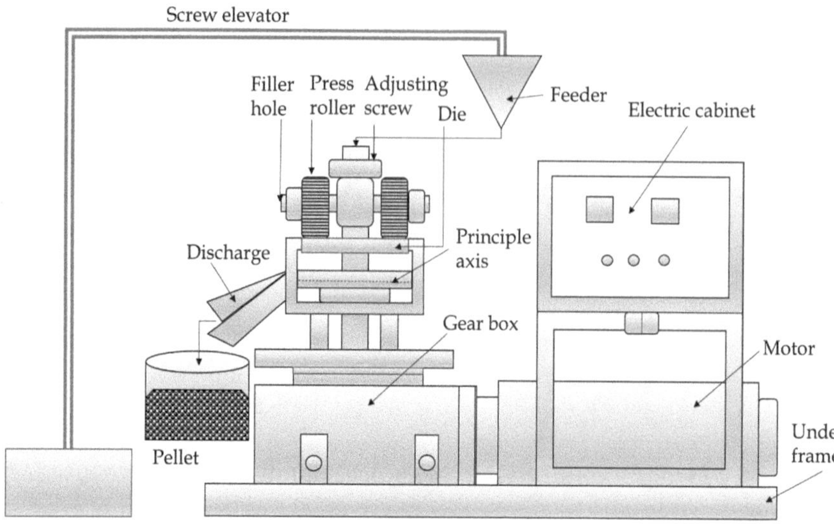

Figure 9.16 A small-scale screw-type pellet mill.

- In the case of a screw press, a platen is attached to the end of the screw, whereas in the case of a hydraulic press, a ram is used, which compresses the powder.
- The mills are generally suitable for producing potassium bromide pellets used in infrared spectroscopy applications.

IMPORTANT POINTS TO REMEMBER

- Motion by a tumbling action caused by the balance between gravity and centrifugal force.
- The rotation speed of the drum granulator can be fixed at 18 rpm to achieve the best mixing and granulation performance. There should be a small hole of diameter 10 mm at the 1 o'clock direction of the granulators to add water.
- The pharmaceutical industry uses vertical axis mixers operating at speeds ranging from 500 to 1500 rpm for mixers less than 30 cm in diameter and 50–200 rpm for mixers larger than 1 m in diameter.
- PCS tablets are prepared using a mixture of PCS granules, silicified microcrystalline cellulose, crospovidone, and magnesium stearate.

- Rotating-disk fluid bed granulator provides good heat and mass transfer because it provides proper mixing, mechanical simplicity, combining the drying and granulation stages, sufficient fines to fill void spaces between granules for better compaction, and continues the bed more or less at a uniform temperature.
- The centrifugal granulation technique is quite promising due to its advantages of simple and compact structure, effortless control, and low energy consumption.
- Prilling is a method to produce reasonably uniform spherical granules from molten solids, robust solutions, or slurries (e.g., nitrogen fertilizers/urea prilling).
- Extrusion is a compressive metal-forming process. In this process, a piston or a plunger is used to apply a compressive force to the metal piece.
- The molding press produces thermoplastic resin, thermoset polymers, etc.

SELF-TESTING QUESTIONS

I. Multiple Choice Questions

Q1. What is the basic principle of a tumbling granulator?
- (a) Compressing materials under high pressure to form granules
- (b) Spraying a liquid binder onto a powder to form granules
- (c) Mixing materials to form granules
- (d) Tumbling and agitating materials in a rotating drum to form granules

Q2. The colloidal particles are used for granulation in a tumbling granulator. This is due to__________________.
- (a) it fills the voids between fine particles to increase cohesion between particles
- (b) it fills the voids between fine particles to decrease cohesion between particles
- (c) it increases the voids between fine particles to increase cohesion between particles
- (d) none of these

Q3. Which of the following materials provides heat during the sintering process?
- (a) Sodium hydroxide
- (b) Coke breeze
- (c) Magnetite
- (d) Gypsum

Q4. The motion of the particles is governed by some form of agitator rotating at speed on a vertical or horizontal axis. This is executed in which of the following granulators?
(a) Fluidized bed granulator
(b) Centrifugal granulator
(c) Prilling granulator
(d) Shear-induced mixer granulator

Q5. Operating the horizontal axis granulation the suitable rotational speed for fertilizer granulation is________________.
(a) 50 rpm
(b) 100 rpm
(c) 200 rpm
(d) 300 rpm

Q6. Vertical Schugi high-shear continuous granulator is generally used for detergents and agricultural chemicals which are operating at around ________________.
(a) 1000 rpm
(b) 2000 rpm
(c) 3000 rpm
(d) 3500 rpm

Q7. The pharmaceutical industry uses vertical axis mixers larger than 1 m in diameter.
Which is operated at a speed range________________.
(a) 0–50 rpm
(b) 50–200 rpm
(c) 250–300 rpm
(d) 350–400 rpm

Q8. The fluidized bed granulation process involves________________.
(a) the suspension of particles
(b) the coagulation of particles
(c) the elutriation of particles
(d) the compaction of particles

Q9. At fluidized conditions, the particles collide with each other, and then the formation of the liquid bridges leads to ________________.
(a) the coating of granules
(b) the dissolving of granules
(c) the acquisition of granules
(d) the suspension of granules

Q10. The liquid bridge adheres the particles to each other in a fluidized bed granulator in by ________________.
(a) hydrostatic suction
(b) surface tension at the interface
(c) both hydrostatic suction and surface tension at the interface
(d) none of these

Q11. Hot air is blown upward between the turntable and the granulation area in a centrifugal granulator. The air causes the cores ___________.

(a) to roll
(b) to make a solid bridge
(c) to make coalescence of the particle
(d) to discharge the particle

Q12. In a centrifugal granulator, when the flow rate is very low, the film breaks into rivulets. As the flow rate increases, a smooth, uniform film is formed in the form of circumferential rings. The process is called ___________

(a) solid stressing
(b) solid layering
(c) solid sublimating
(d) solid mixing

Q13. Prilling is a method to produce reasonably uniform spherical granules from ___________.

(a) molten solids
(b) amorphous solids
(c) liquid vapor
(d) binder solution

Q14. In which operation solidify the liquid drops individually by cooling counter currently?

(a) Distillation
(b) Adsorption
(c) Leaching
(d) Prilling

Q15. Extrusion machine works on the principle of___________.

(a) application of a compressive force to the metal piece
(b) application of an impact force to the metal piece
(c) application of a shearing force to the metal piece
(d) application of a centrifugal force to the metal piece

Q16. In which machine the fine powders are forced between two counter-rotating rolls and pressed the raw materials into a solid compact?

(a) Molding press
(b) Rotary tablet press
(c) Pellet press
(d) Roller compaction machine

Q17. Which of the following is NOT a benefit of fluidized bed granulation?

(a) Improved flow properties of granules
(b) Uniform granule size distribution
(c) Reduced dust generation during processing
(d) Reduced processing time

Q18. Which binder is used in wet massing and fluid bed granulation?
- (a) Polyvinylpyrrolidone
- (b) Polyvinylchloride
- (c) Polyethylene
- (d) Khaya gum

Q19. Match the following granulator methods with product size?

Method	Product size (mm)
A. prilling	i. 0.3–3
B. Spray method	ii. 0.5–20
C. Centrifugal granulator	iii. 0.7–2
D. Tumbling	iv. 0.05–0.5

- (a) A – (iii), B – (iv), C – (i), D – (ii)
- (b) A – (iii), B – (ii), C – (i), D – (iv)
- (c) A – (ii), B – (iv), C – (i), D – (iii)
- (d) A – (iii), B – (i), C – (iv), D – (ii)

Q20. The roller compactor is used suitably for the preparation of directly ________________
- (a) compressible excipients
- (b) incompressible excipients
- (c) granules without binder
- (d) none of these

Answers to the multiple choice questions

Q1. Ans. (d)
Q2. Ans. (a)
Q3. Ans. (b)
Q4. Ans. (d)
Q5. Ans. (a)
Q6. Ans. (c)
Q7. Ans. (b)
Q8. Ans. (a)
Q9. Ans. (c)
Q10. Ans. (c)
Q11. Ans. (a)
Q12. Ans. (b)
Q13. Ans. (a)
Q14. Ans. (d)
Q15. Ans. (a)
Q16. Ans. (d)
Q17. Ans. (d)
Q18. Ans. (a)
Q19. Ans. (a)
Q20. Ans. (a)

II. Objective Type Questions

1. Write the common categories of granulators that are mainly used for the granulation process.
2. Give typical examples of the application of each type of commonly used granulator.
3. Explain the working principle of tumbling granulator.

4. Explain the iron ore granulation in drum granulators.
5. Explain the working principle of a mixer granulator with an example.
6. How fluidized bed granulator works. Describe it with an example.
7. List different types of fluidized bed granulators and their principle.
8. How is the granulation of gellan gum (microbial hydrocolloid) in a fluidized bed granulator done?
9. Explain centrifugal granulation with a schematic diagram and example.
10. How does prilling equipment work? Give an example.
11. Explain the working principle of the extrusion machine. What are the different configurations of extrusion machines?
12. Explain briefly the major types of roller compaction machines.
13. How does a tablet press work? What are the different types of tablet press machines?
14. How a rotary tablet press is classified? How is it different from a pellet mill or press?
15. How does a molding press work? What are the different types of molding presses?

REFERENCES

Capes, V. C. E., *Handbook of Powder Technology: Particle Size Enlargement*, Vol. 1, Elsevier Scientific Publishing Comp., 1980.

Dev, M. J., R. G. Warke, G. M. Warke, G. B. Mahajan and R. S. Singhal, Fluidized bed granulation of gellan gum: Investigations of binder effect on physical, structural and rheological properties, *Powder Technol.*, 415, 118199, 2023.

Dubey, B. N. and E. J. Windhab, Iron encapsulated microstructured emulsion-particle formation by prilling process and its release kinetics, *J. Food Eng.*, 115 (2), 198–206, 2013.

Ennis, B. J. and J. D. Litster, Particle size enlargement, in: R. Perry, D. Green, (Eds.), *Perry's Chemical Engineers' Handbook*, McGraw-Hill, pp. 73–117, 1997.

Fujimoto, Y., N. Hirai, T. Takatani-Nakase, and K. Takahashi, Novel tablet formulation of amorphous indomethacin using wet granulation with a high-speed mixer granulator combined with porous calcium silicate, *J. Drug Deliv. Technol.*, 33, 51–57, 2016.

Ghebre-Sellassie, I., *Pharmaceutical Pelletization Technology*, Marcel Dekker Inc., 1989.

Kleinebudde, P., Roll compaction/dry granulation: Pharmaceutical applications, *Eur. J. Pharm. Biopharm.*, 58, 317–326, 2004.

Parikh, D. M., *Handbook of Pharmaceutical Granulation Technology*, 4th Edition, CRC Press, 2021.

Tan, Y., X. Zhu, H. Wang, X. He, B. Ding, and Q. Liao, Centrifugal granulation characteristics of molten blast furnace slag and performance of the granulated particles, *Appl. Therm. Eng.*, 142, 683–694, 2018.
You, Y., J. Guo, Z. Zheng, G. Li, Y. Li, X. Lv, and R. Yang, Comparisons of iron ore granulation and sintering performance between horizontal high-shear granulators and drum granulators, *Powder Technol.*, 402, 117367, 2022.

Section IV

Mixing of Solids

Chapter 10

Mixing of Solids – Introduction

> I have always felt it is my destiny to build a machine that would allow man to fly.
>
> —*Leonardo da Vinci*

10.1 MIXING, BLENDING, AGITATION

- Mixing is a process of randomizing dissimilar particles within a system.
- Mixing smoothly and inseparably together by imparting minimum energy to the bed is called blending.
- The terms "mixing" and "blending" are synonymously used in the pharmaceutical industry.
- Agitation refers to the induced motion of a material in a specific pattern (mainly circular patterns, it may be for single or multiphase). In contrast, mixing is a random distribution into and through one another (especially for multiphase) (McCabe et al., 2017).
- Mixing and agitation are not the same technically.
- There are four different categories on/between which the mixing occurs:
  - Solid–solid mixing
  - Fluid–solid mixing
  - Liquid–liquid mixing (both miscible and immiscible)
  - Fluid–semisolids mixing

10.2 IMPORTANCE OF MIXING

- Mixing (of gas, liquid, and solid in particulates or powdered form) is essential for any mechanical and transport processes in chemical, biochemical, and pharmaceutical industries.
- The factors that influence the solid mixing are characteristics of particles such as size, shape, volume, surface area, density, porosity, flow charge, and so on.

DOI: 10.1201/9781003571025-14

- For the preparation of many types of formulations requires proper mixing of ingredients.
- The production of several dosage forms (like pellets, capsules, etc.) based on dry and wet mixing or blending.
  - Wet mixing is generally considered in the granulation process in the production of tablets and capsules.
  - Dry mixing is done for various ingredients for direct compression, as in tablets.
  - Formulating dry syrups and compound powders, dry blending of the compositions is recommended for making capsules.

10.3 MECHANISM OF MIXING

- The solid particles are mixed based on the following mechanisms depending on the type of mixers and the particles (Richardson et al., 2003):

Convective mixing:	• Groups of particles are moved from one position with a pattern. • Mixing two or more groups of solid particles so that they are dispersed in each other
Diffusion mixing:	• The particles are distributed over a freshly developed interface. • The powdered materials are mixed in a container where the movement of the powdered particles occurs by tilting the material beyond the angle of repose
Shear mixing:	• The momentum exchange of powder particles having different velocities leads to shear mixing. • Shear mixing can improve semimicroscopic mixing and be favorable in both batch and continuous operation.

10.4 FACTORS AFFECTING MIXING

The factors that affect the mixing are described as follows:

- **Nature of surface:** The particle surface should be smooth for effective mixing.
- **Particle size:** Increasing the difference in particle size will lead to segregation and decrease the ease of mixing.

- **Particle shape:** The particle should be spherical to get a uniform mixture.
- **Particle charge:** Some particles, due to electrostatic charge, exert an attractive force that leads to separation.
- **Proportion of material:** Mixing is easier if powders are in the same quantities of different types.
- **Relative density:** If the components have a distinct density, the denser material will sink through lighter material.
- **Viscosity:** An increase in viscosity reduces the extent of mixing.
- **Surface tension of liquids:** High surface tension reduces the mixing.
- **Temperature:** Temperature affects the mixing because viscosity changes with increased temperature.
- **Mixing time:** The degree of mixing depends on the mixing time.
- **Agitator type, shape, size, and location:** The shape, size, location, and type of agitator also affect the degree of mixing.
- **Speed/rpm of the impeller:** Mixing at less rpm is more homogeneous than that at high rpm.

10.5 CLASSIFICATION OF MIXERS/BLENDERS

- Particulate solids blending/mixing machines can be classified as:

Active

- The active type uses certain moving parts to assist in the randomization of the solid materials or the mixer machine moves about physically in the mixing process.
- Examples : Ribbon blenders, tumbler blenders, and high shear mixing and multi-mechanism blenders.

Passive

- In a passive mixer system the randomization of the materials is achieved by directed flow of the powder streams by baffles, etc. (i.e., no moving parts), as they move through the mixing device.
- Examples: Baffled and gravity in-bed mixing devices.

Another way of classification of mixing equipment (Harnby et al., 1985)

- The mixing equipment can also be classified based on flow properties and based on the scale of mixing:
- **Based on flow properties:** There are two types of mixing devices based on flow properties such as:
  - **Free-flowing solids:** Example – V-cone blender, double-cone blender
  - **Cohesive solids:** Example – Sigma blender, planetary miser.

- **Based on a scale of mixing:** It is based on batch and continuous operations as:
  - **Batch type (small scale):** Example – Morter and pestle, V-cone blender, double-cone blender, ribbon blender, Sigma blender, planetary paddle miser, fluidized mixer.
  - **Continuous (large scale):** Example – Barrel type mixer, Zigzag type mixer.

10.6 WORKING PRINCIPLES AND APPLICATIONS OF BLENDERS

10.6.1 Ribbon Blenders

10.6.1.1 Working principles

- A ribbon rotates within a static trough or open cylinder, and the particles are conveyed by the moving ribbon (single or double helical), as shown in Figure 10.1.
- During blending operations, the outer ribbon moves materials from the ends toward the center of the trough, and the inner ribbon moves the material toward the outside of the trough (i.e., from the center to the ends).

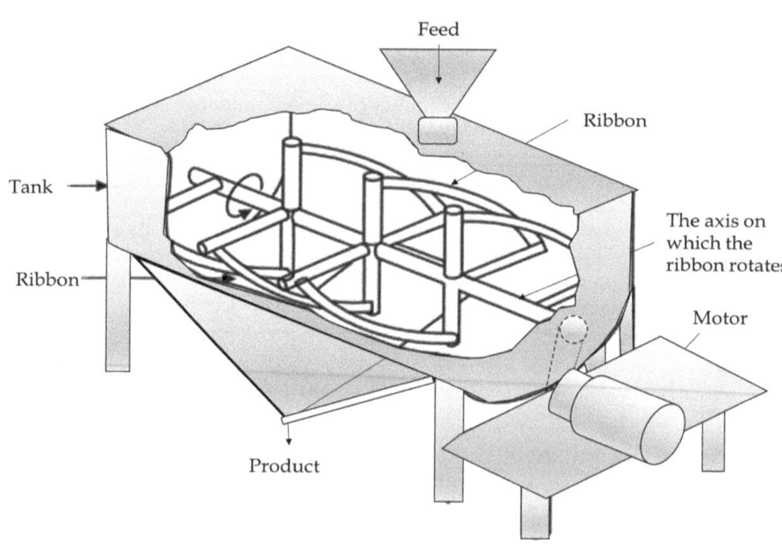

Figure 10.1 Schematic diagram of a ribbon blender.

- Axial movement results from the difference in the peripheral speed of the outer and inner ribbons along the horizontal axis of the blender.
- The mixing efficiency of a ribbon blender is affected by (a) solid particle size and (b) bulk density of the material.

10.6.1.2 Advantages and Disadvantages of Ribbon Blender

Advantages	*Disadvantages*
• It can be operated in both batch and continuous modes • It is recommended for free-flowing and cohesive products	• It cannot be used for fragile materials • It is not suitable for sticky materials

10.6.1.3 Applications

- It is used to blend materials ranging from low-density, finely divided materials that aerate rapidly to fibrous or sticky materials for heating, cooling, coating, and other processes.

10.6.2 Vertical Orbiting Screw Blender

10.6.2.1 Working principles

- It consists of a conical vessel with a rotating screw that lifts material from the base of the conical vessel to the vessel surface, as shown in Figure 10.2. It is also known as Nauta mixer or vertical cone screw blender.

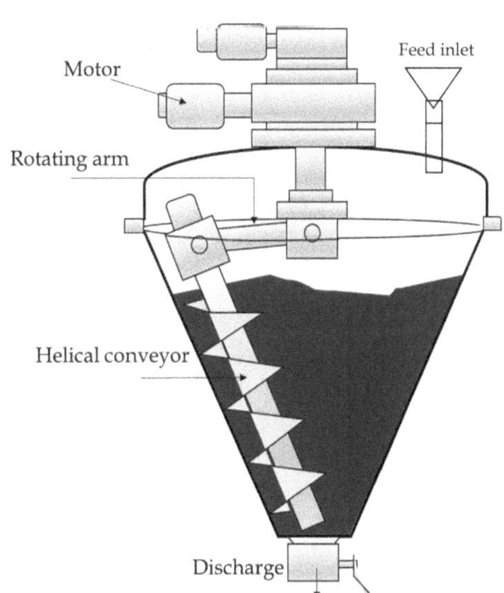

Figure 10.2 Vertical orbiting screw blender.

- The drive system generally consists of two motors: one for the rotation of the main drive and the other for the rotation of the screw.
- The screw rotates at low speed, typically 35–100 rpm.

10.6.2.2 Advantages and Disadvantages of Vertical Orbiting Screw Blender

Advantages	*Disadvantages*
• Mixing is rapid, and the power consumption to produce a uniform blend is thus considerably reduced • Easy to clean after use	• Requires a large free room for installation • It is expensive

10.6.2.3 Applications

- An orbital screw blender is used to mix shear-sensitive and heat-sensitive powders with critical process parameters.

10.6.3 Muller Mixers

10.6.3.1 Working principles

- It (also known as a pan mixer) consists of a flat-bottomed, cylindrical pan-shaped vessel equipped with agitators, large, rotating wheels, and a set of scrapers (called the muller turret), as shown in Figure 10.3.
- The heavy wide wheels shear and mix the materials against the pan bottom (an action similar to that of a mortar and pestle, known as mulling) while the scrapers move the material from the pan's sides and center to flow under the wheels.

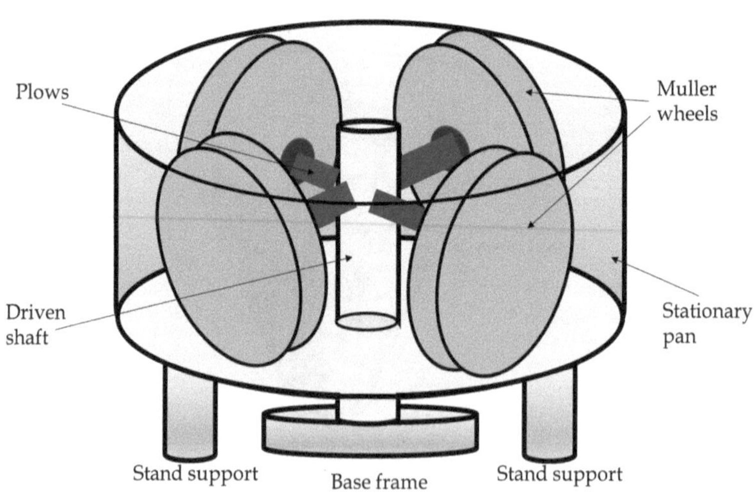

Figure 10.3 Muller mixer.

10.6.3.2 Advantages

- This type of mixer yields uniform consistency of the mixing.
- It provides forces incorporating kneading, shearing, smearing, and blending materials for uniform consistency.
- The mixing takes place without crushing, grinding, or distorting the ingredients.

10.6.3.3 Disadvantages

- It requires high power consumption.
- It has a limited capacity.

10.6.3.4 Applications

- A muller mixer is useful for mixing aggregate breakdown, frictional particle anchorage, and densifying the final mix for mixing solids with liquids to produce pastes.
- It is generally used in foundry, ayurvedic and herbal products, pharmaceuticals, minerals (mica powder grinding), lime mortar grinding, fly ash mixing, rice ash husk mixing, carbon black mixing, snuff, batteries, welding rod flux, abrasives, soaps, putty, paints, etc.

10.6.4 Drum-Type Blenders

10.6.4.1 Working principles

- It consists of a drum or vessel's internal baffles or helical plates.
- The mixing happens by axial rotation of the drum where the internal baffles or helical plates improve the crossflow of solids, as shown in Figure 10.4.

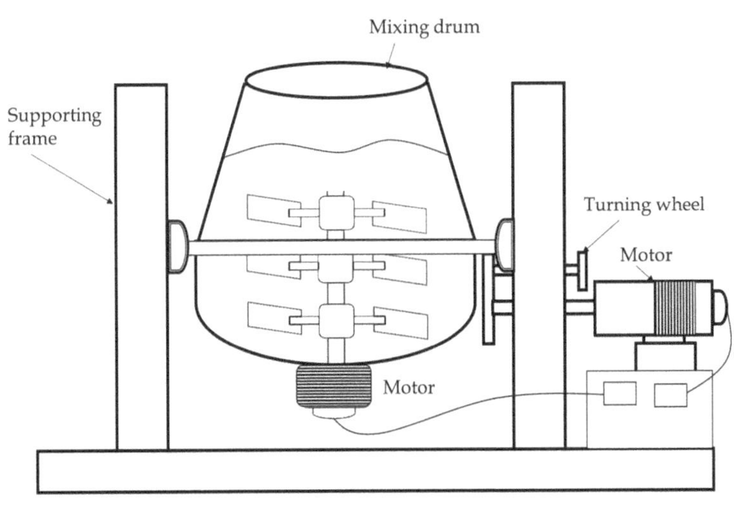

Figure 10.4 Drum-type blenders.

10.6.4.2 Applications

- Drum/barrel mixer is staple machinery used in the pharmaceutical and chemical industries. This equipment is beneficial for mixing ingredients or raw materials to be dispatched in barrels.

10.6.4.3 Advantages

- Highly efficient mixing equipment.
- Unlike other mixers, this mixer has no internal moving parts like blades or impellers.
- There is no dead space, so the material is uniformly blended.
- Dust-free system.

10.6.4.4 Disadvantages

- It is not energy intensive.
- The cleaning is difficult.

10.6.5 Single- and Double-Shaft Paddle Mixer

10.6.5.1 Working principle

- This type of mixer (a twin-rotor or a twin-screw blender) consists of a single shaft with at least two paddles inside a cylindrical shell, as shown in Figure 10.5.
- It usually operates at high tip speeds (30.48–45.72 m/s). All moving particles create a dynamic interaction between subdivided zones of the cylinder.
- The double-shaft paddle mixers operate with high mixing capacity and help transport the free-flowing material at a low shear rate.

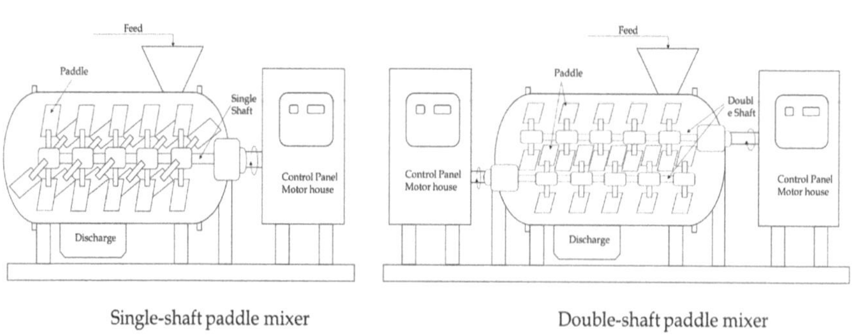

Figure 10.5 Paddle mixer.

10.6.5.2 Applications

- This type of mixer is specifically designed for the mixing procedures of bulk resources like dry powders, granules, and liquids.

10.6.5.3 Advantages

- Long operating life
- Activated by the fully automatic system
- Dirt-free device
- Available with large capacity and short mixing time

10.6.5.4 Disadvantages

- The mixing precision of this mixer is limited.
- Sticky and soft particles are hard to mix.

10.6.6 Turbine Blender/Mixers

10.6.6.1 Working principles

- As shown in Figure 10.6, this type of mixer consists of a circular trough with a housing.
- A spider or a series of legs with plough-shares rotates around through the circular trough, which causes the material to circulate rapidly from the center to the walls, then upward and down to the center.

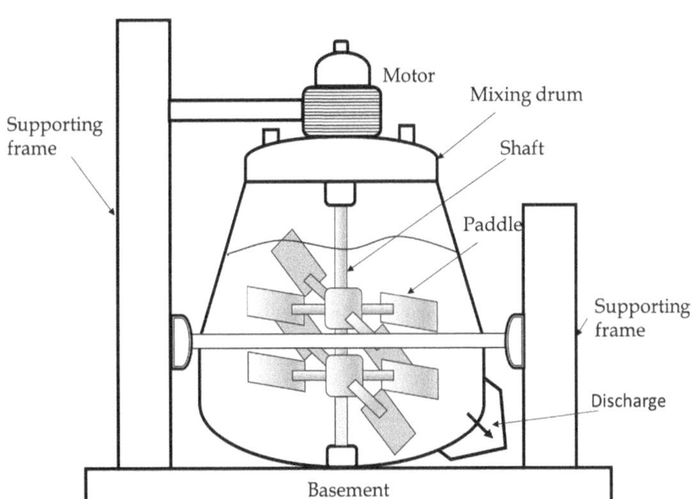

Figure 10.6 Turbine blender/mixers.

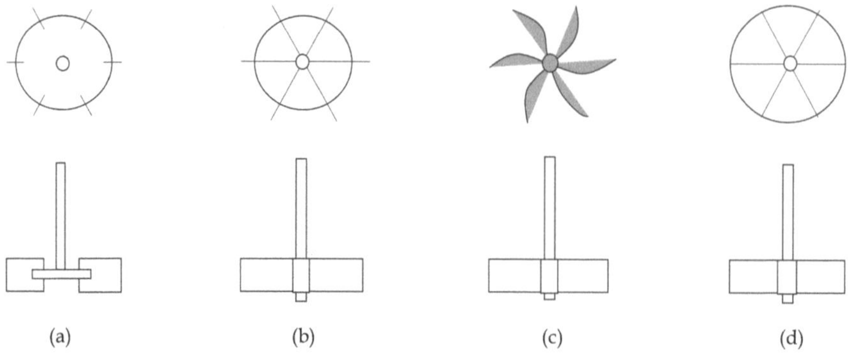

Figure 10.7 Types of turbine. (a) Disk-mounted flat-blade turbine, (b) hub-mounted flat-blade turbine, (c) hub-mounted curve-blade turbine, (d) shrouded turbine impeller.

- A turbine mixer can create a turbulent movement of the fluids due to the combination of centrifugal and rotational motions. These combined motions cause the adequate mixing of low- to medium-viscosity fluids.
- A turbine has a circular disk to which several short blades are attached. The diameter of turbines is approximately 0.13–0.67 to that of the diameter of the vessel, typically 0.33:1.
- The blades may be straight, pitched, curved, or disk type as shown in Figure 10.7. The turbine rotates at a slower speed, usually 50–200 rpm, than the propellers.

10.6.6.2 Advantages

- Suitable in mixing high viscous solutions with a wide range of viscosities up to 7.0 Pa s and up to 60% solids.
- Generate high radial flow efficiency of mixing.
- Efficient for emulsification with turbine shear.

10.6.6.3 Disadvantages

- They are not preferred for solvents of viscosity more than 20 cP.
- Air entrapment may cause oxidation of the material being mixed.
- Turbines have less pumping rate.

10.6.6.4 Applications of turbine mixer

- The mixer is used in chemical reactions and extraction operations.
- It is also used in preparing emulsions, suspensions, and syrups.

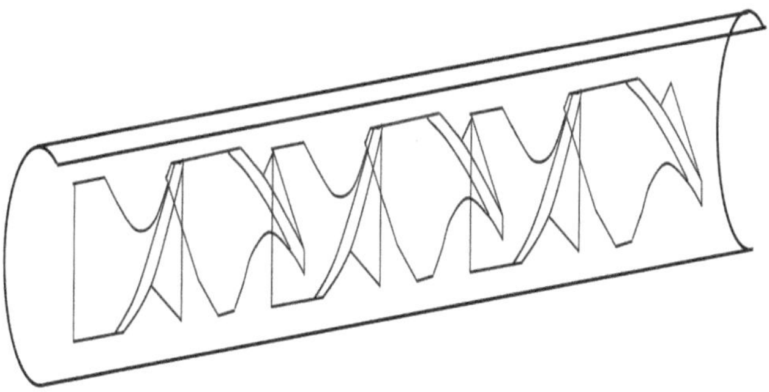

Figure 10.8 Static mixers.

10.6.7 Static Mixers

10.6.7.1 Working principles

- It consists of a cylindrical chamber having chaos-inducting mixing elements located along the axis of the mixer.
- There are at least two elements in a static mixer, which are placed one behind the other as a string.
- The mixing elements (strand) are constructed to give an axial twist to the flow pattern, and each successive part alternately has a left- and a right-hand twist, as shown in Figure 10.8.
- This construction creates turbulence that enhances rapid mixing.

10.6.7.2 Advantages

- The mixer includes reduced residence time requirements and improved material processing performance.

10.6.7.3 Disadvantages

- The material buildup may cause restriction and clogging, which require more energy to get the desired degree of mixing within the device.
- It requires accurate feeding of components in time which may be expensive.

10.6.7.4 Applications

- The mixer is typically used in applications requiring high-intensity mixing, short contact time, and/or continuous processing.
- They are suitable for blending solids such as grains and plastic chips.

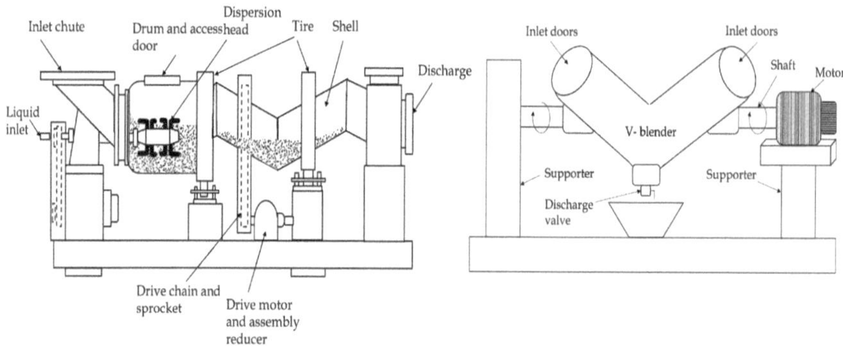

Figure 10.9 Zigzag or V-shaped blender.

10.6.8 Zigzag or V-Shaped Blender

- This consists of a number of V-shaped blenders (called legs) joined together. The unit consists of two cylinders arranged in a "V" shape with an angle, as shown in Figure 10.9.

10.6.8.1 Working principles

- It operates based on tumbling action when it rotates, without internal scrapers, stirring devices, screens, vanes, or paddles. During the motion of the rolling part, the particles are not crushed, smeared, or subjected to excessive mechanical work.
- As the mixer rotates, the V part is inverted, and the material splits into two portions, dividing and recombining continuously, which yields ordered mixing by mechanical means as shown in Figure 10.9.
- The material is loaded approximately 50%–60% of the total volume. The rotation speed is maintained at 35 for a 20-kg capacity and 15 rpm for a 1-ton capacity.
- Sufficient residence time within the mixer is required for the homogeneous mixing of the material.

10.6.8.2 Advantages

- Suitable because of minimum attrition.
- This type of equipment yields the mixing of large capacities of material.
- Cleaning, loading, unloading, and maintenance are easier.

10.6.8.3 Disadvantages

- It needs high headspace for installation.
- It is not suitable for the wider particle size distribution due to a lack of enough applied shear.
- For free-flowing powders, serial dilution is required to add low-dose active ingredients.

10.6.8.4 Applications

- It is widely used for mixing materials like pharmaceuticals, flour for confectionery, fish meal, refined or white flour, whole wheat flour, flour for coating, dextrins, enzymes, coffee, cocoa, milk powder, chocolate, gelatin, crème caramel, and dessert powder, baby food, custard powder, starches, seeds, pigments, colorants, cosmetics, powdered or granulated plastics, fiberglass, etc.

10.6.9 Sigma Blade Mixer

- It consists of two sigma-shaped blades fitted horizontally in each tough of the bowl, as shown in Figure 10.10. These blades are connected to a fixed-speed drive.
- The shapes of blades resemble the Greek letter sigma (Σ) and hence are called sigma blade mixers. The clearance between the blades and the vessel walls is as low as 2 mm. The low clearances produce high shear within the material.
- The blades are rotated through heavy-duty drive systems consisting of a motor, gearbox, couplings, and gears. The top speed of the sigma mixer is generally limited to 60 m/min.
- The mixer trough has a jacket for the circulation of hot or cold media to maintain the required temperature conditions within the mixer.
- The equipment is also attached to the perforated blades to break lumps and aggregates.

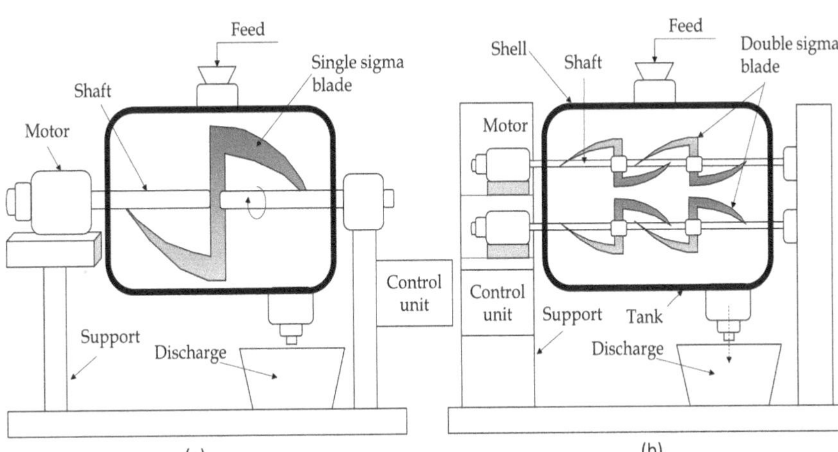

Figure 10.10 Sigma blade mixer. (a) Single-bladed (b) double-bladed.

10.6.9.1 Working principle

- The main mechanism of mixing is shear. The intermeshing of sigma blades creates high shear and kneading action.
- The materials are loaded through the top of the trough to typically 40%–65% of the mixer's total volumetric capacity. The entire mixing process is carried out in a closed trough because of the dust.
- The blades move at different speeds (one about twice the other) using the drive system that includes the motor, gear reducer, couplings, gears, bearings, and seals. The material moves up and down and shears between the blades and the trough wall.
- The discharge of the material is either by tilting the mixing vessel, through the bottom discharge valve or through a discharge screw. The homogeneous mixture is obtained within 10–30 minutes with a mixing homogeneity of up to 99%.

10.6.9.2 Advantages of Sigma blade mixer

- Minimum dead space during mixing.
- Close tolerances between the blade, the sidewall, and the bottom of the mixer shell make mixing efficient.
- The Sigma blade mixer leaves fewer dead spots during mixing operations.
- It is ideal for mixing and kneading highly viscous mass (up to as 10 million centipoises) and sticky products.
- The provision of perforated blades makes it suitable for breaking lumps and aggregates.
- The losses of volatile solvents during mixing can be prevented by closing the trough chamber and maintaining a low temperature.

10.6.9.3 Disadvantages of Sigma blade mixer

- It works at a fixed speed.
- Not energy intensive, though it provides good mixing.
- The power consumption is very high compared to other types of mixers.
- Requires more space for installation.
- It requires proper blade speed adjustments.
- There may be dead spots in the mixing tank.

10.6.9.4 Application of Sigma blade mixer

- It is a commonly used mixer for high-viscosity materials in preparing emulsions and syrups; manufacture of tablets, capsules, and pill masses; mixing of adhesives and chemicals; chewing gums, food and confectionery products; hot melts, inks, and pigment products; soaps and detergents; sugar pastes; ointments; etc.

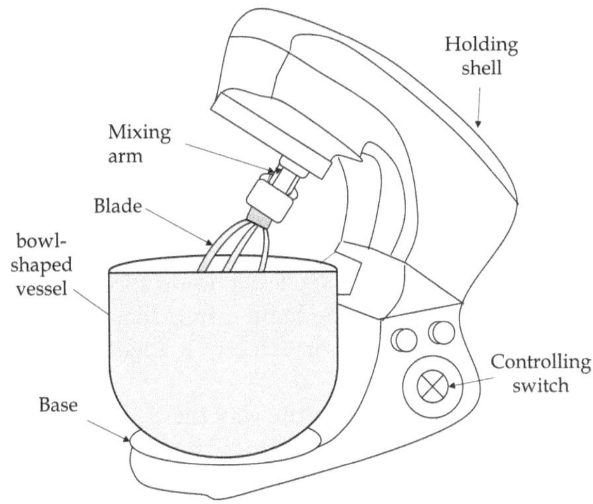

Figure 10.11 Schematic diagram of kneader.

10.6.10 Kneader

- It consists of a mixing arm that moves around its axis and the central axis of a bowl-shaped vessel to reach every spot of the vessel, as shown in Figure 10.11.
- The blade is mounted from the top or bottom of the bowl.
- The mixing shaft is driven by planetary gear and is usually built with variable-speed drive.

10.6.10.1 Working principle

- Mixing occurs by applying shear between the moving blade and the stationary wall.
- The mixing powder gets an upward movement to achieve tumbling action as the plates in the blades are fixed on a slope.

10.6.10.2 Applications

- Efficient for a breakdown of agglomerates.
- It works at low speed where dry blending and fast wet granulation are required.

10.6.10.3 Advantages

- It can be operated with variable speed.
- It is suitable for the wet granulation process.

10.6.10.4 Disadvantages

- Heat buildup within the powder mix due to shear action.
- Not energy efficient.
- Limitation for continuous operation.

10.6.11 Planetary Mixer

- It consists of a vertical cylindrical shell or bowl. The shell or bowl is covered and is provided with nozzles, a liquid spray arrangement viewing ports. The material can be loaded into the mixer either through the nozzles on the top cover or directly loaded into the mixer bowl (Figure 10.12).
- The name "planetary" refers to the way the blades of the mixer rotate around a central point. The mixer rotates the blades on an axis similar to how a planet orbits around the sun.
- Most planetary mixers have two or three blades rotating in different directions as they whip, knead, and mix ingredients.
- This unique design allows for a more thorough mixing than other mixers, making it an essential tool for any baker or chef.
- The drive system consists of a motor and a gearbox that drives the planetary head.

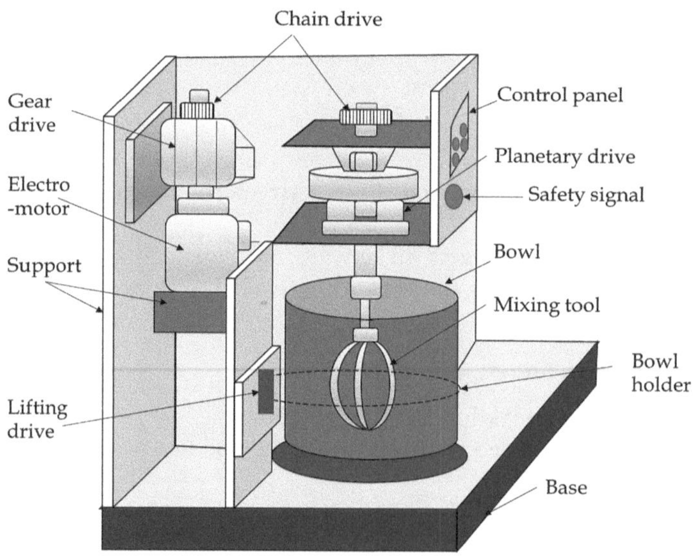

Figure 10.12 Planetary mixer.

10.6.11.1 Working principle

- The materials are mixed by flowing up and down and around the inner cylinder based on the blades' rotation.
- The material to be mixed is loaded into a mixing bowl or shell. The blades rotate on their axis when they orbit the mixing bowl on a common axis. Therefore, the mixer has no dead spots, and high shear is applied. After mixing, the material is discharged through a bottom valve or by manual scooping of the material from the bowl.

10.6.11.2 Advantages

- Reaches the mixing effect in a very short time.
- Almost uniform mixing is attained through a mass shift of ingredients.
- Low maintenance cost and high efficiency.
- The use of this mixer can achieve a higher density of a product.
- Simple construction, operation, and relatively lower cost.
- No dead spot in the mixing.
- The rotation speed of blades can be varied.
- Used for the wet granulation process.
- High mixing efficiency.

10.6.11.3 Disadvantages

- Dumpy ingredients and pastes are difficult to blend.
- Require high power.
- Heat buildup within the powder mix.

10.6.11.4 Applications

- Planetary mixers are commonly used to mix the dough, frosting, pastes, and other thick mixtures.
- It is mainly used for mixing paste, high-viscosity, high-density materials dissolving, mixing, kneading and polymerization, suspending solids in a liquid, blending, emulsifying liquids, dissolving polymers, powders, and granular materials in a liquid base, etc.
- They are often used to chop or grind meat, vegetables, and fruits. Some models of planetary mixers include attachments such as shredders, grinders, and pasta makers.

10.6.12 Fluid Mixers or Fluidized Mixers

- It consists of a mixing silo or a conveying tube with an inverted conical deflector at the top for spreading material, as shown in Figure 10.13.
- Compressed air or gas is introduced intermittently at high velocities through nozzles present at the bottom or side of the silo.

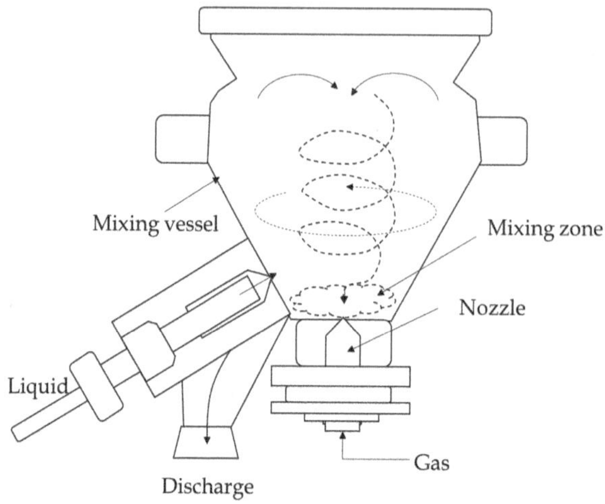

Figure 10.13 Fluidized mixers.

10.6.12.1 Working principle

- The solid particles rise due to the drag force of the air injected or liquid, and an increase in the air or liquid velocity causes agitation in the bed, resulting in the formation of bubbles and fluid internal circulation, which causes blending to take place.

10.6.12.2 Advantages

- Cleaning and maintenance are easier.
- There are no moving parts for which contamination of material is minimal.
- Less handling of materials is required with a pneumatic blender.
- The blender is insensitive to heat.

10.6.12.3 Disadvantages

- Sticky material cannot be suitable to fluidize and mix.
- High energy is required to fluidize for mixing.
- The stagnant zone below the air or liquid distributor at the bottom.

10.6.12.4 Applications

- This is used for mixing dry or free-flow powders of sizes and densities that are similar among all the mixed components.

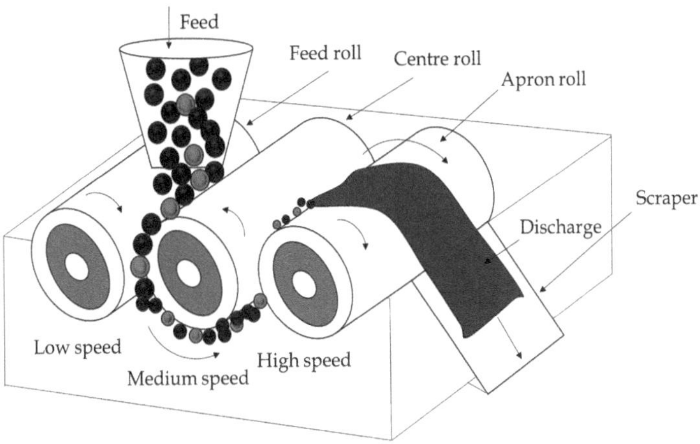

Figure 10.14 Triple roller mill.

10.6.13 Triple Roller Mill

- It consists of three parallel rollers of equal diameters made of stainless steel mounted on a rigid framework horizontally (Figure 10.14).
- The gap between the first two rollers is more than the gap between the last two.
- A hopper is placed in between the first two rollers, and a scrapper is attached to the last roller.
- The first roller rotates at a lower speed compared to the second. Similarly, the second roller's speed is less than that of the third roller.

10.6.13.1 Working principle

- The roller makes high shear which causes the crushing of semisolids and disperses the product homogeneously.

10.6.13.2 Applications

- It is used to mix, refine, disperse, or homogenize viscous materials such as printing inks, electronic thick film inks, high-performance ceramics, cosmetics, plastisol, carbon/graphite, paints, pharmaceuticals, chemicals, glass coatings, dental composites, pigment, coatings, adhesives, sealants, foods, etc.

10.6.13.3 Advantages

- Suitable for small to large batches.
- Well control of temperature.
- It is possible to avoid product contamination due to metal abrasion.
- Uniform dispersion is obtained without loss of material.

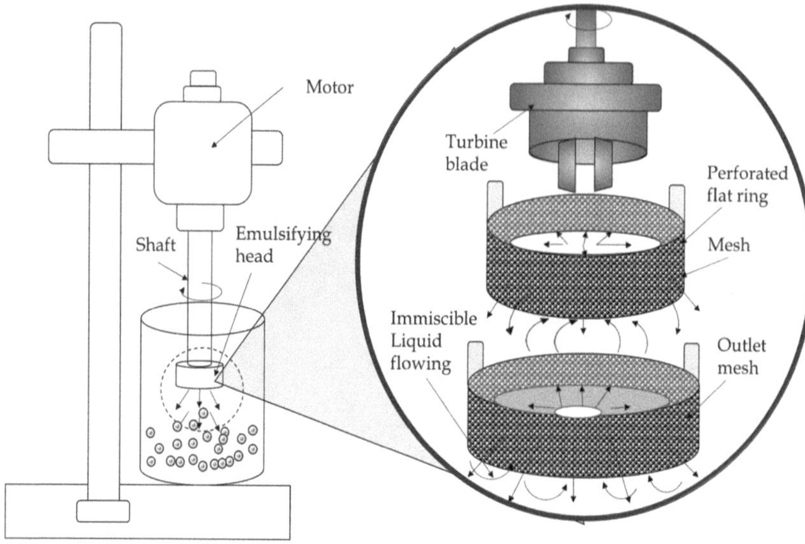

Figure 10.15 Silverson mixer–emulsifier.

10.6.13.4 Disadvantages

- Complex design and expensive

10.6.14 Silverson Mixer–Emulsifier

- It consists of long supporting columns and a central portion, as shown in Figure 10.15. The central portion consists of a shaft that is connected to the motor at one end and the other to the head.
- The head carries turbine blades. The blades are surrounded by a mesh, further enclosed by a cover with openings.

10.6.14.1 Working principle

- It produces shearing forces and turbulence through the use of high-speed rotors.
- During the rotation of the rotors, material circulation occurs through the head by the suction produced in the inlet at the bottom of the head, which ensures rapid breakdown of the dispersed liquid into smaller globules.

10.6.14.2 Applications

- Used for the preparation of emulsions and creams of fine particle size.

10.6.14.3 Advantages

- Wide range of applications possible (from a few milliliters to several thousand liters).
- Suitable for batch and continuous operations by incorporating into a pipeline through which the immiscible liquids flow.

10.6.14.4 Disadvantages

- There is a chance for clogging of the pores of the mesh.

10.6.15 Colloidal Mill

- It consists of the rotor-stator arrangement placed in a cylindrical vessel, as shown in Figure 10.16.
- The rotor and stator are placed closed enough to create a narrow passage where the material size is reduced.

10.6.15.1 Working principle

- The rotor-stator arrangement produces a high shear stress at a speed of 2,000–18,000 rpm applied on the fluid. This stress results in disrupting and breaking down the structure and making emulsion.

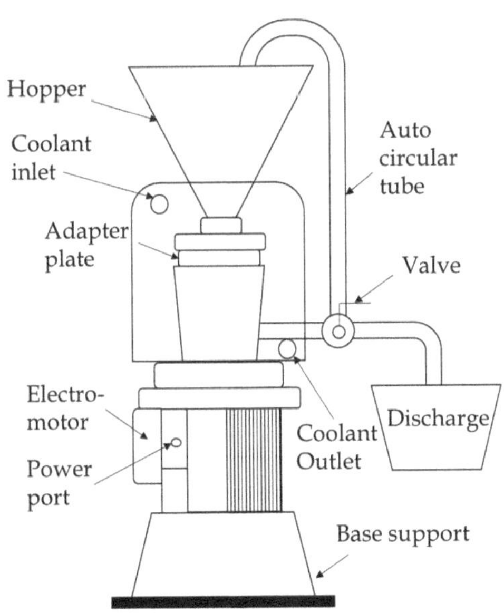

Figure 10.16 Colloidal mill.

- The equipment breaks down materials by forming a dispersion of materials in a liquid.
- When a liquid is suspended in another immiscible liquid, the machine is used alternatively to reduce the size of its droplet.

10.6.15.2 Applications

- It is widely used in homogenizing, emulsifying, dispersing, mixing, and comminuting liquids to highly viscous products.

10.6.15.3 Advantages

- More efficient compared to others.
- Suitable for a wide range of slurry for mixing.
- Simple construction and easy to operate.
- Capacity is high.

10.6.15.4 Disadvantages

- High consumption of energy.
- Lack of wide application in solids.
- No fine grinding.

10.6.16 Ultrasonic Homogenizer

- It consists of a pump driven by a motor. It is connected to an inlet and outlet tubes (Figure 10.17).
- Its head consists of a flat jet for the liquid inlet. A thin blade vibrates at its natural frequency.

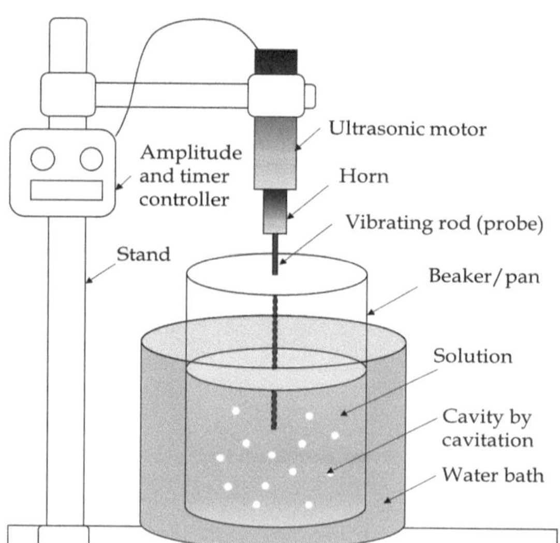

Figure 10.17 Ultrasonic homogenizer.

10.6.16.1 Working principle

- It works by ultrasonic vibrations. Under ultrasonic vibrations, alternate regions of compression and rarefaction are produced in the liquid.
- In rarefaction regions, cavities are formed, which collapse in the compression regions. This generates great forces for emulsification.

10.6.16.2 Applications

- It is used for mixing, dispersing, and emulsification.
- It is also used to produce small particles or droplets in a liquid or slurry to improve the uniformity and stability of the mixture.

10.6.16.3 Advantages

- It can be used for both batch processes and continuous modes.
- It can produce dispersed globules up to one-micron size.
- It is recommended for thermolabile substances since heat is not generated during mixing.

10.6.16.4 Disadvantages

- It is suitable only for low viscous liquids.

IMPORTANT POINTS TO REMEMBER

- Mixing smoothly and inseparably together by imparting minimum energy to the bed is called blending.
- The factors that influence the solid mixing are characteristics of particles such as size, shape, volume, surface area, density, porosity, flow charge, etc.
- In vertical orbiting screw blender the screw rotates at low speed, typically 35–100 rpm.
- Muller mixer is generally used in foundry, ayurvedic and herbal products, pharmaceuticals, minerals (mica powder grinding).
- In single and double-shaft paddle mixer usually operates at high tip speeds 30.48–45.72 m/s.
- The turbine rotates at a slower speed, usually 50–200 rpm, than the propellers.

- Turbine blender/mixers is suitable in mixing high viscous solutions with a wide range of viscosities up to 7.0 Pa s and up to 60% solids.
- The rotation speed of zigzag or V-shaped blender is maintained at 35 for a 20-kg capacity and 15 rpm for a 1-ton capacity.
- The top speed of the sigma mixer is generally limited to 60 m/min.
- The name "planetary" refers to the way the blades of the mixer rotate around a central point.
- The rotor-stator arrangement produces a high shear stress at a speed of 2000 to 18000 rpm applied on the fluid.

SELF-TESTING QUESTIONS

I. Multiple-Choice Questions

Q1. Mixing is a process to ____________________ within a system.
 (a) Transporting of dissimilar particles
 (b) Randomize dissimilar particles
 (c) Fixing of dissimilar particles
 (d) None of these

Q2. Mixing smoothly and inseparably together by imparting minimum energy to the bed is referred to as____________.
 (a) Blending
 (b) Twisting
 (c) Stressing
 (d) Agitation

Q3. Agitation refers to the induced motion of a material in a ___________.
 (a) Random distribution
 (b) Specific pattern
 (c) Uneven distribution
 (d) None of these

Q4. Mixing is______________________, into and through one another.
 (a) Random distribution
 (b) Specific pattern
 (c) Uneven distribution
 (d) None of these

Q5. The factors that influence the solid mixing are characteristics of particles such as size, shape, volume, surface area, density porosity, flow charge, etc. The statement is True or False?
 (a) True
 (b) False

Q6. Mixing two or more groups of solid particles so that they are dispersed in each other is called________________.
(a) Convective mixing
(b) Diffusion mixing
(c) Shear mixing
(d) None of these

Q7. The mixing where the particles are distributed over a freshly developed interface is known as____________.
(a) Convective mixing
(b) Diffusion mixing
(c) Shear mixing
(d) None of these

Q8. The powdered materials are mixed in a container where the movement of the powdered particles occurs by tilting the material beyond the angle of repose. Which mixing mechanism is referred to in this statement?
(a) Convective mixing
(b) Shear mixing
(c) Diffusion mixing
(d) None of these

Q9. In which case can mixing improve semi-microscopic mixing and be favorable in batch and continuous operation?
(a) Convective mixing
(b) Shear mixing
(c) Diffusion mixing
(d) None of these

Q10. The momentum exchange of powder particles having different velocities leads to ______.
(a) Convective mixing
(b) Shear mixing
(c) Diffusion mixing
(d) None of these

Q11. Temperature affects the mixing because ___________.
(a) Surface tension changes with temperature
(b) Thermal conductivity changes with temperature
(c) The viscosity increases with the temperature
(d) None of these

Q12. What should be the following option to fit a uniform mixture?
(a) Particle size should be spherical.
(b) Particle size should be as much as non-spherical.
(c) The particle should be sticky enough.
(d) The particle aspect ratio should be significantly large.

Q13. Which of the following options is correct?
- (a) Mixing at higher rotational speed is more homogeneous than at less rotational speed
- (b) Mixing at less rotational speed is more homogeneous than at a higher rotational speed
- (c) None of these options is correct
- (d) Both options are correct.

Q14. Which of the following is not a batch-type blender?
- (a) Ribbon mixer
- (b) Muller mixer
- (c) Turbine mixer
- (d) Pug mill

Q15. The _______type of mixer consists of a vertical cylinder with blades mounted on a central shaft.
- (a) Pug mills
- (b) Static mixers
- (c) Free fall impeller blender
- (d) Continuous blender

Q16. In ______, mixing particles are distributed over a freshly developed interphase.
- (a) Convective
- (b) Diffusion
- (c) Shear
- (d) None of the above

Q17. Which mechanisms are suitable for powder mixing?
- (a) Convective mixing only
- (b) Diffusion mixing only
- (c) Shear mixing only
- (d) All of the above

Q18. In the ______ system, the randomization of the materials is achieved by the directed flow of the powder systems.
- (a) Active mixer
- (b) Passive mixer
- (c) (a) & (b)
- (d) None of the above

Q19. The ___________uses certain moving parts to assist in the randomization of the solid materials, or the mixer machine moves about physically in the mixing process.
- (a) Active mixer
- (b) Passive mixer

(c) (a) & (b)
(d) None of the above

Q20. Which type of mixer system uses baffles to direct the flow of powder streams?
(a) Active mixer system
(b) Passive mixer system
(c) Both active and passive mixer systems
(d) None of the above

Answers to the Multiple-Choice Questions

Q1. Ans. (b)	Q8. Ans. (c)	Q15. Ans. (c)
Q2. Ans. (a)	Q9. Ans. (b)	Q16. Ans (b)
Q3. Ans. (b)	Q10. Ans. (b)	Q17. Ans. (d)
Q4. Ans. (a)	Q11. Ans. (c)	Q18. Ans (b)
Q5. Ans. (a)	Q12. Ans. (a)	Q19. Ans (a)
Q6. Ans. (a)	Q13. Ans. (b)	Q20. Ans. (b)
Q7. Ans. (b)	Q14. Ans. (d)	

II. Objective Type Questions

1. How can mixing, blending, and agitation be distinguished?
2. Explain different mechanisms of mixing.
3. What are the factors that affect mixing?
4. Classify different types of mixers or blenders.
5. Based on flow properties how can you classify the mixing devices?
6. Give some examples of mixing devices based on batch and continuous operations.
7. Explain the working principle of the turbine mixer.
8. Name different types of turbines.
9. What is the working principle of in-bin blenders?
10. What is a static mixer? How is it advantageous?
11. Give examples of the use of zigzag blenders.
12. What are the operating conditions of the sigma blade mixer?
13. Explain the working principle of the planetary mixer.
14. What are the advantages of a fluidized bed mixer? How does it work?
15. Give examples of applications of triple roller mills.

REFERENCES

Harnby, N., M. F. Edwards and A. W. Nienow, (Eds.), *Mixing in the Process Industries*, Butterworth, 1985.

McCabe, W. L., J. Smith, and P. Harriot, *Unit Operations of Chemical Engineering*, 7th Edition, McGraw-Hill, International Edition, 2017.

Richardson, J. F., J. H. Harker, and J. R. Backhurst, *Chemical Engineering, Vol. 2, Particle Technology and Separation Processes, A Volume in Chemical Engineering Series*, 5th Edition, Elsevier, 2003.

Chapter 11

Degree of Mixing and its Assessment

Scientists investigate that which already is; engineers create that which has never been.

—*Albert Einstein*

11.1 ASSESSMENT OF DEGREE OF MIXING

- The performance of the industrial mixer is determined by the degree of mixing (called mixing effectiveness or mixing index) of the material after mixing.
- The mixing index is determined by analyzing spot samples collected from the mixer.
- The mixer acts on separate materials to mix them nearly always in a random fashion.
- The degree of mixing can be assessed based on the random distribution of the material.

11.2 ASSESSMENT OF DEGREE OF MIXING BY A KINETIC MODEL

- The mixing states of the particle system are defined by the percentage (by volume) of particles in a mixture in a mixed condition in the system at a mixing time. For example, a mixing degree of 33% means that 1/3 of the volume of particles is in a state of mixed condition in the system at time t. The law of mixing of solids appears to follow the first order.

$$M = A(1 - e^{-kt}) \tag{11.1}$$

where M = degree of mixing after time t, A and k = constants.

DOI: 10.1201/9781003571025-15

Table 11.1 Degree of mixing with time

Time (seconds)	0.00	30	60	120	150	180	210
Degree of mixing (M)	0.00	0.71	0.92	0.99	1.00	1.00	1.00

- The parameters $A = 1$ and $k = \infty$ signify the degree of mixing of solids and the mixing rate constant at perfect homogeneous mixed conditions (ideal mixing) at a time t. The mixing rate constant k is determined from the slope of the plot of log $(A–M)$ versus t.
- The parameters A and k depend on:
  - the mixer geometry
  - the physical characteristics of the powders and
  - the proportion of the material being mixed.

Example 11.1

A mixing state of particles in a mixing system with time is shown in the following Table 11.1. Find the mixing rate constant.

Solution:

The rate equation is $M = A(1 - e^{-kt})$
Rearranging and taking logarithm on both sides, one can get

$$-\ln(1 - M/A) = kt$$

Time, t (seconds)	0.00	30	60	120	150	180	210
Degree of mixing (M)	0.00	0.71	0.92	0.99	1.00	1.00	1.00
$-\ln(1 - M/A)$	0.00	1.23	2.47	4.93	6.17	7.40	8.63

The slope of graph of $-\ln(1 - M/A)$ vs. t yields the mixing rate constant $(k) = 0.0411$

11.3 ASSESSMENT OF DEGREE OF MIXING BY STATISTICAL ANALYSIS

- Consider a solid mixer, as shown in Figure 11.1, in which some tracer material is added, which will not react with the other materials in the sample.
- Let the overall average fraction of tracer in the mixture is μ
- Take N small samples randomly from various locations in the mixture
- Determine the fraction of tracer x_i in each small sample
- The average value of the measured concentration is $\overline{x}$
- If N is very large $\overline{x}$ will be equal to μ; otherwise, they may be significantly different.

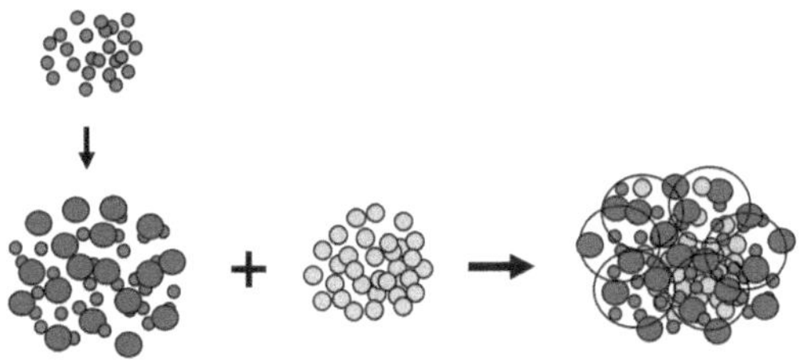

Figure 11.1 Mixing of tracer material with a sample of solid blend.

- If solids were perfectly mixed, every measured value of x_i would be equal to $\overline{x}$
- If mixing is not complete, the measured value of x_i will differ from $\overline{x}$, and their standard deviation about the average value of $\overline{x}$ will measure the degree or quality of mixing.
- This standard deviation is estimated by

$$S = \sqrt{\frac{\sum_{i=1}^{N} (x_i - \overline{x})^2}{N-1}} \tag{11.2}$$

$$\overline{x} = \frac{1}{N} \sum_{i=1}^{N} x_i \tag{11.3}$$

- The value of S is a relative measure of mixing valid only for testing a specific material in a mixer.
- Its significance depends on the amount of tracer in the mix.
- The standard deviation S diminishes toward zero as mixing proceeds. So low value of S means good mixing.

11.4 MIXING INDEX

- Various statistical measures that are independent of the amount of tracer
- Harnby et al. (1997) introduce a mixing index to quantify the degree of mixing
- **The mixing index for pastes** is defined as

$$I_P = \frac{\sigma_0}{S} = \sqrt{\frac{(N-1)\mu(1-\mu)}{\sum_{i=1}^{N}(x_i - \bar{x})^2}} \tag{11.4}$$

- The index is the reciprocal of the ratio of s to the standard deviation at zero mixing (σ_0).
- Before mixing has begun, the material in the mixture exists as two layers, one containing no tracer and one containing all tracers. Under this condition,

$$\sigma_0 = \sqrt{\mu(1-\mu)} \tag{11.5}$$

Example 11.2

A paste mixture containing moisture was mixed in a large muller mixer with an average 15% tracer after a certain time of mixing 8 random samples were taken and analyzed the mass fraction of tracer material (x_i) and found as from the data as shown as 0.00, 0.03, 0.06, 0.12, 0.15, 0.18, 0.21, 0.25. Estimate the mixing index.

Solution:

The overall average fraction of tracer in the mixture is μ=0.15
Number of samples=8
From the given data, $\sum_{i=1}^{N}(x_i - \bar{x})^2 = 0.00669$
Therefore, the mixing index is

$$I_P = \frac{\sigma_0}{S} = \sqrt{\frac{(N-1)\mu(1-\mu)}{\sum_{i=1}^{N}(x_i - \bar{x})^2}} = \sqrt{\frac{(8-1)(0.15)(1-0.15)}{0.0069}} = 11.35$$

11.4.1 The Mixing Index for Granular Materials

- With granular solids, the mixing index is based not on conditions at zero mixing but on the standard deviation that would be observed with a completely random, fully blended mixture (McCabe et al., 2017).
- With pastes, assuming the analyses are perfectly accurate, this value is zero. With granular solids, it is not zero.
- Consider, for example, a completely blended mixture of salt and sand grains from which N number of samples, each containing n particles, are taken.
- Suppose the fraction of sand in each sample is determined by counting particles of each kind.
- Let the overall fraction, by the number of particles, of sand in the total mix be μ_p.

- If n is small (say, about 100), the measured fraction x_i of sand in each sample will not always be the same, even when the mix is completely and perfectly blended; there is always some chance that a sample drawn from a random mixture will contain a more significant (or smaller) proportion of one kind of particle than the population from which it is taken.
- Thus, for any given number of samples, there is a theoretical standard deviation for a completely random mixture. This standard deviation σ_R is given by

$$\sigma_R = \sqrt{\frac{\mu(1-\mu)}{n}} \tag{11.6}$$

- Then the mixing index of granular materials is defined as (McCabe et al., 2017)

$$I_{\text{GS}} = \frac{\sigma_R}{S} = \sqrt{\frac{\mu(1-\mu)(N-1)}{n\sum_{i=1}^{N}(x_i-\overline{x})^2}} \tag{11.7}$$

- The mixing index I_{GS} measures how far mixing has proceeded toward equilibrium.

11.4.2 Mixing Index at Zero Time

- The mixing index at zero time is defined by (McCabe et al., 2017)

$$I_{\text{GS},0} = \frac{\sigma_R}{\sigma_0} = \frac{1}{\sqrt{n}} \tag{11.8}$$

where σ_0 is the standard deviation at zero mixing and σ_R is the equilibrium standard deviation for complete mixing.

Example 11.3

In an experiment of mixing granules of sand and salt, 50 random samples are collected from the output of mixing. From each sample, 100 numbers of particles of sand are considered to analyze the mixing index of mixing granules. The experimental data on the mixing of sand and salt particles yields an average fraction of tracer, $\mu = 0.20$. From the sample analysis, it was found that $\sum_{i=1}^{N}(x_i-\overline{x})^2 = 0.00045$. Under these conditions, find the mixing index of mixing of granules and mixing index at zero mixing time.

Solution:

Average fraction of tracer, $\mu=0.20$.
Numbers of particles of sand are considered $(n)=100$
Number of random samples $(N)=50$.
From the sample analysis it was found that $\sum_{i=1}^{N}(x_i-\bar{x})^2=0.00045$

Therefore, the mixing index of mixing of granules

$$I_{GS}=\frac{\sigma_R}{S}=\sqrt{\frac{\mu(1-\mu)(N-1)}{n\sum_{i=1}^{N}(x_i-\bar{x})^2}}=\sqrt{\frac{(0.20)(1-0.20)(50-1)}{100(0.00045)}}=13.33$$

Mixing index at zero mixing time, $I_{GS,0}=\frac{\sigma_R}{\sigma_0}=\frac{1}{\sqrt{n}}=\frac{1}{\sqrt{100}}=0.10$

11.4.3 Lacey Mixing Index

- The degree of mixing can also be assessed by the Lacey mixing index (1954).
- The Lacey mixing index is defined as

$$\text{Lacey mixing index } (I_L)=\frac{\sigma_0^2-s^2}{\sigma_0^2-\sigma_R^2} \tag{11.9}$$

- For a two-component system, the theoretical upper and lower limits of mixture variance are defined as:

$$\text{(a) Upper limit (complete segregation)}: \quad \sigma_0^2=\mu(1-\mu) \tag{11.10}$$

$$\text{(b) Lower limit (randomely mixed)}: \quad \sigma_R^2=\frac{\mu(1-\mu)}{n} \tag{11.11}$$

where μ and $(1-\mu)$ are the fractions of the two components determined from samples, and n is the number of particles in each sample.
- A Lacey mixing index of **zero** represents complete segregation, and a value of **unity** represents thorough mixing. The mixing index mainly lies within the range of **0.75–1.0.**

11.4.4 Poole Mixing Index

- The mixing index suggested by Poole et al. (1964) to have a better representation of the degree of mixing is defined as

$$\text{Poole mixing index } (I_{Po})=\frac{s}{\sigma_R} \tag{11.12}$$

- This index approaches unity and indicates completely random mixing.

Example 11.4

Mixture of powdered herbs and minerals used in the preparation of ayurvedic medicine. In an experiment of mixing of herbs and minerals, 30 random samples are collected from the output of their mixing. From each sample, 5×10^6 numbers of particles of the herb are considered to analyze the mixing index of their mixing to prepare the medicine. The experimental data on the mixing yields an average fraction of tracer, μ=0.15. From the sample analysis, it was found that $\sum_{i=1}^{N}(x_i-\bar{x})^2=0.085$. Under these conditions, find the mixing index of mixing of herbs and minerals by the Lacey mixing theory and Poole et al. mixing theory.

Solution:

The mixing index of mixing of herbs and minerals by the Lacey mixing theory is

$$I_L=\frac{\sigma_0^2-s^2}{\sigma_0^2-\sigma_R^2}=\frac{\mu(1-\mu)-\dfrac{\sum_{i=1}^{N}(x_i-\bar{x})^2}{N-1}}{\mu(1-\mu)-\dfrac{\mu(1-\mu)}{n}}=\frac{0.15(1-0.15)-\dfrac{0.085}{30-1}}{0.15(1-0.15)-\dfrac{0.15(1-0.15)}{5\times10^6}}=0.977$$

The mixing index of mixing of herbs and minerals by the Poole et al. mixing theory is

$$I_{Po}=\frac{s}{\sigma_R}=\frac{\sqrt{\dfrac{\sum_{i=1}^{N}(x_i-\bar{x})^2}{N-1}}}{\sqrt{\dfrac{\mu(1-\mu)}{n}}}=\frac{\sqrt{\dfrac{0.085}{30-1}}}{\sqrt{\dfrac{0.15(1-0.15)}{5\times10^6}}}=3.39\times10^{-4}$$

11.5 RATE OF MIXING BASED ON MIXING INDEX

- The mixing rate is proportional to the driving force for getting the material mixed.
- For short mixing times, the rate of change of I_{GS} is expressed as (McCabe et al., 2017)

$$\frac{dI_{GS}}{dt}=k(1-I_{GS}) \tag{11.13}$$

where k is a constant. The equilibrium value of I_{GS} is 1. The driving force for mixing at any time is $1-I_{GS}$. After integrating within limits, one can get

Table 11.2 Fraction of sand in a spot sample (x_i) at a fixed time of 60 seconds

Mixing Time (seconds)	*Fraction of Sand in a Spot Sample (x_i)*
60	0.60, 0.61, 0.75, 0.72, 0.65, 0.62, 0.66, 0.69, 0.85, 0.63

$$\int_0^t dt = \frac{1}{k}\int_{I_{GS,0}}^{I_{GS}} \frac{dI_{GS}}{1-I_{GS}} \tag{11.14}$$

or

$$t = \frac{1}{k}\ln\frac{1-I_{GS,0}}{1-I_{GS}} \tag{11.16}$$

Or after substitution of $I_{GS,0}$ from Equation (11.8)

$$t = \frac{1}{k}\ln\frac{1-1/\sqrt{n}}{1-I_{GS}} \tag{11.17}$$

Example 11.5

Experimental data on the rate of mixing of sand and salt particles in a fluidized bed is given in Table 11.2 for a mixing time of 60 seconds. In the given run, the initial charge to the reactor was 300 gm of salt on top of 500 gm of sand. The average number of particles in the spot sample was 400. How long will it take to reach the mixing index of 0.9?

Solution:

Overall average fraction of salt in the mixture, $\mu = 300/(300+500) = 0.38$

$$N = 10, n = 400$$

$$\bar{x} = \frac{1}{N}\sum_{i=1}^{N} x_i = 0.678$$

$$\sum_{i=1}^{N}(x_i - \bar{x})^2 = 0.05416$$

$$I_{GS} = \frac{\sigma_R}{s} = \sqrt{\frac{\mu(1-\mu)(N-1)}{n\sum_{i=1}^{N}(x_i-\bar{x})^2}} = \sqrt{\frac{0.38(1-0.38)(10-1)}{400(0.05416)}} = 0.0979$$

$$t = \frac{1}{k}\ln\frac{1-1/\sqrt{n}}{1-I_{GS}} = 60 \text{ which implies, } k = 8.618\times10^{-4}$$

Therefore, the time takes to reach the mixing index of 0.9

$$t = \frac{1}{8.618\times10^{-4}}\ln\frac{1-1/\sqrt{400}}{1-0.9} = 2612.175 \text{ seconds} = 43.54 \text{ minutes}$$

IMPORTANT POINTS TO REMEMBER

- The degree of mixing can be assessed based on the random distribution of the material.
- The mixing states of the particle system are defined by the percentage (by volume) of particles in a mixture in a mixed condition in the system at a mixing time.
- The law of mixing of solids appears to follow the first order.
- The standard deviation S diminishes toward zero as mixing proceeds. So low value of S means good mixing.
    - The mixing index (I_p) is unity at the start of any batch mixing.
    - The higher the mixing index (I_p) value, the mixing is higher.
    - The mixing index (I_p) would become infinite at a long mixing time.
- The mixing rate, as measured by the rate of change of I_p with time, varies significantly with the kind of mixer and the nature of the mixed materials.
- The mixing index falls within ranges of 10–150.
- The mixing index I_{GS} measures how far mixing has proceeded toward equilibrium.
- A Lacey mixing index of zero represents complete segregation, and a value of unity represents thorough mixing. The mixing index mainly lies within the range of 0.75–1.0.
- The Poole mixing index approaches unity and indicates completely random mixing.
- The driving force for mixing at any time is $1-I_{GS}$.

SELF-TESTING QUESTIONS

I. Multiple-Choice Questions

Q1. The mixing time for ideal mixing is________.
(a) Infinite
(b) 10
(c) 0
(d) 100

Q2. Which of the following statement(s) is (are) incorrect?
(a) The mixing index falls within ranges 10–150.
(b) In any batch mixing, the mixing index (I_p) is unity at the start.
(c) The mixing index (I_p) would become infinite at long mixing time.
(d) Lower the value of the mixing index (Ip) the mixing is higher.

Q3. The driving force for mixing of solids at any time is__________ where the mixing index I_{GS} is a measure of how far mixing has proceeded toward equilibrium.
(a) $1-I_{GS}$
(b) $(1+I_{GS})/(1-I_{GS})$
(c) $1+I_{GS}$
(d) $1/1-I_{GS}$

Q4. Lacey mixing index is given by__________.
(a) $\dfrac{\sigma_0^2-\sigma_R^2}{\sigma_0^2-\sigma^2}$
(b) $\dfrac{\sigma_R^2-\sigma^2}{\sigma_R^2-\sigma_0^2}$
(c) $\dfrac{\sigma_0^2-\sigma^2}{\sigma_0^2-\sigma_R^2}$
(d) $\dfrac{\sigma^2-\sigma_0^2}{\sigma^2-\sigma_R^2}$

Q5. Lacey mixing index of zero represents__________.
(a) Complete segregation
(b) Completely random mixture
(c) Both (a) and (b)
(d) None

Q6. Consider a solid mixer in which a tracer material is added, which does not react with the other materials in the sample. Take a number of small samples (N) randomly from various locations in the mixture. The fraction of tracer is x_i in each small samples i. What would be the overall average fraction of tracer in the mixture if the standard deviation at zero mixing is negligible.
(a) 0
(b) –1
(c) 1
(d) Infinity

Q7. Two statements are given below:
Statement 1: In any batch mixing, the mixing index at start is unity.
Statement 2: Higher the value of mixing index, higher the extent of mixing.

Which option is correct in favor of the statements?
(a) Statement 1 is TRUE but Statement 2 is false
(b) Statement 2 is TRUE but Statement 1 is false
(c) Both statements 1 & 2 are TRUE
(d) Both statements 1 & 2 are FALSE

Q8. Which one of the following values of the Lacey mixing index represents "complete segregation"?
(a) Zero
(b) Unity
(c) Infinite
(d) None of the above

Q9. Which of the following relation is correct?

(a) $I_{GS,0} = \frac{\sigma_R}{\sigma_0}$

(b) $I_{GS,0} = \frac{\sigma_0}{\sigma_R}$

(c) $I_{GS,0} = \sigma_R \times \sigma_0$

(d) $I_{GS,0} = \frac{\sigma_R}{s}$

Q10. What is the Poole mixing index when it approaches completely random mixtures?
(a) Infinity
(b) 1
(c) 0
(d) ½

Q11. In any batch mixing, the mixing index is __________ at the start.
(a) Zero
(b) Unity
(c) ∞
(d) It depends on other parameters.

Q12. A Lacey mixture index of zero would represent __________.
(a) Complete segregation.
(b) Completely random mixture.
(c) It does not provide sufficient discrimination between mixtures.
(d) None of the above.

Q13. Poole mixing index approaches __________ for completely random mixtures.
(a) 0
(b) 0.5

(c) 1
(d) ∞

Q14. Mixing index for pastes is defined as________________.
(a) Ratio of standard deviation at zero mixing to variance which is the value of relative measure of mixing
(b) Ratio of variance to the standard deviation at zero mixing
(c) Ratio of variance to the standard deviation at complete mixing
(d) None of these

Q15. In any batch mixing, the mixing index at the beginning is______________.
(a) Unity
(b) Zero
(c) Infinity
(d) None of the above

Q16. A Lacey mixture index of zero represents____________________.
(a) Completely random mixture
(b) Complete segregation
(c) It does not provide sufficient discrimination between mixtures
(d) None of the above

Q17. For completely random mixing, Poole mixing index approaches_____________.
(a) 0.5
(b) 1
(c) 0
(d) Infinity

Q18. Degree of mixing at infinite time as per first-order mixing characteristics is____________.
(a) 1.0
(b) 2.0
(c) Infinity
(d) 0

Q19. The standard deviation tends to zero indicates___________.
(a) Incomplete mixing
(b) Perfect mixing
(c) Does not indicate any mixing
(d) None of these

Q20. The mixing index (I_p) would become ________at a long mixing time.
(a) Infinite
(b) Zero
(c) One
(d) None

Answers to the multiple-choice questions

Q1. Ans. (a)
Q2. Ans. (d)
Q3. Ans. (a)
Q4. Ans. (c)
Q5. Ans. (a)
Q6. Ans. (c)
Q7. Ans. (c)
Q8. Ans. (a)
Q9. Ans. (a)
Q10. Ans. (b)
Q11. Ans. (b)
Q12. Ans. (a)
Q13. Ans. (c)
Q14. Ans. (a)
Q15. Ans. (a)
Q16. Ans. (b)
Q17. Ans. (b)
Q18. Ans. (a)
Q19. Ans. (b)
Q20. Ans. (a)

II. Objective Type Questions

1. How is a degree of mixing assessed?
2. Write the kinetic equation of the degree of mixing.
3. The kinetic parameters of the degree of mixing depend on which factors?
4. How mixing index for paste is defined? How is it related to standard deviation at zero mixing?
5. How mixing index for granular solids is defined? How is it different from that of paste?
6. What do you understand about the mixing index at zero time?
7. What is the Lacey mixing index? What are the upper and lower limits of the Lacey mixing index?
8. Define the Poole mixing index. What is its significance?
9. Explain the rate of mixing.
10. Derive the mixing time from the rate of mixing.

III. Numerical Problems

Q1. A moisture paste mixture was mixed in a large muller mixer with an average 20% tracer. After a certain time of mixing, 15 random samples were taken and analyzed. What is its mixing index (I_p)? (Consider $\sum (x_i - \bar{x})^2 = 0.002$). [Ans. 33.466]

Q2. If the overall average fraction of tracer particles in a solid mixture is 0.1. Find the standard deviation. [Ans. 0.3]

Q3. In N number solid–solid mixture samples, the number of particles in each sample is 100. If the average fraction of tracer in the mixture is 0.5, what would be the lower variance limits for random mixing? [Ans. 0.0025]

Q4. For granular solid mixture, the mixing index (I_{GS}) is__________ if μ=0.1, n=150, the number of samples (N)=10, and $\sum (x_i - \bar{x})^2 = 0.0012$. [Ans. 2.323]

Q5. The experimental data on the rate of mixing of sand and solid particles and how long it will take for the given sample to reach the mixing index (I_{GS}) to 0.9 if the given conditions are: The average number of particles in spot sample is, n=300 and the average fraction of tracer is, μ=0.1, number of samples, N=20, and mixing time is 120 seconds. $\sum (x_i - \bar{x})^2 = 0.02$. [Ans. 448.622 seconds]

Q6. A silty soil containing (14%) moisture was mixed in a large muller mixer with (10 wt.%) of a tracer containing dextrose and picric acid. After 3 minutes of mixing, 12 random samples were taken from the mix and analyzed calorimetrically for tracer material. The measured concentrations in the samples were, in weight percent tracer, 10.58, 9.63, 7.88, 10.02, 11.18, 10.98, 11.66, 9.54, 9.38, 10.37, 10.62, and 10.98. Under this condition, find the value of standard deviation 's' for this case. [Ans. 0.01012]

Q7. A silty soil containing (14%) moisture was mixed in a large muller mixer with (10 wt.%) of a tracer containing dextrose and picric acid. After 3 minutes of mixing 12 random samples were taken from the mix and analyzed calorimetrically for tracer material. The measured concentrations in the samples were, in weight percent tracer, 10.58, 9.63, 7.88, 10.02, 11.18, 10.98, 11.66, 9.54, 9.38, 10.37, 10.62, and 10.98. Find the mixing index. [Ans. 29.64]

Q8. If the standard deviation values at zero mixing, standard deviation at randomly mixed, and 's' are 0.02, 0.01, and 0.012, respectively, then what is the value of the Lacey mixing index (I_L). [Ans. 0.8533]

Q9. A mixing experiment takes 60 seconds to reach a mixing index of 0.085. How long will it take to reach the mixing index of 0.90? (Take mixing index at zero time, $I_{GS,0}$=0.03). [Ans. 38.92]

Q10. The compressive strength of a cement-stabilized clay sample from a small two-arm kneader varies with the mixing index according to the relation: S=65.87 In (I_p)−12.47. From the relation with the mixing time as shown in Figure 11.2, what mixing time would be required with each of these soils to give a product with a compressive strength of 160 lbf/in^2? [Ans. 0.35 minutes]

Q11. A soil containing 14% moisture was mixed in a large muller mixer with 20 wt.% of a tracer containing maltose and uric acid. After 5 minutes of mixing, 10 random samples were taken from the mix and analyzed calorimetrically for tracer material. The measured concentrations of tracer in the sample were (in weight): 10, 9, 8, 11, 12, 6, 7, 8, 9, and 13. Calculate the mixing index $I_{P.}$ [Ans. 0.181]

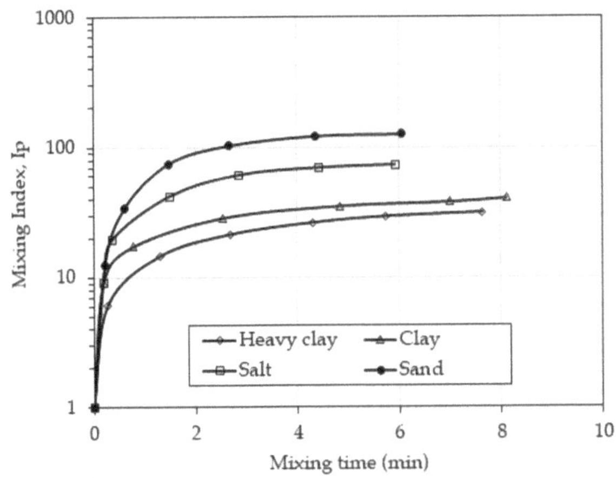

Figure 11.2 Variations of mixing index with mixing time.

Q12. A silty soil containing 14% moisture was mixed in a large muller mixer with 10.00 wt.% of a tracer consisting of dextrose and picric acid. After 5 minutes of mixing, 10 random samples were taken from the mix and analyzed calorimetrically for tracer material. The measured concentrations in the sample were, in weight percent tracer, 10.00, 9.50, 8.75, 11.25, 11.90, 6.50, 10.50, 7.75, 9.15, and 11.75. Calculate the mixing index I_P. [Ans. 17.174]

Q13. Table 11.3 gives experimental data on the rate of mixing of sand and salt particles in a fluidized bed. In the given run, the initial charge to the reactor was 250 gm of salt on top of 300 gm of sand. The average number of particles in the spot sample was 500.

How much time will it take for the mixing index to reach 0.9? [Ans. 7.68 minutes]

Q14. The compressive strength of a cement-stabilized clay sample from a small two-arm kneader varies with the mixing index according to the relation: $S = 65.87 \ln I_p - 12.47$. From the relation with the mixing time as shown in Figure 11.2, what mixing time would be required with each of these soils to give a product with a compressive strength of 150 lbf/in^2? [Ans. 0.27 minutes]

Q15. The compressive strength of a cement-stabilized clay sample from a small two-arm kneader varies with the mixing index according to the relation: $S = 65.87 \ln I_p - 12.47$. After prolonged mixing in this mixer, find the maximum strength obtainable with the soil. [Ans. –12.47]

Table 11.3 Fraction of sand in a spot sample at a fixed time 50 seconds

Mixing Time (seconds)	*Number Fraction of Sand in a Spot Sample*
50	0.66, 0.64, 0.77, 0.70, 0.67, 0.58, 0.60, 0.65, 0.87, 0.60

Q16. A random mixture consists of components A and B, in proportions 60% and 40% by mass, respectively. The spherical particles A and B have 2500 and 3660 kg/m^3 particle densities, respectively. The cumulative undersize mass distributions of the two components are shown in Table 11.4. If samples of 1 g are withdrawn from the mixture, what is the expected value for the standard deviation of the composition of the samples? [Ans. 0.0165]

Q17. A loose soil containing 20% moisture was mixed in a large muller mixer with 5 wt.% of a tracer containing dextrose and picric acid. After 10 minutes of mixing, 12 random samples were taken from the mix and analyzed calorimetrically for tracer material. The measured concentrations in the sample were, in wt.% tracer, 5.24, 4.30, 2.94, 5.24, 6.08, 5.03, 6.91, 4.72, 4.20, 5.76, 5.97, and 5.55. What is the mixing index? [Ans. 20.96]

Q18. A loose soil containing 15% moisture was mixed in a large muller mixer with 15 wt.% of a tracer containing dextrose and picric acid. After 5 minutes of mixing, 10 random samples were taken from the mix and analyzed calorimetrically for tracer material. The measured concentrations in the sample were, in wt.% tracer, 15.64, 14.35, 13.74, 14.24, 16.08, 15.93, 16.89, 14.72, 14.70, and 15.06. What is the standard deviation? [Ans. 0.0098]

Q19. Experimental data on the rate of mixing of sand and salt particles in an air-fluidized bed is given in the Table 11.5. In each run, the initial charge to the reactor was 284.4 g of salt on top of 400 g of sand. The average number of particles in each spot sample was 750.

Based on this problem, answer the following questions as follows:

Qi. What is the mixing index after 1 run? [Ans. 0.0963]

Qii. What is the mixing index after 500 runs? [Ans. 0.961]

Qiii. When will the mixing index be 0.5? [Ans. 553 seconds]

Qiv. What is the minimum number of runs for the mixing index to be at least 0.5? [Ans. 11]

Table 11.4 Size distribution of particles A and B

Size x	*2045*	*1676*	*1405*	*1204*	*1003*	*853*	*699*	*599*	*500*	*422*
$F_A(x)$	1.0	0.80	0.50	0.32	0.19	0.12	0.07	0.04	0.02	0
$F_B(x)$	0	0	1.0	0.88	0.38	0.44	0.21	0.08	0	0

Table 11.5 Fraction of sand in a spot sample at a fixed time 54 seconds

Run No.	*Mixing Time (seconds)*	*Number Fraction of Sand in Spot Samples*
1	54	0.54, 0.67, 0.49, 0.25, 0.26, 0.68, 0.64, 0.62, 0.32, 0.26, 0.74, 0.77, 0.49, 0.33, 0.21, 0.73, 0.67, 0.44, 0.65, 0.42

Table 11.6 Fraction of sand in a spot sample at a fixed time 45 seconds

Run No.	*Mixing Time (seconds)*	*Number Fraction of Sand in Spot Samples*
1	45	0.64, 0.66, 0.49, 0.25, 0.26, 0.68, 0.64, 0.52, 0.32, 0.26, 0.74, 0.77, 0.49, 0.33, 0.21, 0.63, 0.7, 0.54, 0.65, 0.32

Q20. A loose soil containing 14% moisture was mixed in a large muller mixer with 10 wt.% of a tracer consisting of dextrose and picric acid. After 3 minutes of mixing, 12 random samples were taken from the mix and analyzed calorimetrically for tracer material. The measured concentrations in the sample were, in wt.% tracer, 10.24, 9.30, 7.94, 10.24, 11.08, 10.03, 11.91, 9.72, 9.20, 10.76, 10.97, and 10.55. Calculate the mixing index and the standard deviation *s*. [Ans. 0.0104]

Q21. Experimental data on the rate of mixing of sand and salt particles in an air-fluidized bed is given in Table 11.6. How long will it take for the mixing index to reach 0.95? In each run, the initial charge to the reactor was 254.4 g of salt on top of 300 g of sand. The average number of particles in each spot sample was 500.
[Ans. 1.32 hours]

Q22. The compressive strength of a cement-stabilized clay sample from a small two-arm kneader varies with the mixing index according to the relation.

$$C = 50.07 \ln I_p - 15.29$$

a. From the Figure 11.2 shown for clay and heavy clay, what mixing time would be required with each of these soils to give a product with a compressive strength of 125 lbf/in^2? [Ans. 0.804 minutes]
b. What would be the maximum strength obtainable with each soil after prolonged mixing in this mixer? [Ans. –15.29]

NOMENCLATURE

A	constant [-]
I_{GS}	mixing index of granular solids [-]
$I_{GS,0}$	mixing index granular solids at zero time [-]
I_p	mixing index for pastes
I_{po}	mixing index suggested by Poole et al. (1964)
I_L	Lacey mixing index [-]
k	rate constant [1/s]
M	degree of mixing after time t [-]
n	small sample number [-]
N	number of sample [-]
S	standard deviation [-]
t	time [s]
x_i	mass fraction of sample I [-]
$\bar{x}$	mean of mass fraction [-]

GREEK LETTERS

μ	mean of sample [-]
σ_0	standard deviation at zero mixing [-]
σ_R	theoretical standard deviation for a completely random mixture [-]

SUBSCRIPTS

0	zero mixing time
GS	granular solids
i	index number
L	Lacey
P	paste
Po	Poole
R	Random

REFERENCES

Harnby, N., The mixing of cohesive powders, in: N. Harnby, M.F. Edwards, A.W. Nienow, (Eds.), *Mixing in the Process Industries*, Butterworth-Heinemann, 1997.

Lacey, P. M. C., Developments in the theory of particulate mixing, *J. Appl. Chem.*, 4, 257, 1954.

McCabe, W. L., J. Smith, and P. Harriot, *Unit Operations of Chemical Engineering*, 7th Edition, McGraw-Hill, International Edition, 2017.

Poole, K. R., R. F. Taylor, and G. P. Wall, Mixing powders to fine-scale homogeneity: Studies of batch mixing, *Trans. Inst. Chem. Eng.*, 42, T305, 1964.

Section V

Transport and Storage of Solids

Chapter 12

Pneumatic Transport of Solid

The engineer has been, and is, a maker of history.

—James Kip Finch, American Engineer and Educator.

12.1 TRANSPORTATION OF SOLID

The term "flow," when applied to the transport of granular materials, is a wrong statement because it is a fact that only fluids can flow; bulk solids slide, roll, or fall under the action of gravity (Stepanoff, 1969). The only way that bulk solids can be pumped from one location to another is when the particles are suspended in either flowing gas (pneumatic conveying) or flowing liquid (hydraulic transport) (Klinzing, 2018).

- The common equipment used to transport the solid are:
  - **Gravity chutes:** The solids which fall under the action of gravity.
  - **Air slides:** Particles are transported suspending partially suspended in a channel by the upward flow of air through a porous distributor.
  - **Belt conveyors:** The solids are conveyed or transported horizontally, or at small angles to the horizontal, on a continuous moving belt.
  - **Screw conveyors:** A rotating helical impeller transports the solids along a pipe or channel.
  - **Bucket elevators:** The particles are carried vertically in buckets attached to a continuously moving vertical belt.
  - **Vibrating conveyors:** The particles are moved by asymmetric vibration over a surface. During the onward stroke of the surface, the particles are carried forward in contact with it. At the same time, at the high acceleration in the reverse stroke, the surface slips under the particles.
  - **Pneumatic hydraulic transport:** The particles are transported in a stream of air/water through a standpipe.

DOI: 10.1201/9781003571025-17

12.2 FLOW REGIMES OF PNEUMATIC TRANSPORT

- There are two broad categories of flow regimes of the pneumatic transport of particulate solids (Woodcock and Mason, 1987):
  - **Dilute (or lean) phase flow**
  - **Dense phase flow**

12.2.1 Dilute-Phase Transport

- Dilute phase flows as shown in Figure 12.1, are observed at high gas velocities (greater than 20 m/s), low solids concentrations (less than 1% by volume), and low-pressure drops per unit length of transport line (typically less than 5 mbar/m).
- The regime is limited to short routes and continuous transport of solids at rates of less than 10 t/h.
- This is the only regime capable of operation under negative pressure.

12.2.2 Dense Phase Transport

- The dense phase flow, as shown in Figure 12.2, is distinguished by low gas velocities (1–5 m/s), high solids concentrations (greater than 30% by volume), and high-pressure drops per unit length of pipe (typically greater than 20 mbar/m).
- In this flow regime, particles are not fully suspended, and there is much interaction between the particles.

12.2.3 Dilute vs. Dense Phase

- Dense-phase flow can be distinguished from dilute-phase flow based on the following variables:
  - solids/air mass flow rates
  - solids concentration

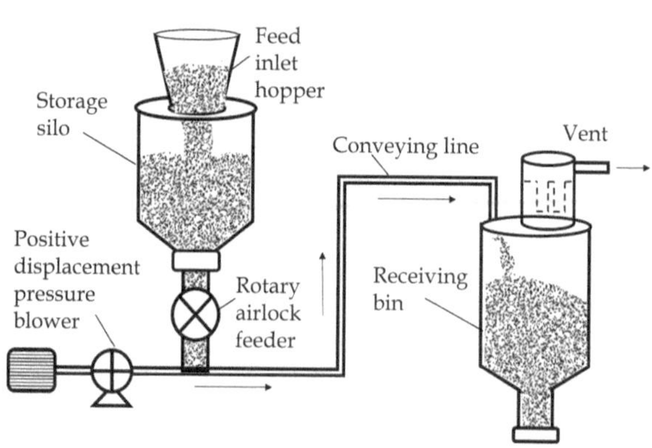

Figure 12.1 Dilute-phase transport.

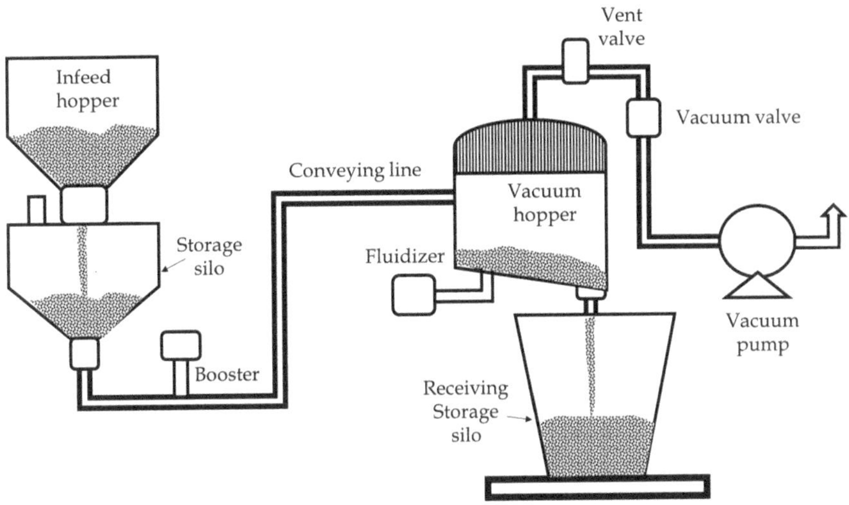

Figure 12.2 Dense-phase transport.

- cross-section of the pipe
- flow patterns
- Choking and saltation velocities are used to mark the boundaries between dilute-phase transport and dense-phase transport in vertical and horizontal pipelines, respectively. (choking and saltation velocity were described in a later section)

12.3 TYPES OF PNEUMATIC SYSTEMS

The various types of pneumatic systems are as follows:

- Low-pressure pneumatic conveying systems:
  - Positive-pressure systems
  - Negative-pressure (vacuum) systems
  - Combined negative/positive-pressure systems
- High-pressure systems:
  - Single-blow tank systems
  - Twin blow tanks and continuously operating systems
  - Long-distance conveying
- Low-velocity conveying and the use of supplementary air feeds:
  - Plug forming systems
  - Plug-limiting systems
  - Air-injection and booster systems

12.4 LOW-PRESSURE PNEUMATIC CONVEYING SYSTEMS

12.4.1 Positive-Pressure Systems

- Air (or other gas) at a velocity of 15–30 m/s is blown along a pipeline to transport the solids, as shown in Figure 12.3.
- Fans or blowers of a maximum pressure of under one bar are generally used.
- The Low-pressure pneumatic conveying systems transport a wide range of pulverized, granular, and fibrous materials.

12.4.2 Negative-Pressure (Vacuum) Systems

- The particle is transported through the suction pipe under vacuum pressure (at about 0.7 bar, in practice), as shown in Figure 12.4. The domestic vacuum cleaner is an example of such a system.
- They can achieve high tonnages or convey over long distances compared to positive-pressure systems.

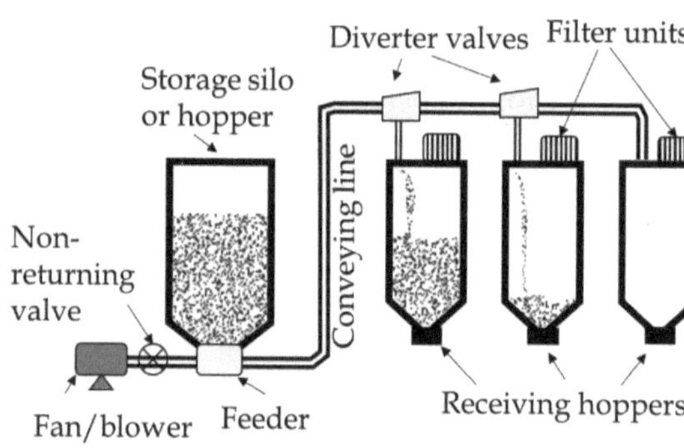

Figure 12.3 Positive-pressure system.

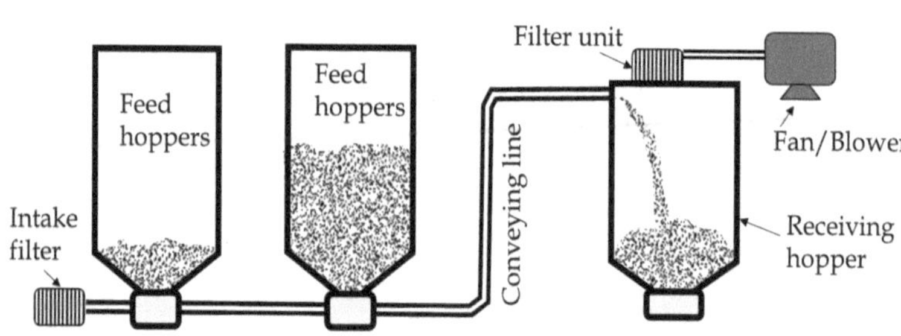

Figure 12.4 Negative-pressure (vacuum) system.

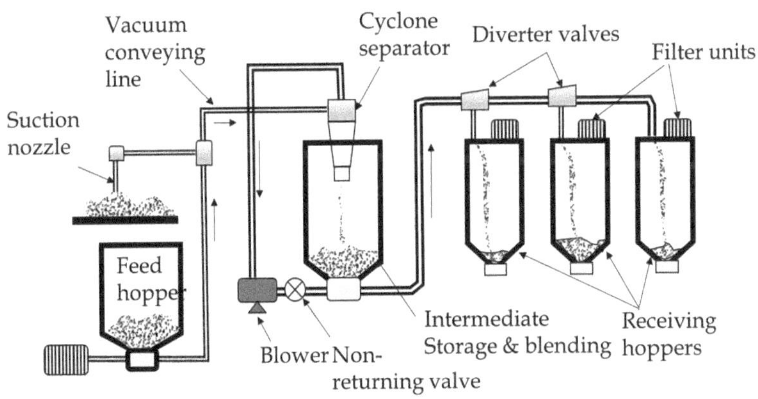

Figure 12.5 Combined negative/positive-pressure systems.

12.4.3 Combined Negative/Positive-Pressure Systems

- The system is used in industry when a bulk solids are collected from different locations and then redistributed to several delivery points.
- This system comprises vacuum pick-up and delivery to an intermediate storage hopper. This is also called the "pull-push" or "suck-blow" system, as shown in Figure 12.5.

12.5 HIGH-PRESSURE PNEUMATIC CONVEYING SYSTEMS

- This is basically a dense-phase conveying. Reciprocating or screw-type compressors are required to generate high pressures (7–8 bar) for transporting the material, as shown in Figure 12.6.
- The blow tank is one of the important components of such a high-pressure pneumatic conveying system. This blow tank provides the means for transporting materials into the pipeline.
- Blow tanks operating in parallel consist of two identical blow tanks, which are generally placed alongside each other. While one is being filled, the other is being discharged to improve the cycling efficiency.
- Twin blow tanks that operate in series have one blow tank mounted on top of another. The top vessel acts as a lock hopper, and this allows conveying on a continuous basis from the blow tank below. This particular design is because of headroom problems, which is designed by positioning the lock hopper alongside the blow tank.

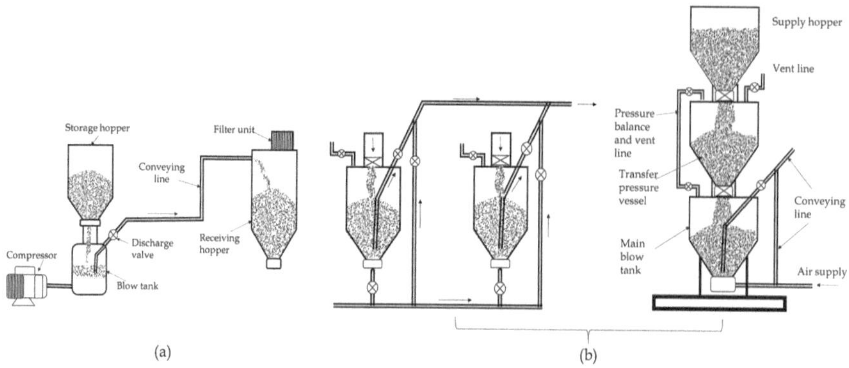

Figure 12.6 High-pressure pneumatic conveying systems. (a) Single-blow tank feeder, (b) Twin blow tank – arranged in parallel and in series.

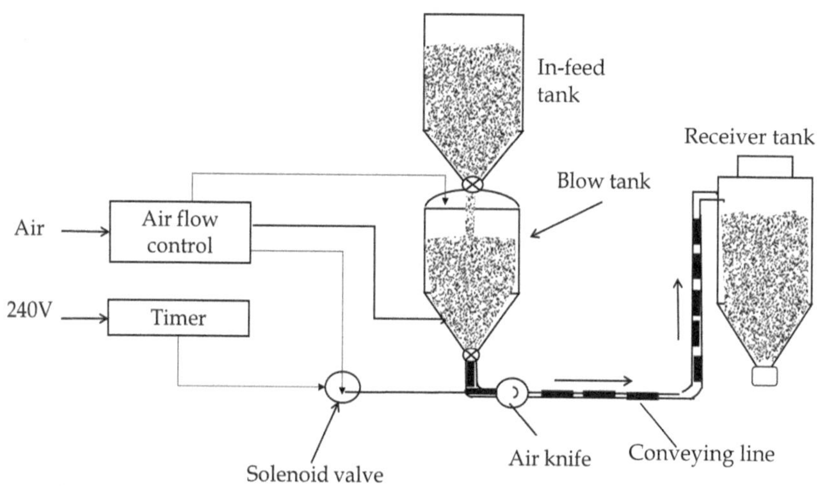

Figure 12.7 Low-velocity pneumatic conveying systems.

12.6 LOW-VELOCITY PNEUMATIC CONVEYING SYSTEMS

- The system is operated at low velocity (around half of the conventional dense-phase conveying speed) to minimize product degradation and erosive wear of the conveying line and system components.
- This type of conveying is done by dividing the bulk solid into a series of short plugs separated from each other by "air cushions", as shown in Figure 12.7. That is why the pressure required to convey them is much less than that needed to move a single plug of equivalent length.

12.7 THE CHOKING VELOCITY IN VERTICAL TRANSPORT

- The gas velocity at which the gas no longer entrains the solid for transport is called choking velocity.
- To convey a bulk solid in the dilute phase, the minimum chocking velocity in the vertical part of the line must be maintained.
- The choking velocity (denoted by U_{CH}) is the lowest velocity at which the dilute-phase transport line is operated at the solids feed rate (G), which can be calculated by Equations (12.1) and (12.2) (Rhodes, 1990).

$$\frac{U_{CH}}{\varepsilon_{CH}} - U_t = \frac{G}{\rho_p\left(1-\varepsilon_{CH}\right)} \tag{12.1}$$

$$\varepsilon_{CH} = \left(\frac{\left(U_{CH}/\varepsilon_{CH} - U_t\right)^2 \rho_f^{0.77}}{2.25D} + 1\right)^{-1/4.7} \tag{12.2}$$

where ε_{CH}=the voidage in the pipe at the choking velocity U_{CH} (m/s), ρ_p=the particle density (kg/m^3), ρ_f=the gas density, G=the mass flux of solids (mass flowrate per unit cross-sectional area of pipe), U_T=the free fall, or terminal velocity of a single particle in the gas, D=the diameter of the pipe. (Note that the constant is dimensional and that SI units must be used).
- Both the Equations (12.1) and (12.2) must be solved simultaneously by trial and error to calculate U_{CH} and ε_{CH}.

Example 12.1

A powder of particle of density 1530 kg/m^3 and size 75 μm is to be transported through the pipeline of a certain size of length 20 m vertically. It is required to transport the powder by a gas at a choking velocity. The volume fraction of the solid and its mass flow rate are to be maintained at 13% and 72.3 kg/m^2.s at its choking conditions. Calculate the chocking velocity at which the solids can be transported. The particles are considered almost spherical in shape. The viscosity of air at room temperature is 18.6 × 10^{-6} Pa.s. What should be the pipe diameter of the transport line?

Solution:

Assume the choking velocity and terminal velocity are 0.55 and 0.269 m/s. It is given that ε_p=0.13, therefore, ε_{CH}=1−0.13=0.87

Therefore, the relative velocity, $u = \dfrac{U_{CH}}{\varepsilon_{CH}} - U_T = 0.363$ m/s

$$\mathrm{Re}_p = \frac{\rho_f u d_p}{\mu_f} = \frac{(1.2)(0.363)(75\times10^{-6})}{18.6\times10^{-6}} = 1.757$$

The drag coefficient between Stoke's and turbulent condition ($0.1<\mathrm{Re}_p<500$), Haider and Levenspiel (1989) correlation can be used, which is expressed as

$$C_D = \frac{24}{\mathrm{Re}_P} + 3.3643\mathrm{Re}_P^{0.3471} + \frac{0.4607\mathrm{Re}_P}{\mathrm{Re}_P + 2682.5}, \quad \text{for} \quad \varphi_s = 1$$

where $\mathrm{Re}_p = \dfrac{\rho_f u d_p}{\mu_f}$, φ_s = sphericity

Therefore, the drag coefficient as per the Equation

$$C_D = \frac{24}{\mathrm{Re}_P} + 3.3643\mathrm{Re}_P^{0.3471} + \frac{0.4607\mathrm{Re}_P}{\mathrm{Re}_P + 2682.5} = 17.75$$

Therefore, the terminal velocity

$$U_T = \sqrt{\frac{4gd_p}{3C_D}\left(\frac{\rho_p - \rho_f}{\rho_f}\right)} = 0.269 \text{ m/s (Checked)}$$

The mass flow rate of the solid is to be controlled as per the equation

$$\frac{U_{CH}}{\varepsilon_{CH}} - U_T = \frac{G}{\rho_p(1-\varepsilon_{CH})}$$

Since the mass flow rate is controlled at 72.3 kg/m².s, Equation (12.1) is satisfied at the choking velocity. Also, the void fraction at this chocking velocity is to be satisfied as per Equation (12.2). So

$$\varepsilon_{CH} = \left(\frac{\left(U_{CH}/\varepsilon_{CH} - U_T\right)^2 \rho_f^{0.77}}{2.25D} + 1\right)^{-1/4.7} = 0.872 \text{ (Checked)}$$

It implies pipe diameter $(D) = 0.075$ m.

12.8 FLOW PATTERNS IN HORIZONTAL PNEUMATIC CONVEYING

- There are different patterns in the case of dense phases, such as (a) continuous dense-phase flow, (b) discrete plug flow, (c) plug flow, and (d) dune flow, whereas in dilute phase, the pattern is observed as dilute-phase flow.
- The different flow patterns of pneumatic conveying of the solid particles in the horizontal direction in the case of dilute and dense-phase conditions are shown diagrammatically in Figure 12.8.

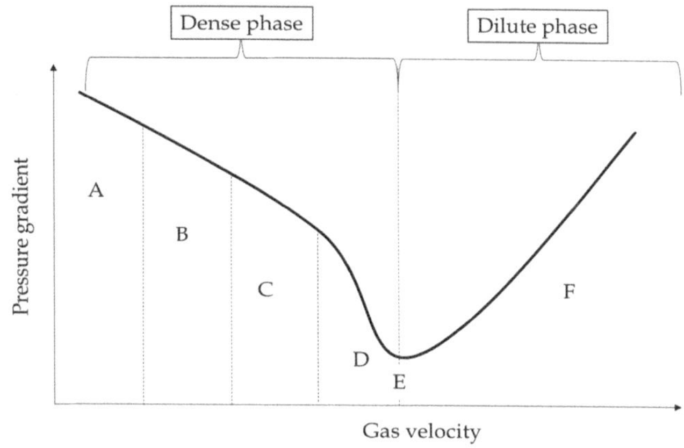

Figure 12.8 Flow patterns in horizontal pneumatic conveying. A, continuous dense-phase flow; B, plug flow; C, discrete plug flow; D, dune flow; E, saltation flow; F, dilute-phase flow (Rhodes, 2008).

- All these flow patterns exhibit different pressure gradients, as shown in Figure 12.8.
- The transition of these dense and dilute-phase flows is recognized based on saltation flow.
- **The continuous dense-phase flow pattern:** The solids occupy the entire pipe. This flow pattern requires very high gas pressures, and it is limited to short, straight pipe lengths and granular materials with high permeability. It results in product degradation by attrition. The pattern is characterized by low gas velocities (1–5 m/s), high solids concentrations (greater than 30% by volume), and high-pressure drops per unit length of pipe (typically greater than 20 mbar/m). In dense-phase transport, particles are not fully suspended, and there is much interaction between the particles.
- **Discrete plug flow:** In this pattern, plugs of solids occupy the full pipe cross-section of the pipe.
- **Dune flow:** A layer of solids settled at the bottom of the pipe moves along in the form of rolling dunes. As the gas velocity increases, particles are conveyed in suspension above a layer of settled solids. Particles may be deposited and re-entrained from this layer. As the gas velocity is decreased, the thickness of the layer of settled solids increases, which results in dune flow.
- **Plug flow:** A hybrid of discrete plug flow and dune flow in which the rolling dunes completely fill the pipe cross-section, but no discrete plugs exist.
- **Saltating flow:** It is encountered at gas velocities below the saltation velocity. Saltating flow occurs at a velocity too close to the saltation velocity and is therefore unstable.

- **Dilute-phase flow:** The dilute-phase flow pattern is a continuous flow of solid with high voidage and with relatively low pressures. It is mostly seen in the advent of Fluid Bed Catalytic Cracking. The pattern is mostly characterized by high gas velocities (greater than 20 m/s), low solids concentrations (less than 1% by volume), and low-pressure drops per unit length of transport line (typically less than 5 mbar/m). In this flow pattern, the solid particles behave as individuals, fully suspended in the gas, and fluid-particle forces dominate.

12.9 THE SALTATION VELOCITY IN HORIZONTAL TRANSPORT

- The saltation velocity is the conveying velocity below which the conveyed solids start to settle at the bottom of horizontal pipes (Rhodes, 2008).
- The minimum conveying velocity in all parts of the line must be more than the saltation velocity to convey the solids (such as powder or granulates) in a horizontal pipeline.
- The saltation velocity marks the boundary between the dilute-phase flow and dense-phase flow in horizontal pneumatic transport, which can be predicted from the correlation suggested by (Rizk, 1973):

$$\frac{M_p}{\rho_g U_{\text{salt}} A} = \left(\frac{1}{10^{(1440 d_p + 1.96)}}\right)\left(\frac{U_{\text{salt}}}{\sqrt{gD}}\right)^{1100 d_p + 2.5} \tag{12.3}$$

where, the left-hand side term of Equation (12.3) is called the solid loading in the suspension defined by

$$\text{Solid Loading} = \frac{\text{Mass flowrate of solids}}{\text{Mass flowrate of gas}} = \frac{M_p}{\rho_g U_{\text{salt}} A} \tag{12.4}$$

where $\frac{U_{\text{salt}}}{\sqrt{gD}}$ is called Froude number at saltation state, U_{salt}=the superficial gas velocity at saltation, M_p=the mass flow rate of solids, D=the pipe diameter, d_p=the particle diameter. All units are in SI system.

- The choking velocity happens in vertical lines, whereas the saltation velocity happens in horizontal lines.
- The saltation velocity is higher than the choking velocity, thus, designing the system to run above the saltation velocity will allow to avoid crossing the choking velocity in vertical pipes.

Example 12.2

A powder with a particle diameter of 150 μm is to be transported through a pipe of diameter 5 cm to a section where it will be used for chemical synthesis. The transportation was done at a velocity just above the saltation velocity. If the mass flux of the solid cannot be more than 50 kg/m².s, what would be the saltation velocity above which the transportation was done? Calculate the solid loading of suspension for the transport system.

Solution:

Given: $d_p = 150\,\mu\text{m} = 150\times10^{-6}\,\text{m}$, $M_p/A = 50\ \text{kg/m}^2.\text{s}$, $D = 0.05\,\text{m}$.

Substituting the values in Equation (12.3) and solving by trial and error method or by other numerical method, one can get

$$\frac{50}{(1.2)U_{\text{salt}}} = \left(\frac{1}{10^{(1440(150\times10^{-6})+1.96)}}\right)\left(\frac{U_{\text{salt}}}{\sqrt{(9.81)(0.05)}}\right)^{1100(150\times10^{-6})+2.5}$$

which implies, $U_{\text{salt}} = 8.379$ m/s

12.10 PRESSURE DROP ACROSS A LENGTH OF THE TRANSPORT LINE

- Pressure drop across a length of transport line has in general six components:
  Component (1): pressure drop due to gas acceleration;
  Component (2): pressure drop due to particle acceleration;
  Component (3): pressure drop due to gas-to-pipe wall friction;
  Component (4): pressure drop related to solid-to-pipe friction;
  Component (5): pressure drop due to the static head of the solids;
  Component (6): pressure drop due to the static head of the gas.
- If the gas and the solids are already accelerated in the line, then components (1) and (2) should be omitted from the calculation of the pressure drop; if the pipe is horizontal, components (5) and (6) can be omitted.

12.10.1 Pressure Loss in Transport in Horizontal Pipe

- **Component 1:** pressure drop due to gas acceleration

$$\Delta P_{a,f,H} = \frac{\rho_f \varepsilon_H U_{fH}^2}{2} \tag{12.5}$$

- **Component 2:** pressure drop due to particle acceleration

$$\Delta P_{a,p,H} = \frac{\rho_p(1-\varepsilon_H)U_{\mathrm{pH}}^2}{2} \tag{12.6}$$

where U_f=superficial fluid velocity, U_p=particle velocity, ε=porosity, ρ=density, d_p=particle diameter. Subscript H refers to horizontal.

- The horizontal particle velocity can be calculated from the well-known correlation developed by Hinkle, which is expressed as

$$U_{\mathrm{pH}} = U(1-0.0638d_p^{0.3}\rho_p^{0.5}) \tag{12.7}$$

- **Component** (3): In dilute transport, the gas-to-wall friction is often assumed independent of the presence of the solids, and so the Fanning equation for the gas flow can be used as

$$\Delta P_{\mathrm{f\text{-}w},H} = \frac{2f_f\rho_f U_f^2 L_H}{D} \tag{12.8a}$$

where f_f=Fanning friction factor for fluid (here gas) flow only, U=Superficial fluid velocity, L_H=length of pipe (horizontal), D=diameter of pipe, ρ_f=fluid density (here gas density), the subscript H refers to the horizontal sections.

$$\begin{aligned} f_f &= \frac{16}{\mathrm{Re}_f} \quad \text{if } \mathrm{Re}_f\left(=\frac{\rho_f U_f D}{\mu}\right) < 2100 \\ &= 0.125\left[0.0112 + \frac{1}{\mathrm{Re}_f^{0.3185}}\right] \quad \text{if } 2100 \le \mathrm{Re}_f < 4000 \\ &= \frac{0.079}{\mathrm{Re}_f^{0.25}} \quad \text{if } \mathrm{Re}_f \ge 4000 \end{aligned} \tag{12.8b}$$

- **Component 4:** The pressure drop due to solid-to-pipe wall friction is

$$\Delta P_{\mathrm{s\text{-}w},H} = \frac{2f_p\rho_p(1-\varepsilon_H)U_{\mathrm{pH}}^2 L_H}{D} \tag{12.9}$$

- The particle velocity through the pipe can be found from the solid flux, which is related to (Rhodes, 2008)

$$U_{\mathrm{pH}} = \frac{G}{\rho_p(1-\varepsilon_H)} \tag{12.10}$$

$$f_p = \frac{3}{8}\frac{\rho_f}{\rho_p}\frac{D}{d_p}C_D\left(\frac{U_f - U_p}{U_p}\right)^2 \tag{12.11}$$

$$\begin{aligned} &\text{if } \mathrm{Re}_p < 1, && C_D = 24/\mathrm{Re}_p \\ &\text{if } 1 < \mathrm{Re}_p < 500, && C_D = 18.5\,\mathrm{Re}_p^{-0.6} \\ &\text{if } 500 < \mathrm{Re}_p < 2\times 10^5, && C_D = 0.44 \end{aligned} \tag{12.12}$$

$$\mathrm{Re}_p = \frac{\rho_f (U_f - U_p) d_p}{\mu_f} \tag{12.13}$$

- Components (5) and (6) can be omitted for the horizontal pipe. Therefore, the total pressure drop for horizontal pipe

$$\begin{aligned} \Delta P_{H,\text{total}} &= \Delta P_{a,f} + \Delta P_{a,s} + \Delta P_{\text{f-w}} + \Delta P_{\text{s-w}} + \Delta P_{\text{sh},s} + \Delta P_{\text{sh},f} \\ &= \underbrace{\frac{\rho_f \varepsilon_H U_{fH}^2}{2}}_{\text{Component 1}} + \underbrace{\frac{\rho_p (1-\varepsilon_H) U_{\text{pH}}^2}{2}}_{\text{Component 2}} + \underbrace{\frac{2 f_f \rho_f U^2 L_H}{D}}_{\text{Component 3}} + \underbrace{\frac{2 f_p \rho_p (1-\varepsilon_H) U_{\text{pH}}^2 L_H}{D}}_{\text{Component 4}} \\ &\quad + \underbrace{0}_{\text{Component 5}} + \underbrace{0}_{\text{Component 6}} \end{aligned} \tag{12.14}$$

- Component (5): pressure drop due to the static head of the solid ($\Delta P_{\text{sh},s}$); Component (6): pressure drop due to the static head of the fluid ($\Delta P_{\text{sh},g}$).

12.10.2 Pressure Loss in Transport in a Vertical Pipe

- For vertical pipe, the total pressure drop due transport of solids through the vertical pipe can be expressed as (Konno and Saito, 1969). Component (1): pressure drop due to gas acceleration ($\Delta P_{a,f,v}$); component (2): pressure drop due to particle acceleration ($\Delta P_{a,s,v}$) are considered as zero. Therefore,

$$\begin{aligned} \Delta P_{v,\text{total}} &= \Delta P_{a,f,v} + \Delta P_{a,s,v} + \Delta P_{\text{f-w},v} + \Delta P_{\text{s-w},v} \\ &\quad + \Delta P_{\text{sh},s} + \Delta P_{\text{sh},f} + \Delta P_{\text{others}} \\ &= \underbrace{0}_{\text{Component 1}} + \underbrace{0}_{\text{Component 2}} + \underbrace{\frac{2 f_f \rho_f U^2 L_v}{D}}_{\text{Component 3}} + \underbrace{0.057 G L_v \sqrt{\frac{g}{D}}}_{\text{Component 4}} \\ &\quad + \underbrace{\rho_p (1-\varepsilon_v) g L_v}_{\text{Component 5}} + \underbrace{\rho_f \varepsilon_v g L_v}_{\text{Component 6}} + \Delta P_{\text{others}} \end{aligned} \tag{12.15}$$

where

$$G = \rho_p(1-\varepsilon_v)U_{pv} \tag{12.16}$$

$$U_{pv} = \frac{U}{\varepsilon_v} - U_T \tag{12.17}$$

- Subscript v refers to the vertical pipe. From these two Equations (12.16) and (12.17) yield

$$\varepsilon_v^2 U_T - \left(U_T + U + \frac{G}{\rho_p}\right)\varepsilon_v + U = 0 \tag{12.18}$$

where the particle terminal velocity, U_T can be calculated by

$$U_t = \sqrt{\frac{4gd_p}{3C_D}\left(\frac{\rho_p - \rho_f}{\rho_f}\right)}$$

where drag coefficient can be calculated as follows:

$$\begin{aligned}
&\text{if } \mathrm{Re}_p < 1, && C_D = 24/\mathrm{Re}_p \\
&\text{if } 1 < \mathrm{Re}_p < 500, && C_D = 18.5\,\mathrm{Re}_p^{-0.6} \\
&\text{if } 500 < \mathrm{Re}_p < 2\times10^5, && C_D = 0.44
\end{aligned}$$

$$\mathrm{Re}_p = \frac{\rho_f(U/\varepsilon_v - U_t)d_p}{\mu} \tag{12.19}$$

[Note that in this region, the terminal velocity is independent of the fluid viscosity and proportional to the square root of the particle diameter.]

12.11 POWER CONSUMPTION FOR PNEUMATIC TRANSPORT

- The power consumption can be calculated from the pressure drop and the volumetric flowrate of the solid, which are transferred as

$$P = \dot{Q}_s \Delta P = \frac{M_p}{\rho_p}\Delta P \tag{12.20}$$

where $\dot{Q}_s$=the volumetric flowrate of the solid, M_p=the mass flowrate of the solid, ΔP=the pressure drop, andρ_p=the solid density.

IMPORTANT POINTS TO REMEMBER

- Dilute-phase flows are observed at gas velocities greater than 20 m/s, low solids concentrations less than 1% by volume, and low-pressure drops per unit length of transport line typically less than 5 mbar/m.
- The dense-phase flow happens at low gas velocities (1–5 m/s), high solids concentrations (greater than 30% by volume), and high-pressure drops per unit length of pipe (typically greater than 20 mbar/m).
- Choking and saltation velocities are used to mark the boundaries between dilute-phase transport and dense-phase transport in vertical and horizontal pipelines, respectively.
- The particle is transported through the suction pipe under vacuum pressure (at about 0.7 bar), in practice.
- Reciprocating or screw-type compressors are required to generate high pressures (7–8 bar) for transporting the material.
- The velocity of the gas at which the gas no longer entrains the solid for transport is called choking velocity.
- The saltation velocity marks the boundary between the dilute-phase flow and dense-phase flow in horizontal pneumatic transport.
- The saltation velocity is the conveying velocity below which the conveyed solids start to settle at the bottom of horizontal pipes.
- The choking velocity happens in vertical lines, whereas the saltation velocity happens in horizontal lines.

SELF-TESTING QUESTIONS

I. Multiple-Choice Questions

Q1. For transferring solids downwards from a vessel at low pressure to a vessel at a higher pressure, which of the following systems is to be used?
 (a) Standpipes
 (b) Gravity cutes
 (c) Vibrating conveyors
 (d) Elevators

Q2. The dilute-phase flow, in its most recognizable form, is characterized by ________.
 (a) Gas velocities < than 20 m/s, solids concentrations > 1% by volume and pressure gradient > 5 mbar/m.
 (b) Gas velocities > than 1 m/s, solids concentrations < 20% by volume and pressure gradient < 5 mbar/m.

(c) Gas velocities > than 5 m/s, solids concentrations < 1% by volume and pressure gradient < 20 mbar/m.
(d) Gas velocities > than 20 m/s, solids concentrations < 1% by volume and pressure gradient < 5 mbar/m.

Q3. The case of solid transport vertically distinguishes the dilute and dense flow pattern by which of the following options of threshold condition?
(a) The superficial velocity of the particle
(b) Choke velocity of the fluid
(c) Saltation velocity
(d) Both choke and saltation velocities

Q4. To assess frictional pressure drop in the transport of solids through the pipeline, which of the following components is (are) not considered?
(a) pressure drop due to fluid and solid accelerations
(b) pressure drop due to gas-to-pipe wall friction
(c) pressure drop related to solid-to-pipe friction
(d) pressure drop due to the static head of the solids and fluids

Q5. The gas velocity at which it is insufficient to maintain the solids in suspension and the solids begin to settle out in the bottom of the pipe.
(a) Critical velocity
(b) Saltation velocity
(c) Terminal velocity
(d) Instant velocity

Q6. It is the lowest velocity at which this dilute-phase transport line can be operated at the solid feed rate. Identify it.
(a) Transition velocity
(b) Choking velocity
(c) Escape velocity
(d) None of these

Q7. The pneumatic transport of particulate solids is broadly classified into flow regimes as ________________.
(a) Dense-phase flow
(b) Dilute-phase flow
(c) Lean phase flow
(d) All of these

Q8. Particles are not fully suspended and there is ________ between the particles in dense-phase transport.
(a) Much interaction
(b) Not much interaction
(c) Zero interaction
(d) None of these

Q9. __________ pressure pneumatic conveying systems can handle a wide range of pulverized granular and fibrous materials.

(a) High
(b) Zero
(c) Low
(d) None of these

Q10. The choking velocity is defined as ____________________.

(a) It is the lowest velocity at which this dilute-phase transport line can be operated at the solid feed rate.
(b) It is the lowest velocity at which this dense-phase transport line can be operated at the solid feed rate.
(c) It is the maximum velocity at which this dilute-phase transport line can be operated at the solid feed rate.
(d) It is the lowest velocity at which this dense-phase transport line can be operated at the solid feed rate.

Q11. The true about saltation velocity is ______________.

(a) The saltation velocity is the conveying velocity below which the solids that are conveyed start to settle at the bottom of horizontal pipes.
(b) To convey the solids in a horizontal pipeline such as a powder or granulates, the minimum conveying velocity in all parts of the line.
(c) The saltation velocity marks the boundary between the dilute-phase flow and dense-phase flow in horizontal pneumatic transport.
(d) All of the above.

Q12. Pneumatic conveying is done under which of the mentioned conditions?

(a) High pressure
(b) Vacuum
(c) Fluidization
(d) Any of the above

Q13. In dense-phase transport, particles are ______ and there is ________ between the particles.

(a) not fully suspended, much interaction
(b) not fully suspended, no interaction
(c) fully suspended, much interaction
(d) fully suspended, no interaction

Q14. What is the mechanism of a pneumatic transport system?

(a) Use of gas to transport particulate solids through a pipeline
(b) Use of hydraulics to transport particulate solids through a pipeline

(c) Use of a liquid to transport particulate solids through a pipeline
(d) There is no such thing as a pneumatic transport system

Q15. Which of the following is not true about dilute-phase pneumatic transport systems?
(a) It is characterized by high gas velocities
(b) It is characterized by low solid concentrations
(c) It is the only system capable of operating under negative pressure
(d) It is a long-route transport system

Q16. Which of the following is not true about dense-phase pneumatic transport systems?
(a) It is characterized by low gas velocities
(b) It is characterized by high solid concentrations
(c) It is characterized by low-pressure drops per unit length of pipe
(d) In this phase, particles are not fully suspended and there is much interaction between the particles

Q17. The gas velocity at which it is insufficient to maintain the solids in suspension and the solids begin to settle out in the bottom of the pipe is termed as ___________.
(a) Choking velocity
(b) Saltation velocity
(c) Suspension terminal velocity
(d) Superficial velocity

Q18. Froude number at saltation is related to the pipe diameter as-
(a) Inversely proportional to D
(b) Directly proportional to D
(c) Inversely proportional to $D^{1/2}$
(d) Directly proportional to $D^{1/2}$

Q19. Match the following ___________________________.

i. Belt conveyors gravity	p. the solids fall under the action of
ii Pneumatic transport impeller	q. the solids are moved by a rotating
iii. Gravity chutes	r. solid transport by the flow of fluid
iv. Screw conveyor moving belt	s. solids conveyed on a continuous

(a) (i) – (p); (ii) – (q); (iii) – (r); (iv) – (s)
(b) (i) – (s); (ii) – (r); (iii) – (q); (iv) – (p)
(c) (i) – (s); (ii) – (r); (iii) – (p); (iv) – (q)
(d) (i) – (r); (ii) – (q); (iii) – (p); (iv) – (s)

Q20. What is the unit of solid loading?
(a) kg/s
(b) m3/s
(c) $kg/m^2.s$
(d) dimensionless

Answers to the multiple-choice questions

Q1. Ans. (a)
Q2. Ans. (d)
Q3. Ans. (b)
Q4. Ans. (d)
Q5. Ans. (b)
Q6. Ans. (b)
Q7. Ans. (d)
Q8. Ans. (a)
Q9. Ans. (c)
Q10. Ans. (a)
Q11. Ans. (d)
Q12. Ans. (d)
Q13. Ans. (a)
Q14. Ans. (a)
Q15. Ans. (d)
Q16. Ans. (c)
Q17. Ans. (b)
Q18. Ans. (c)
Q19. Ans. (c)
Q20. Ans. (d)

II. Objective Type Questions

1. What are the common equipment used to transport the solid?
2. What are the basic components of a pneumatic transport system?
3. Write the advantages and disadvantages of pneumatic transport.
4. Explain the two flow regimes of the pneumatic transport of particulate solids.
5. How dense-phase flow can be distinguished from dilute-phase flow?
6. State the various types of pneumatic systems.
7. Describe the positive and negative-pressure systems of solid transport.
8. Explain the high-pressure pneumatic conveying systems.
9. Explain the working principle of low-velocity pneumatic conveying systems.
10. What is choking velocity? How is it related to the solid flux?
11. Explain the flow patterns in the case of dense phase such as (a) continuous dense-phase flow, (b) plug flow, (c) discrete plug flow, and (d) dune flow.
12. What is saltation velocity? What is the significance of this velocity for solid transport?
13. Explain the different pressure components of solid transport in a transport pipeline.
14. Which of the pressure components are not dominant in a horizontal pipeline?
15. How can the power consumption be calculated from the pressure drop and the volumetric flowrate of the solid?

III. Numerical Problems

Q1. It is supposed to transport a powder through a pipeline vertically at a fluid velocity equal to the terminal velocity of a particle. Under this condition, find the volume fraction of the fluid in the pipe. [Ans. 0.5]

Q2. A powder is to be transported through the pipeline of a specific size of length 50 m horizontally. Transporting the solid of size (diameter) of 100-μm m by gas at a velocity three times the particle's terminal velocity is required. The frictional pressure between the particle and the pipe wall should not exceed 500 Pa/m. The friction factor of the particle is to be considered as 0.002. The density of the particle is 1530 kg/m³. The terminal velocity of the particle is 0.3 m/s. The dilution of the mixture is maintained at a concentration of 20% by volume. What should the pipe diameter be under this condition to transport the material? [Ans. 0.070 m]

Q3. A horizontal pipeline transports zinc oxide powder from one bin to another at a distance of 20 m by airflow. The pressure drop due to gas acceleration during the flow follows the following equation $\Delta P_H = 0.12U_{fH}^2$. Under this condition, what would be the volume fraction of air to be maintained so there will be no uneven situation? The density of air is to be considered as 1.2 kg/m³. [Ans. 0.20]

Q4. A powder is to be transported through the pipeline of a certain size of length 10 m vertically. It is required to transport the powder of a particle of size (diameter) of 100 μm by a gas at a choking velocity 1.25 times the particle's terminal velocity. The density of the particle is 1530 kg/m³. It was observed that the volume fraction of the solid is 25% at choking conditions. The mass flow rate of the solid is controlled at 153 kg/m².s. What is the terminal velocity of the particle? [Ans. 0.6 m/s]

Q5. To transport a powder through a pipeline at a mass flux of 200 kg/m².s. It is seen that the frictional force resisting powder transport is 10 N. If the density of the particle is 1530 kg/m³, what will be the power required to transport the solid through the pipeline? [Ans. 2 N.m/s]

Q6. Calculate total pressure loss in a horizontal pipe for positive-pressure dilute phase (10% solid by vol.) pneumatic transport system to transport 2150 kg/h of sand of (density 2500 kg/m³ and mean particle size 100 μm) between two points in a plant separated by 10 m vertical distance and 30 m horizontal distance using ambient air of flowrate 179.47 lpm. Assume that six 90° bends are required and the allowable pressure loss is 0.55 kPa for each bend. The inner pipe diameter is to be 78 mm. The viscosity of the fluid is 18.4×10^{-6} Pa.s. [Ans. 19.84 kPa]

Q7. Calculate total pressure loss in a vertical pipe of diameter 0.075 m for positive-pressure dilute phase (15% solid by vol.) pneumatic transport system to transport of zeolite of (density 720 kg/m^3 and mean particle size 150 μm) between two points in a plant separated by 15 m vertical distance using ambient air of flowrate 387.84 lpm. The viscosity of the fluid is 18.4×10^{-6} Pa.s. [Ans. 16.14 kPa]

Q8. A 5 m long vertical standpipe of inside diameter 0.5 m transports solids at the flux of 600 kg/m^2.s from an upper vessel held at a pressure of 1.25 bar, to a lower vessel held at 1.5 bar. The particle density of the solids is 1600 kg/m^3, and the surface volume mean particle size is 200 mm. Assuming that the voidage is 0.5 and is constant along the standpipe. (Properties of gas in the system: density, 1.5 kg/m^3; viscosity 2×10^{-5} Pa.s.). Determine relative particle velocity. [Ans. 0.75 m/s]

Q9. A long vertical standpipe of 5 m of inner diameter 0.05 m is used to transport solids at flux of 433.08 kg/m^2.s by airflow from a lower vessel held at a pressure of 1.66 bar to an upper vessel held at 1.25 bar. The particle density of the solids is 1600 kg/m^3, and the surface volume mean particle size is 150 μm. Assuming that the voidage is 0.5 and is constant along the standpipe. (Properties of gas in the system: density, 1.2 kg/m^3; viscosity 1.84×10^{-5} Pa.s.). Determine the air flowrate in lpm. Consider that transport happens in turbulent flow conditions. [Ans. 175.40 lpm]

Q10. Calculate the choking velocity in a vertical transport system given that the voidage in the pipes is 0.75, the particle density is 2100 kg/m^3, the mass flux is 1000 kg/m^2.s, and the terminal velocity of a single particle in the gas is 0.25 m/s. [Ans. 1.616 m/s]

Q11. Calculate the Froude number at saltation given the superficial gas velocity at saltation is 5 m/s and the pipe diameter is 9.81 m. [Ans. 0.51]

Q12. A powder of a particle diameter of 105 μm is to be transported through a pipe of 4 cm diameter to a section where it will be used for chemical synthesis. During transportation, it was observed that the minimum Froude no. was 0.50 when saltation started. Calculate the solid loading of suspension for the transport system. [Ans. 1.26×10^{-3}]

Q13. The saltation velocity in a horizontal transport is 10 cm/s. The pipe diameter is 75 cm and the particle diameter is 0.04 cm. What is the value of solid loading? [Ans. 1.78×10^{-7}]

Q14. There is a vertical transport of solids to utilize the same in a chemical process in industry. The density of the particle is 3200 kg/m^3, and the mass flux of the solid is 150 kg/m^2. The terminal velocity of a single particle in gas is 1515 cm/s. The voidage of the pipe at chocking velocity is 0.4. What is the value of choking velocity? [Ans. 609.125 cm/s]

Q15. Solids are transported vertically in an industry to be utilized in a chemical process. The density of the particle is 3200 kg/m^3, and the mass flux of the solid is 150 kg/m^2. The terminal velocity of a single particle in gas is 1515 cm/s. The voidage of the pipe at chocking velocity is 0.4. What is the density of the gas if the diameter of the pipe is 0.5 m? [Ans. 370 kg/m^3]

Q16. For a chemical synthesis with a gas, catalyst particles are transported vertically in an industry. The density of gas in the transport is 550 kg/m^3. The velocity of gas is 0.5 m/s, and the porosity is 0.6. What is the pressure drop due to gas acceleration? [Ans. 41.25 Pa]

Q17. A solid is transported in a horizontal pipe. The density of solid in the transport is 2550 kg/m^3. The superficial velocity of gas is 0.75 m/s, and the porosity is 0.6. The particle diameter is 0.0025 m. What is the pressure drop due to particle acceleration? [Ans. 62.475 Pa]

Q18. The total pressure drop across the length of the horizontal transport line is 2500 Pa. The pressure drop due to gas acceleration is 200 Pa. The pressure drop due to particle acceleration is 450 Pa. The pressure drop due to gas-to-pipe friction is 20 Pa. What is the value of pressure drop due to solid-pipe friction? [Ans. 1830 Pa]

NOMENCLATURE

A	cross-sectional area of pipe [m^2]
C_D	drag coefficient [-]
D	diameter of the pipe [m]
d_p	particle diameter [m]
f_f	Fanning friction factor for fluid flow only [-]
G	mass flux of solids [kg/m^2.s]
f_p	Fanning friction factor for solid flow only [-]
L_H	length of horizontal pipe [m]
L_v	length of vertical pipe [m]
M_p	mass flowrate of solid [kg/s]
P	power [W]
$\Delta P_{a,p,H}$	pressure drop due to gas acceleration in horizontal pipe [N/m^2]
$\Delta P_{s,w,H}$	pressure drop due to solid-to-pipe wall friction [N/m^2]
$\Delta P_{H,total}$	total horizontal pressure drop [N/m^2]
$\Delta P_{v,total}$	total vertical pressure drop [N/m^2]
U_{CH}	choking velocity [m/s]
U_T	free fall, or terminal velocity of a single particle [m/s]

U_{pH} particle velocity in horizontal pipe [m/s]
U_{sal} superficial gas velocity at saltation [m/s]
U_t terminal velocity [m/s]
U_f superficial fluid velocity [m/s]
U_p particle velocity [m/s]
$\dot{Q}_s$ volumetric flowrate of the solid [m^3/s]
Re_p Reynolds number of particles [-]

GREEK LETTERS

ε_{CH} voidage in the pipe at choking condition [-]
ρ_p particle density [kg/m^3]
ρ_f density of gas [kg/m^3]
ε_H porosity or void fraction in horizontal pipe [-]
ρ_f Density of fluid [kg/m^3]

SUBSCRIPTS

CH choke
D drag
f fluid
H horizontal
p particle
s solid
sal saltation
s-w solid to wall
t terminal
v vertical

REFERENCES

Haider, A. and O. Levenspiel, Drag coefficient and terminal velocity of spherical and nonspherical particles, *Powder Technol.*, 58, 63–70, 1989.
Klinzing, G. E., A review of pneumatic conveying status, advances and projections, *Powder Technol.*, 333, 78–90, 2018.
Konno, H. and S. Saito, Pneumatic conveying of solids through straight pipes, *J. Chem. Engng Japan*, 2, 211–217, 1969.
Rhodes, M., *Introduction to Particle Technology*, John Wiley & Sons Ltd., 2008.
Rhodes, M. J., Pneumatic conveying, in: M.J. Rhodes, (Ed.), *Principles of Powder Technology*, Wiley, Chap. 7, 1990.

Rizk, F., *Pneumatische Förderung von Kunststoffgirranulaten in horizontalen Rahrleitungen unter Berucksichtigung des Gewichtseinflusses in Zusammenhang mit Gut-und Rohrwerkstoffeigenschaften, insbesondere im optimalen For der-bereich*, Dr-Ing Dissertation, Technische Hochschule Karlsruhe, 1973.
Stepanoff, A. J., *Gravity Flow of Bulk Solids and Transportation of Solids in Suspension*, John Wiley & Sons, 1969.
Woodcock, C. R. and J. S. Mason, *Bulk Solids Handling: An Introduction to the Practice and Technology*, Springer, 1987.

Chapter 13

Transport by Conveyor and Elevator Systems

> As engineers, we were going to be in a position to change the world – not just study it.
>
> —*Henry Petroski, American engineer and author specializing in failure analysis.*

13.1 TRANSPORTATION OF SOLID BY CONVEYORS

- As per the requirements in industrial processes, transporting solids has led to developing a wide range of equipment. The previous chapter gives a brief introduction to conveying and elevator systems. This chapter highlights important conveying and elevator systems for sild transport with their principles, advantages, disadvantages, and applications.

13.2 TRANSPORTATION BY GRAVITY CHUTE

- It is a gravity conveyor in the form of an inclined plane, trough, or framework that depends on sliding friction to control the rate of descent, as shown in Figure 13.1a.

13.2.1 Working Principle

- It uses the force of gravity to move materials over a downward path.
  - As per the diagram shown in Figure 13.1b, applying Newton's second law, the force balance can be written as:
  - Along the plane force balance:

$$mg\sin\theta - F = ma \tag{13.1}$$

  - Perpendicular to the plane, force balance:

DOI: 10.1201/9781003571025-18

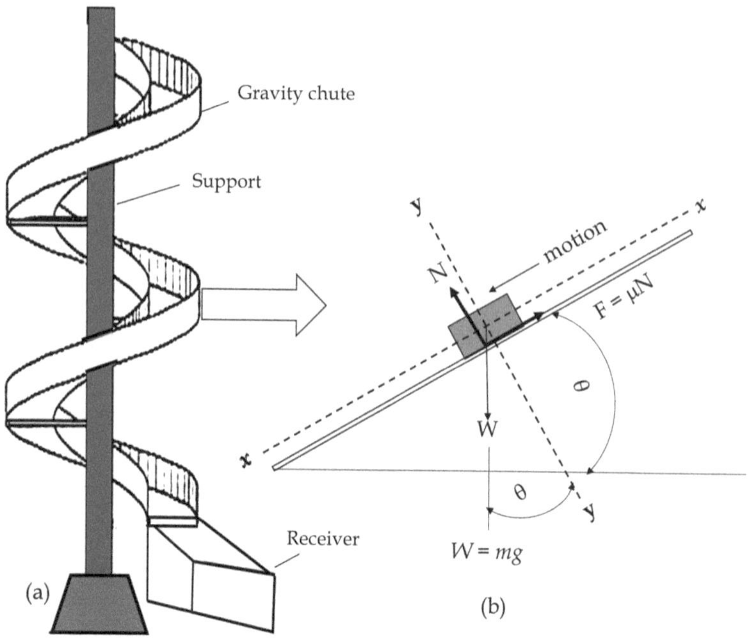

Figure 13.1 Schematic of forces acting on gravity chute. *N*, normal force perpendicular to *x*-direction; *F*, force acting in *x*-direction; *W*, weight of particle; *m*, mass of particle; μ, dynamic coefficients of friction; ϑ, angle of inclination.

$$N = mg\cos\theta \tag{13.2}$$

$$F = \mu N = \mu mg\cos\theta \tag{13.3}$$

- By substitution of *F* from Equation (13.3) in Equation (13.1):

$$a = g(\sin\theta - \mu\cos\theta) \tag{13.4}$$

- If the initial velocity is V_i, the exit velocity V_e of the particle can be calculated as follows:

$$V_e = V_i - at = V_i - [g(\sin\theta - \mu\cos\theta)]t \tag{13.5}$$

Example 13.1

A 10 kg steel object is falling through a gravity chute at an initial velocity of 1.5 m/s, and it takes 10 seconds to fall on to the ground. What should be the exit velocity at which the object falls? What is the frictional force acting on the object? The gravity chute is made of steel plate. The dynamic coefficient of friction between materials is 0.42. The angle of inclination of the chute in each turn is 22°.

Solution:

The initial velocity (V_i) of the object is 1.5 m/s.
The time to reach the ground (t) is 10 seconds.
The dynamic coefficient of friction between materials is 0.42.
The angle of inclination of the chute in each turn is 22°.
Therefore, using the Equation (13.5), the exit velocity at which the object falls is

$$V_e = V_i - at = V_i - [g(\sin\theta - \mu\cos\theta)]t$$

$$= 1.5 - [(9.81)(\sin 22° - (0.42)\cos 22°)](10) = 2.95 \text{ m/s}$$

The frictional force acting on the object is

$$F = \mu N = \mu mg\cos\theta = (0.42)(10)(9.81)\cos 22° = 38.20 \text{ N}$$

13.3 TRANSPORTATION BY BELT CONVEYORS

- The belt conveyors consist of drums (driving and non-driving), belts, drive, idler rollers, belt tensioners, and belt cleaners, as shown in Figure 13.2.
- Belts may be flat or fitted with a pattern of cleats or flights molded into their surface to reduce the tendency for the conveyed material to slip. The conveyors are supported on idler rollers (trough and return).
- The drive may be at either end of the conveyor. Belt tensioners respond to changes in the belt's operating conditions. Belt cleaner minimizes material build-up on snub pulleys and return idlers.

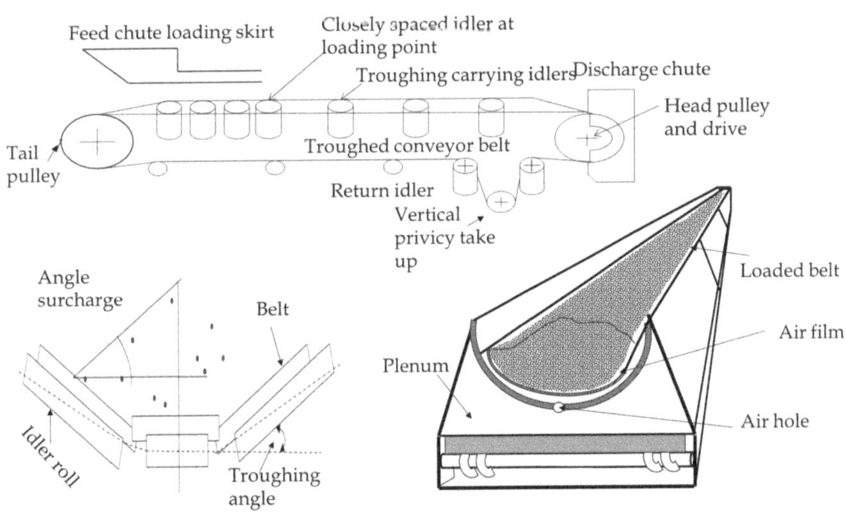

Figure 13.2 Belt conveyor.

13.3.1 Working Principle

- As a driving motor rotates the pulleys (drum), the belt will move with the roller due to frictional resistance. Thereby, the material placed on the belt will be transported from one place to another.

13.3.2 Types of Belt Conveyors

- The belt conveyors can be classified into some basic types as per the convenience of transport of material, which are shown in pictorial form in Figure 13.3.
  - **Roller belt conveyor:** The surface of belt is made up of roll. Used for sorting, assembling, inspecting, and airport baggage handling.
  - **Flat belt conveyor:** Uses a series of powered pulleys to move the belt. Transporting items within a facility.
  - **Modular belt conveyor:** Made of countless interlocked pieces, used to carry food products around corners.
  - **Cleated belt conveyor:** Designed with vertical cleats or barriers. Used to carry loose materials up a steep incline.
  - **Curved belt conveyor:** Uses a curved frame to carry items around corners. Used in bag handling systems.
  - **Inclined belt conveyor:** Made with a center drive, gear motor, and take-up with a single or double nose. Suitable for boosting gravity flow systems.
  - **Sanitary, washdown belt conveyor:** Specially designed to withstand sanitary procedures, applied in the food and pharmaceutical industries for harsh washing and sterilization.

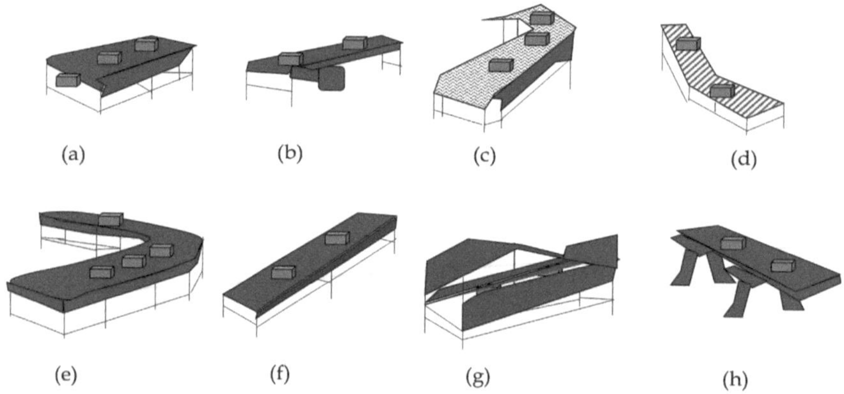

Figure 13.3 Different types of belt conveyors. (a) Roller bed conveyor belt, (b) flat belt conveyor, (c) modular belt conveyor, (d) cleated belt conveyor, (e) curved belt conveyor, (f) inclined belt conveyor, (g) sanitary and washdown conveyors, (h) specialty conveyor belt.

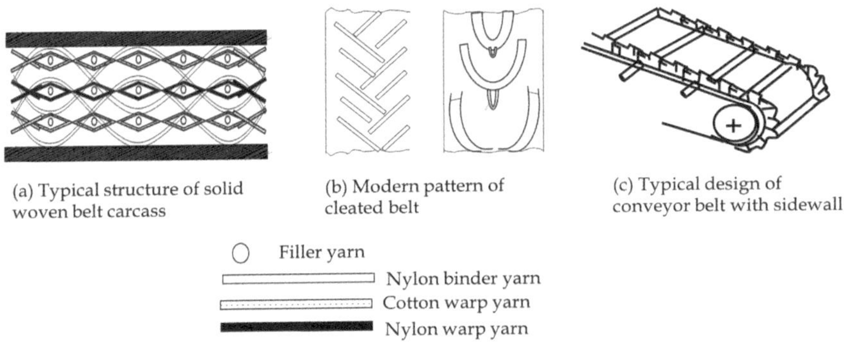

Figure 13.4 Some basic construction structures of the belt conveyors.

- There are several other kinds of conveyors tailored for different applications. Some of these are shown in Figure 13.4.
  - **Fiberglass conveyor belts:** Made of fiberglass. They are useful for tasks that require extreme temperatures.
  - **Metal nub conveyor belts:** The belt is fitted with metal nubs to better handle complex machines and their parts.
  - **Narrow-width belts:** The belt is the best choice for handling small items.
  - **Back-lit conveyor belts:** They come with a light fixture on the belt to aid with quality control.
  - **Vacuum conveyor belts:** The belt creates suction to keep light products (sheets of paper, leaflets, etc.) despite inclines and high speeds.
  - **Magnetic conveyor belt:** The belt can transport materials that contain iron through equally spaced electromagnets that are placed within the bed of the conveyor.
  - **Sandwich belt conveyors:** Use two face-to-face belts to securely carry items at steep inclines or even up a vertical run.

13.3.3 Some Important Design Points

- The main design components of the belt conveyors are belt width, speed, capacity, and power requirement.
- The factors that affect the belt conveyor's capacity are the belt's speed, shape, and width, method of feeding, and solids bulk density and particle size.
- The basic needs for power input to the belt conveyor are as follows:
  - to move the load,
  - to move the belt itself,
  - to overcome friction in the idlers,
  - to operate trippers, inclined belts, and
  - to elevate the material.

- Some basic construction structures of the belt conveyors are shown in Figure 13.4.
- The temperature of the material to be transported should not be above 120°C.
- The materials may be conveyed at angles of inclination of up to 22° with standard belts.
- To reduce the tendency for the conveyed material to slip, the belt should have patterns of cleats of height between 15 and 25 mm or flights molded into its surface.
- The idler with concentrator rolls should be with an outer roll angle of 30°–35°.

13.3.4 Drive Arrangement

- The different types of drive arrangements are shown in Figure 13.5. The potency of the conveyor drive depends on the following factors:
  - the tension between the 'tight side' and the "slack side" of the belt
  - the friction between the belt and the drive drum
  - the angle of wrap, or arc of contact, of the belt to the drum
- Typical angles of the wrap as per drives are 180° for single drum, 180°–240° for snubbed drum, and 360°–480° for Tandem or dual drive (Woodcock and Masson, 1987).

13.3.5 Surcharge Angle, Maximum Lumps

- For most materials, the surcharge angle is 5°–15° less than the angle of repose, as shown in Figure 13.6.
- The maximum lump size would be as follows: one-fifth of the belt width for surcharge angles of 20° or one-tenth of the belt width for surcharge angles up to 30°.

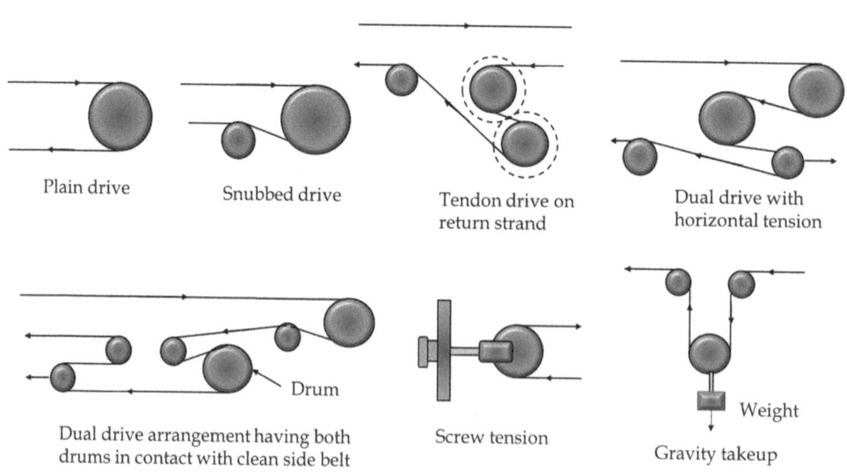

Figure 13.5 Different types of drive arrangement.

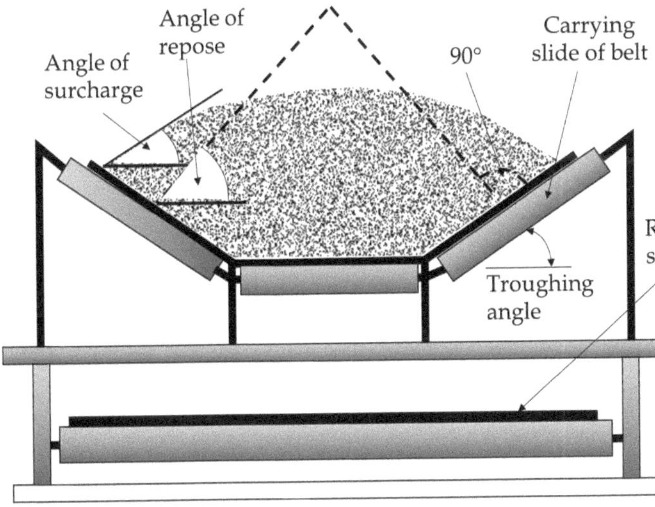

Figure 13.6 Surcharge angle and angle of repose.

Table 13.1 Correlations for belt speed (in m/s) as a function of belt width (in mm)

Type of Material	*Correlation for Belt Speed (in m/s) as a Function of Belt Width (in mm)*
For fine and free-flowing non-abrasive material	$v = 0.938 + 4.38\times10^{-3}B - 1.24\times10^{-6}B^2$
For fine and mildly abrasive or lumpy material	$v = 0.859 + 3.72\times10^{-3}B - 1.05\times10^{-6}B^2$
For granular abrasive or lumpy mildly abrasive material	$v = 0.709 + 3.27\times10^{-3}B - 9.27\times10^{-7}B^2$
For granular and very abrasive or lumpy moderately abrasive material	$v = 0.441 + 3.08\times10^{-3}B \quad 9.14\times10^{-7}B^2$
For lumpy and very abrasive material	$v = 0.354 + 2.70\times10^{-3}B - 8.14\times10^{-7}B^2$

$v = 1.755$ for belts used with belt-propelled trippers, $v = 0.747$ for belts used with plows, and $v = 0.372$ for picking and feeding belts.

13.3.6 Belt Speed

- The applied belt speed (v) (in m/s) can be calculated from the following correlations as a function of belt width (B) (in mm) for different conveyed materials, as shown in Table 13.1:

Example 13.2

A belt of 600 mm in width operates to transport solid materials that are lumpy and very abrasive. What would be the ideal belt speed so that the material cannot be slipped during transport?

Solution:

For the lumpy and abrasive material, the belt speed can be calculated by

$$v = 0.354 + 2.70\times10^{-3}B - 8.14\times10^{-7}B^2$$

$$= 0.354 + 2.70\times10^{-3}(600) - 8.14\times10^{-7}(600)^2 = 1.974 \text{ m/s}$$

13.3.7 Belt Carrying Capacity and Width

- The carrying capacity (m_s) (in kg/s) of a belt conveyor is defined as

$$\dot{m}_s = \rho_b k_s A v \tag{13.6}$$

where ρ_b=the bulk density of the conveyed material, A=the average cross-sectional area of the material stacked on the horizontal moving belt, v=the belt speed, and k_s=a 'slope factor' to take into account the decrease of the load stream cross-section when the belt operates on a gradient. The shape factor is a function of the angle of inclination of the belt (α, in degree), which can be calculated from the correlation as follows:

$$k_s = 1 + 1.91\times10^{-4}\alpha - 4.82\times10^{-4}\alpha^2 \tag{13.7}$$

- The maximum value of A depends on the material's nature, the belt's width, and the idlers' configuration.
- The surface of the conveyed material is generally considered parabolic, which can be expressed as

$$k_s A = \frac{1}{6}b^2\tan\delta = \left(\frac{1}{6}\tan\delta\right)(b^2) = k_s b^2 \tag{13.8}$$

Therefore, the carrying capacity as per the parabolic surface is

$$\dot{m}_s = A k_s v \rho_b = \left(\frac{1}{6}b^2\tan\delta\right)v\rho_b \tag{13.9}$$

where b is the width of the load stream on the belt, and δ is the surcharge angle, as shown in Figure 13.7.

- Therefore, the shape factor can be calculated by comparing Equations (13.6) and (13.9) as

$$k_s = \frac{1}{6}\tan\delta \tag{13.10}$$

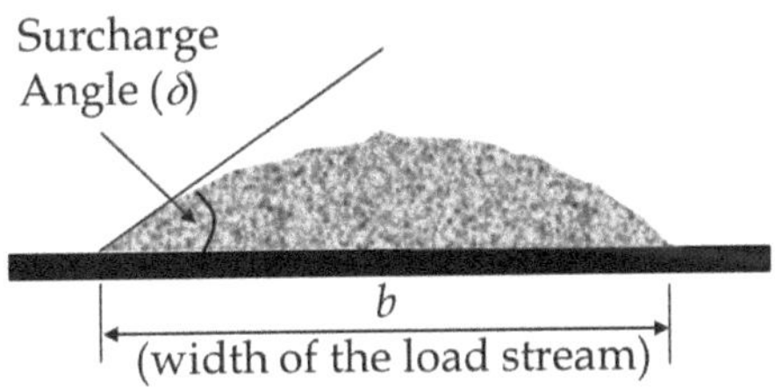

Figure 13.7 Surcharge angle (δ).

13.3.8 Belt Width

- The minimum width (B_{min}) of the belt (recommended in BS 5934/ISO 5048) is:

$$B_{min} = 1.11b + 0.056 \text{ (in meter)} \tag{13.11}$$

- The minimum belt width required to transport material also depends on bulk density (ρ_b), the carrying capacity of belt conveyor ($\dot{m}_s$), shape factor (k_s), and belt speed (v).
- The minimum belt width based on the above factors can also be calculated from the correlation:

$$B_{min} = 1.11\left(\frac{\dot{m}_s}{\rho_b k_s v}\right)^{0.5} + 0.056 \tag{13.12}$$

- The shape factor (k_s) is also related to the troughing angle (β) and surcharge angle (δ) as:

$$k_s = p + q\beta + r\beta^2 \tag{13.13}$$

where

$$p = 3.70 \times 10^{-3}\delta - 0.0166 \tag{13.13a}$$

$$q = 5.75 \times 10^{-3} - 1.40 \times 10^{-5}\delta \tag{13.13b}$$

$$r = 8.02 \times 10^{-7}\delta + 5.70 \times 10^{-5} \tag{13.13c}$$

where all the angles in the Equations (13.13a) to (13.13c) are to be considered in degree.

- The recommended conveyor belt width is (by the UK Mechanical Handling Engineers Association) in mm: 400, 500, 600, 650, 800, 1000, 1200, 1400, 1600, 1800, and 2000.

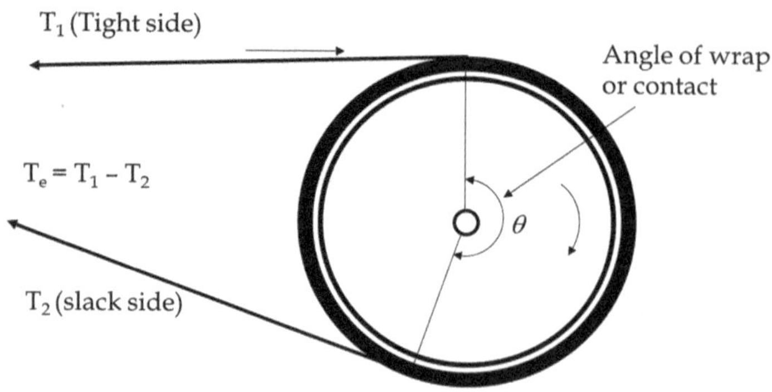

Figure 13.8 The tensions in the belt as it approaches and leaves the driving drum.

13.3.9 Belt Tension

- Effective tension and drive factor (or wrap factor) are useful parameters in belt conveyor design.
- The effective tension (T_e) is defined as the difference ($=T_1-T_2$) between the tensions in the belt as it approaches and leaves the driving drum, as shown in Figure 13.8.
- The 'drive factor' or 'wrap factor' (K_d) is the ratio of the "slack-side" tension T_2 to the effective tension (T_e), which is expressed as

$$K_d = \frac{T_2}{T_e} = \frac{T_2}{T_1 - T_2} \tag{13.14}$$

Also, as per the force balance

$$\frac{T_1}{T_2} = \exp(\mu\theta) \tag{13.15}$$

- From Equations (13.12) and (13.13)

$$T_1 = (1 + K_d)T_2 \quad \text{and} \quad T_2 = K_d T_e \tag{13.16}$$

where

$$K_d = \frac{1}{\exp(\mu\theta) - 1} \tag{13.16a}$$

where ϑ=the angle of wrap as shown in Figure 13.8, μ=the dynamic coefficient of friction between the belt and the drum. Typical values of the friction coefficient are shown in Table 13.2.

Table 13.2 Typical values of friction coefficient (μ) between drive drum and belt (Woodcock and Masson, 1987)

Operating Conditions	*Bare Drum*	*Lagged Drum*
Dry	0.3	0.35
Cleat wet	0.2	0.2 to 0.3
Wet and dirty	0.1	0.2

- In practice, for a single drive, K_d should be 20% greater if the drum is bare and 30% greater if it is lagged, and for a tandem or dual drive, K_d should be about 25% greater for both bare and lagged drums.

13.3.10 Important Note

- The value of the minimum tension should be sufficient to prevent the belt from sagging excessively between the idlers. Sag is usually the lowest point on a belt between two points.
- The tension throughout the length of the belt must always exceed 'sag tension', which can be estimated as follows:
- For the carrying side,

$$T_{sag} = K_s L_{ic}(m_b + m_L)g \tag{13.17}$$

- For the return side

$$T_{sag} = K_s L_{ir} m_b g \tag{13.18}$$

where m_b and m_L are, respectively, the mass per unit length of the belt and of the conveyed material on the belt, L_{ic}=the idler pitch on the carrying side, L_{ir}=the idler pitch on the return side, and K_s=a 'sag factor' which has values 4.2 for 3% sag, 6.25 for 2% sag and 8.4 for 1.5% sag.

13.3.11 Idler Spacing

- It is recommended that the return idler sets on a 3 m pitch, but on the carrying side, the idler spacing would typically be from 0.93 to 1.6 m.

13.3.12 Power Requirements for Transport by Belt Conveyor

- The power required to transport material on a belt conveyor is absorbed by overcoming different resistances. Effective tension T_e is equal to the total resistance (F_R) in the belt that can be expressed as

$$T_e = F_R = \underbrace{F_{fb} + F_{fL}}_{\text{Main resistance}} + \underbrace{F_{st}}_{\text{Slope resistance}} + \underbrace{F_N}_{\text{Secondary resistance}} \tag{13.19}$$

where F_{fb}=Belt friction resistance, F_{fL}=Load friction resistance, F_{st}=Load slope resistance, F_N=Load acceleration resistance. The belt friction and the load friction resistance are called 'main resistance' denoted by F_H.

13.3.13 Belt Friction Resistance (F_{fb})

- The belt friction resistance depends on the total mass of the moving parts. It can be calculated by

$$F_{fb} = \mu_r M_c g \tag{13.20}$$

where, M_c is the total mass of moving parts which can be calculated as

$$M_c = (m_{ic} + m_{ir} + 2m_b \cos\alpha)L \tag{13.21}$$

where m_{ic}=the mass of the rotating parts of the idlers per unit length of belt on the carry side, m_{ir}=the mass of the rotating parts of the idlers per unit length of belt on the return side, m_b=the mass of the belt per unit length overall, α=the angle of elevation, L=the overall length of the conveyor (center-to-center), and μ_r=belt friction coefficient=0.025.

13.3.14 Load Friction Resistance (F_{fL})

- The load friction resistance can be calculated by

$$F_{fL} = \mu_r m_L g L = \mu_r \frac{\dot{m}_s g L}{v} \tag{13.22}$$

where the mass of load (m_L) is defined as

$$m_L = \frac{\dot{m}_s}{v} \tag{13.23}$$

and

$$\dot{m}_s = \frac{1}{6} \rho_b v b^2 \tan\delta \tag{13.24}$$

where L=the conveying distance, μ_r=0.025, v=Linear velocity of belt, m_L=the mass of conveyed material per unit length of the belt, $\dot{m}_s$=carrying capacity (mass of solids per unit time) of a belt, b=the width of the load stream on the belt, and δ=the surcharge angle.

13.3.15 Load Slope Resistance (F_{st})

- It may be positive or negative, depending upon whether the movement of the conveyed material is upwards or downwards, which can be expressed as

$$F_{st,vertical} = \pm m_L g H = \pm \frac{\dot{m}_s g H}{v} \tag{13.25}$$

where H is the net change in vertical elevation which can be positive or negative. The sign+ve is for upward transport, and –ve is for negative transport.

- For horizontal belt conveyors slope resistance,

$$F_{st,\ horizontal} = 0 \tag{13.26}$$

13.3.16 Main Resistance (F_H)

- The main resistance is the sum of the belt and load friction resistance. The main resistance can be expressed as

$$F_H = F_{fb} + F_{fL} = \mu_r M_c g + \mu_r \frac{\dot{m}_s g L}{v} = 0.025 g\left(\frac{M_c}{L} + \frac{\dot{m}_s g}{v}\right) L \tag{13.27}$$

13.3.17 Load Acceleration Resistance, F_N

- The load acceleration resistance can be calculated as

$$F_N = K_{SR} F_H \tag{13.28}$$

where the K_{SR} is called the secondary resistance coefficient, which varies with the conveyor length and can be correlated with the length of the belt as

$$K_{SR} = \frac{32.44}{L^{0.80}}. \tag{13.29}$$

- Therefore, effective tension T_e can be expressed as

$$T_e = F_R = (1 + K_{SR}) F_H + F_{st} \tag{13.30}$$

13.3.18 Motor Power

- The motor power depends on the effective tension and belt velocity, which can be defined as

$$P_m = \frac{T_e v}{\eta} \tag{13.31}$$

where η=efficiency of the motor of the motor/drum transmission (usually around 85%–95%), v=belt speed.

Example 13.3

A belt conveyor of effective overall length 100 m (center-to-center) and width 600 mm at a vertical elevation of 30 from the ground is used to transport a lumpsum load of bulk density 1530 kg/m^3 upwardly at a speed of 1.2 m/s. The motor which was used to run the belt for the operation is 95% efficient. The material placed in the belt aligned with a surcharge angle of 18°. The width of the load stream on the belt should be 90% of the belt width. The mass of the rotating parts of the idlers per unit length of the belt on the carry side and the mass of the rotating parts of the idlers per unit length of the belt on the return side are 15 kg/m each. The overall mass of the belt per unit length is 10 kg/m. The belt friction coefficient is 0.025. Calculate the total power required to transport the material by the belt conveyor.

Solution:

Given:

m_{ic}=the mass of the rotating parts of the idlers per unit length of belt on the carry side=15 kg/m,

m_{ir}=the mass of the rotating parts of the idlers per unit length of belt on the return side=15 kg/m,

m_b=the mass of the belt per unit length overall=10 kg/m,

α=the angle of elevation=30°

δ=surcharge angle of 18°

L=the overall length of the conveyor (center-to-center)=100 m

H=the net change in vertical elevation=$L \times \sin 30°$=50 m

b=The width of the load stream on the belt should be 90% of the belt width=0.90×600 mm=540 mm=0.54 m.

μ_r=belt friction coefficient=0.025.

v=belt speed=1.2 m/s.

η=efficiency of the motor of the motor/drum transmission=0.95

After substitution of the above values in the respective equations as follows one can get:

$$M_c = (m_{ic} + m_{ir} + 2m_b \cos\alpha)L = 4732.05 \text{ kg}$$

$$\dot{m}_s = \frac{1}{6}\rho_b v b^2 \tan\delta = 28.99 \text{ kg/s}$$

$$F_{fb} = \mu_r M_c g = 1160.54\ \text{N}$$

$$F_{fL} = \mu_r m_L g L = \mu_r \frac{\dot{m}_s g L}{v} = 592.53\ \text{N}$$

$$F_{st,vertical} = +m_L g H = +\frac{\dot{m}_s g H}{v} = 11850.67\ \text{N}$$

$$F_H = F_{fb} + F_{fL} = 1753.07\ \text{N}$$

$$K_{SR} = \frac{32.44}{L^{0.80}} = 0.81$$

$$F_N = K_{SR} F_H = 1428.50\ \text{N}$$

$$T_e = F_R = (1 + K_{SR}) F_H + F_{st} = 15032.23\ \text{N}$$

Therefore, the power required to transport the material is

$$P_m = \frac{T_e v}{\eta} = 18,988.08\ \text{W.} = 18.99\ \text{kW}$$

13.4 TRANSPORTATION BY SCREW CONVEYORS

- Screw conveyors primarily consist of a rotating screw shaft that is installed within a trough (Figure 13.9).
- It rotates in a trough, as shown in Figure 13.9, to move a bulk solid continuously from one end of the trough to the other.
- In a screw conveyor, transportation of bulk material is achieved by the rotation of a helical screw, thus moving the material axially in a trough.

13.4.1 Some Design Feature

- For most conveying applications, a single screw with continuous flighting and a constant pitch-to-diameter ratio of unity is the standard geometry used (Figure 13.10).
- For very easy-flowing materials, a pitch equal to 1.5 times the screw diameter may be sufficient to move the bulk material effectively.
- For vertical applications and for screw feeders, the pitch-to-diameter ratio is usually 2/3 to 1/2.
- Feeding lumpy materials, the tapered screw is suitable, usually from a pitch-to-diameter ratio of 2/3 to 1.0.
- The two essential parameters to be established in the design or selection of a screw conveyor for a given application are the screw size and its rotational speed.
- The diameter of the screw conveyor (D_{sc}(in mm)) depends on the maximum lump size (x in mm within 0–150 mm) and lump amount (y in %), which is to be conveyed and can be obtained from the following correlation:

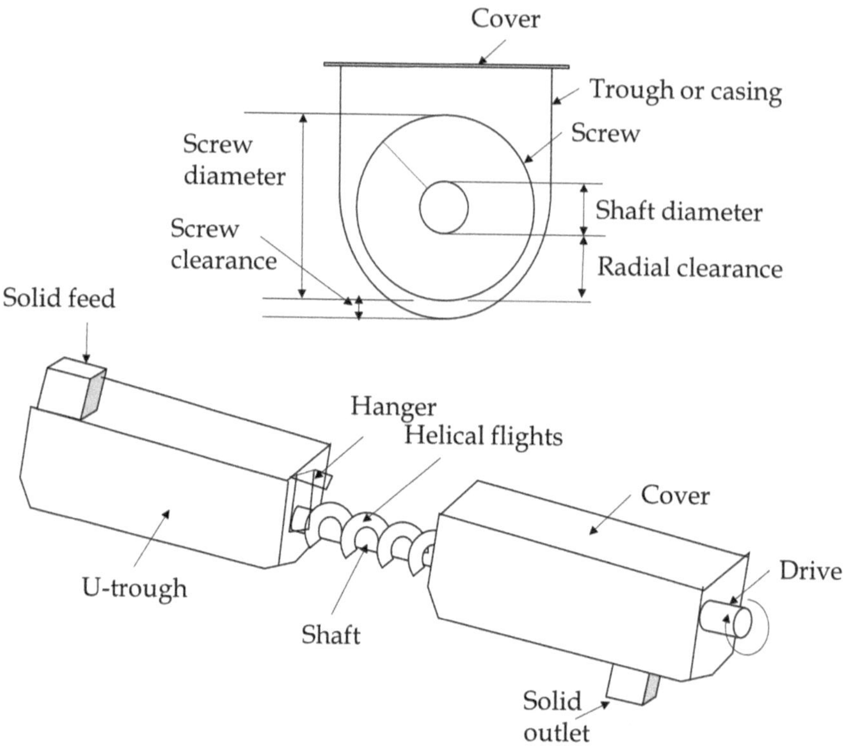

Figure 13.9 Screw conveyor.

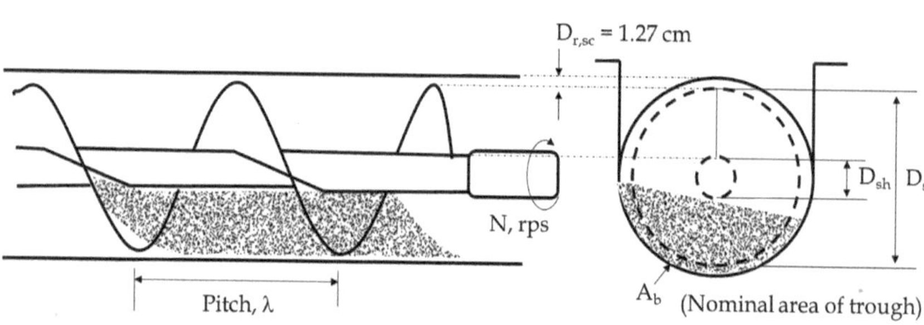

Figure 13.10 Pitch-to-diameter in screw conveyor.

$$D_{sc}\text{(in mm)} = px^q; \quad x = \text{Maximum lump size in mm} \tag{13.32}$$

where

$$p = 3.8979y^{0.2994}; \quad y = \% \text{ of lumps} \tag{13.32a}$$

$$q = 1.045y^{-0.072}; \quad y = \% \text{ of lumps} \tag{13.32b}$$

- The radial clearance of the screw conveyor ($C_{r,sc}$(in mm)) depends on the maximum lump size (x in mm within 0–150 mm) and lump amount (y in %), which is to be conveyed, which can be obtained from the following correlation:

$$C_{r,sc}(\text{in mm}) = mx^n; \quad x = \text{Maximum lump size in mm} \tag{13.33}$$

where

$$m = 0.37y^{0.69}; \quad y = \%\text{ of lumps} \tag{13.33a}$$

$$n = 1.196y^{-0.081}; \quad y = \%\text{ of lumps} \tag{13.33b}$$

13.4.2 Capacity

- The distance that the conveyed material moves forward during one revolution of the screw is approximately equal to the pitch of the screw.
- The average velocity U of this material can be written as

$$U = \lambda N \tag{13.34}$$

where λ is the pitch of the screw and N is the rotational speed in rev/s.
- The volumetric throughput of the screw conveyor is then given by

$$\dot{V}_s = A_b \lambda N \tag{13.35}$$

where A_b is the cross-sectional area of the moving bed of particulate material.
- The mass flow rate can be calculated by

$$\dot{m}_s = \rho_b \frac{1}{4}\pi(D_{sc}^2 - D_{sh}^2)k\lambda N \tag{13.36}$$

where ρ_b=bulk density of the material which is to be conveyed. k=the conveyor loading factor, D_{sc}=trough or screw diameter, D_{sh}=shaft diameter, N=the rotational speed in rev/s.
- Screw conveyor shaft diameter refers to the outer diameter of the shaft. For screw conveyors, the diameter will be one of the Conveyor Equipment Manufacturer's Association (CEMA) standardized sizes; 3.81, 5.08, 6.19, 7.62, and 8.73 cm.
- Conveyor loading factor (k) can be calculated based on trough inclination angle (β in degree) by correlations given in Table 13.3.
- Maximum operating speeds (in r.p.s) for screw conveyors as a function of screw diameter (D_{sc} in mm) can be obtained from the correlations given in Table 13.4.

Table 13.3 Correlations for conveyor loading factor

Materials	*Correlations for Conveyor Loading Factor*
For heavy bulk materials, very abrasive and aggressive (e.g., ash, gravel, minerals)	$k = 0.147 - 0.0028\beta$; β in degree
For average materials, abrasive, graded from grains to small lumps (e.g., salt, sand, and coals)	$k = 0.3 - 0.0059\beta$; β in degree
For screws without hangers, materials that flow easily, slightly abrasive (e.g., flour and cereals)	$k = 0.451 - 0.009\beta$; β in degree

Table 13.4 Correlations for maximum operating speeds (in r.p.s) for screw conveyors

Materials	*Correlations for Maximum Speed* D_{sc} *is in mm and* N_{max} *is in rev/min*
For 15% loading of non-abrasive materials and 30% loading of abrasive materials	$N_{max} = 65.858 - 0.0484D_{sc}$
For 30% loading of non-abrasive materials	$N_{max} = 129.38 - 0.1144D_{sc}$
For 45% loading of non-abrasive materials	$N_{max} = 186.86 - 0.1475D_{sc}$

- In practice, the capacity is also affected by flight capacity and mixing paddle capacity. If both the capacity factor is taken care, the capacity of a screw conveyor may then be expressed, in general as

$$\dot{m}_s = \rho_b \frac{1}{4}\pi(D_{sc}^2 - D_{sh}^2)\frac{k\lambda N}{CF_f \times CF_m} \tag{13.37}$$

where

CF_f is the flight capacity factor depending on the type of flight (as shown in Figure 13.11), which can be calculated from the correlations:

$$CF_f = 6.14 - 7.98k \quad \text{for cut and folded flight} \tag{13.38}$$

$$CF_f = 1.135k^{-0.293} \quad \text{for cut flight} \tag{13.39}$$

$$CF_f = 2.172k^{0.383} \quad \text{for ribbon flight} \tag{13.40}$$

The conveyor loading factor k is generally within the range of 0.05–0.5.

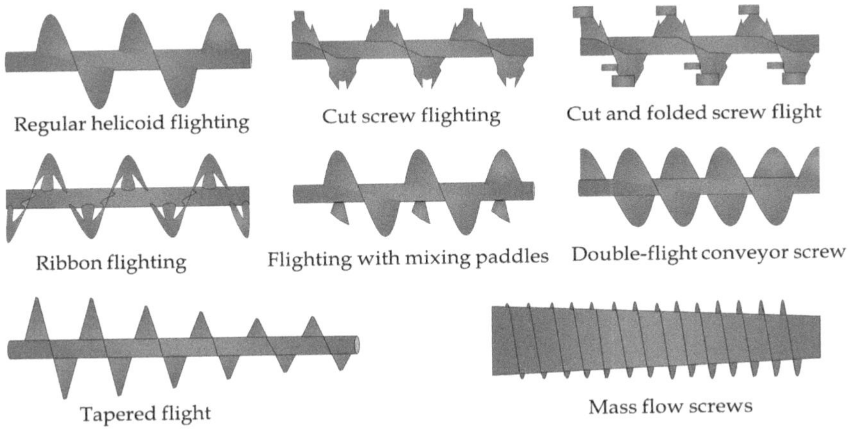

Figure 13.11 Different types of flighting of screw conveyor.

- CF_m is the mixing paddle capacity factor, which can be obtained from the correlation:

$$CF_m = 0.081n_{mp} + 0.997 \tag{13.41}$$

where n_{mp}=number of mixing paddles per pitch within a range of $0 < n_{mp} < 5$.

13.4.3 Screw Conveyor Power

- The total power for the screw conveyor operating is calculated by considering the mass transportation power (P_{mat}) and power required for frictional resistance (P_{frict}), which can be expressed as

$$P_{tot} = P_{mat} + P_{frict} \tag{13.42}$$

where

$$P_{mat} = F_s F_m \rho_b g \dot{V}_s L \tag{13.43}$$

$$P_{frict} = 75.7LND_{sc}^{1.7} \tag{13.44}$$

where F_s=coefficient to account for differences in the types of screw that could be used, F_m=coefficient to account for the nature of the conveyed material, D_{sc}=the screw diameter (m), L is its length (m), and N is the rotational speed (rev/s).

- Motor power for non-elevated screws can be calculated by

$$P_{mot} = \frac{(P_{mat} + P_{frict})F_0}{\eta} \tag{13.45}$$

where η=the drive efficiency (85%–95%). F_0 is the overload factor, which can be calculated from the following correlation,

$$F_0 = 1.64P_{tot}^{-0.294}; \quad P_{tot} \text{ is in kW} \tag{13.46}$$

- The value F_s can be considered unity for a standard helical screw (at all conveyor loadings). The common bulk solids with corresponding values of F_m, which, in general, range from 0.8 to 4.0 (Woodcock and Mason, 1987).
- The power of the driving motor for an elevated screw conveyor is

$$P_{mot} = \frac{(P_{mat} + P_{frict} + P_{st})F_0}{\eta} \tag{13.47}$$

where

$$P_{st} = \dot{m}_s gH \tag{13.48}$$

$$P_{mat} = F_s F_m \rho_b g \dot{V}_s L \tag{13.49}$$

$$P_{frict} = 75.7LND_{sc}^{1.7} \tag{13.50}$$

where H is the vertical elevation of the top end of the screw above the feed point.

13.5 TRANSPORTATION BY VIBRATORY CONVEYORS

- The particles are subjected to an asymmetric vibration and travel in a series of steps over a table.
- Vibratory conveyors use both vibration and gravity to move material.
- The speed of the material transport is determined by the amplitude range and the frequency of the conveyor's vibrations.
- **Normal operating ranges for vibratory conveyor:**
  - The frequency range=3–20 Hz.
  - The amplitude of vibration=50–5 mm.
  - The angle of oscillation is generally around 20°–30°.

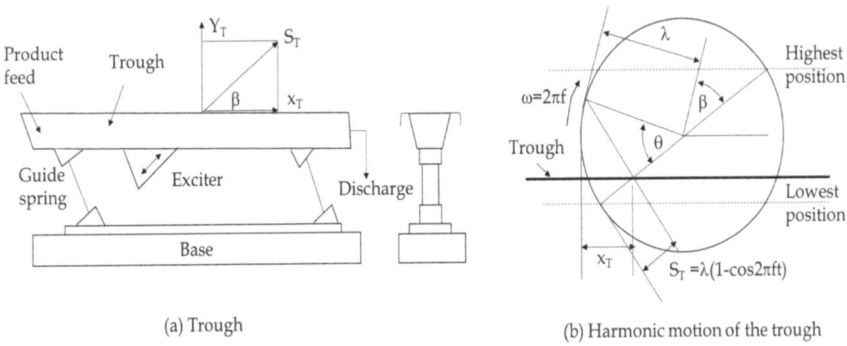

Figure 13.12 Fundamental model of vibratory conveyor.

13.5.1 The Motion of the Trough of the Vibrating Conveyor

- The trough can move only in a direction perpendicular to the fixed guide springs.
- The line of motion of the trough S_T makes an angle β with the horizontal as shown in Figure 13.12.
- This angle, β, termed the 'angle of oscillation' or 'drive angle' is generally around 20°–30°.
- If the trough is oscillating at frequency f and with amplitude λ, its position at any time t is given by

$$S_T = \lambda(1 - \cos 2\pi f t) \tag{13.51}$$

- Dynamic material coefficient or throw factor (Γ):
  - The material will lift off the trough's surface when the trough's acceleration, in the downward vertical direction, will equal the gravitational acceleration g.
  - The throw factor is a parameter that will determine the point at which the flight phase begins when $y_T / g = -1$
  - The maximum value of this parameter (throw factor) is defined by

$$\Gamma = \frac{\ddot{y}_{T\max}}{g} = \frac{\lambda(2\pi f)^2 \sin\beta}{g} \tag{13.52}$$

  - If the positive value of the throw factor $\Gamma < 1$, the bulk solid will not leave the surface of the trough, and forward movement will be little.

- In general, the higher the frequency, the smaller must be the amplitude. The relation is expressed as

$$\lambda = \frac{\ddot{S}_{T\max}}{(2\pi f)^2} = \frac{K \times g}{(2\pi f)^2} \tag{13.53}$$

where

$$K = \frac{\ddot{S}_{T\max}}{g} = \frac{\Gamma}{\sin\beta} \tag{13.54}$$

- The term K is called 'dynamic machine coefficient' or 'machine number', which lies between 1 and 4.

- **The average conveying velocity of the vibrating conveyor**
  - The average velocity of the bulk solid along the trough can be conveniently expressed as

$$u_s = \eta_u \dot{x}_{T\max} = \eta_u \lambda (2\pi f)\cos\beta \tag{13.55}$$

where the horizontal components are:

$$x_T = \lambda(1 - \cos(2\pi f t))\cos\beta \tag{13.56}$$

$$\dot{x}_T = 2\lambda\pi f \sin(2\pi f t)\cos\beta \tag{13.57}$$

$$\dot{x}_{T\max} = 2\lambda\pi f \cos\beta \tag{13.58}$$

where η_u='efficiency of transport'

- **The capacity of vibratory conveyors:**
  - Drive angle, β (in degree) is likely to be in the range 30°–50° as per throw factor (dynamic material coefficient, likely to be within a range of 1.5–2.0).
  - Dynamic machine coefficient, K is likely to be in the range 0.7–10 as per throw factor (dynamic material coefficient)
  - The capacity of a vibrating conveyor can be expressed as

$$\dot{m}_s = \rho_b A u_s \tag{13.59}$$

Example 13.4

A vibratory conveyor of through factor 1.5 and of cross-sectional area 50 cm^2 is operating at a frequency 60 1/s to transport the solid of lump height 0.5 m. The efficiency of the transport is 90%. Calculate the capacity of the vibratory conveyor. The drive angle for the conveyor is 30°. The bulk density of the material is 1350 kg/m^3.

Solution:

Given: amplitude, $\lambda = 0.20\,\text{m}$, frequency, $f = 60$ 1/s, throughput factor $(\Gamma) = 1.5$, the drive angle, β (in degree) $= 30°$, Area of the vibratory conveyor $= 50\,\text{cm}^2$

The dynamic machine coefficient, K is

$$K = \frac{\ddot{S}_{T\max}}{g} = \frac{\Gamma}{\sin\beta} = \frac{1.5}{\sin(30°)} = 3.0$$

The amplitude of the drive operation is

$$\lambda = \frac{\ddot{S}_{T\max}}{(2\pi f)^2} = \frac{K \times g}{(2\pi f)^2} = \frac{3 \times 9.81}{(2\pi(20))^2} = 0.00186$$

The average velocity of the bulk solid along the trough

$$u_s = \eta_u \dot{x}_{T\max} = \eta_u \lambda (2\pi f)\cos\beta = (0.90)(0.00186)(2 \times \pi \times 20)\cos(30°) = 0.183\ \text{m/s}$$

$$\dot{m}_s = \rho_b A u_s = (1350)(50/(100 \times 100))(0.183) = 1.232\ \text{kg/s}$$

13.6 TRANSPORTATION BY BUCKET ELEVATORS

- In some applications, bulk solid materials are elevated in an area where sufficient floor space is unavailable for inclined belts or screw conveyors. For such conditions, a bucket elevator may be a suitable alternative.
- The bucket elevator consists of an endless chain of metal or plastic buckets mounted on a belt or a single or double chain, as shown in Figure 13.13.
- The buckets are fixed to load at one end and discharge at a higher level.
- Buckets may be discharged centrifugally, in which case the particles are thrown from each bucket to a discharge chute.
- The centrifugal action of this type of elevator requires relatively high operating speeds of about 1–3.0 m/s.

13.6.1 Design Feature

- The design feature is strongly influenced by the material's nature, shape, and construction of the bucket.
- The material would be dry, free-flowing, of lump size less than 100 mm, at ambient to moderate temperature, slightly abrasive, and not especially friable.
- There are two types of elevators mostly used. Chain elevators are generally limited to a speed of about 1.3 m/s, whereas belt elevators are operated satisfactorily at speeds up to 2.5 m/s.

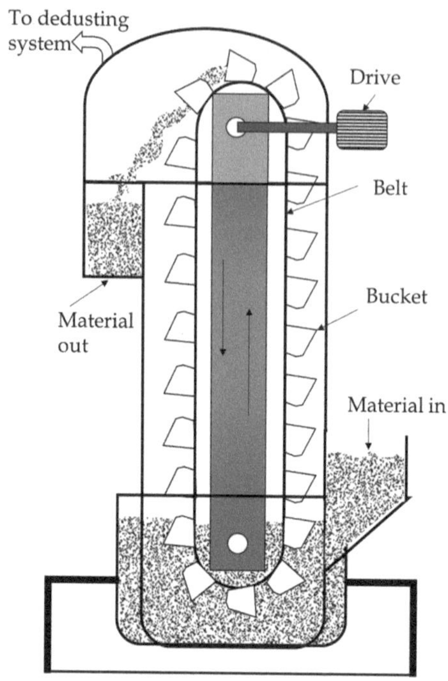

Figure 13.13 Bucket elevators.

13.6.2 Discharge Characteristics of a Bucket Elevator

- When the loaded buckets travel vertically toward the head wheel, the forces acting upon the load are:
  - The gravity force F_G.
  - The accelerative force F_A acts on the load.
  - The resultant force F_R acting on the line of action passes through a fixed-point P, called the pole point, which lies vertically above the center of the head wheel, as shown in Figure 13.14.
- The distance between center of the head wheel and the pole point is

$$L = r\frac{F_G}{F_A} \tag{13.60}$$

where r is the radial distance of the center of mass C of the load in the bucket from the head wheel center.

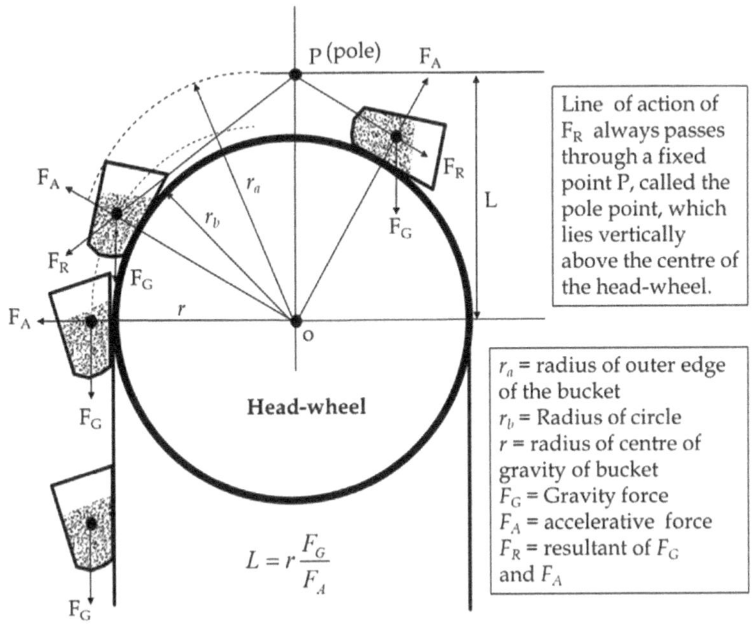

Figure 13.14 Representation of different forces acting during elevation of bucket.

- Now, if m is the mass of bulk solid in the bucket and v is the linear velocity of point C, then,

$$F_G = mg \tag{13.61}$$

$$F_A = \frac{mv^2}{r} \tag{13.62}$$

$$v = 2\pi rN \tag{13.63}$$

- Then, from the above three equations, one can write

$$L = g\left(\frac{r}{v}\right)^2 = \frac{g}{4\pi^2} \cdot \frac{1}{N^2} \tag{13.64}$$

where N is the rotational speed of the head wheel in revolutions per second.

  - If $r_a < L$, the elevator can be classified as a gravity discharge type (Figure 13.15a).

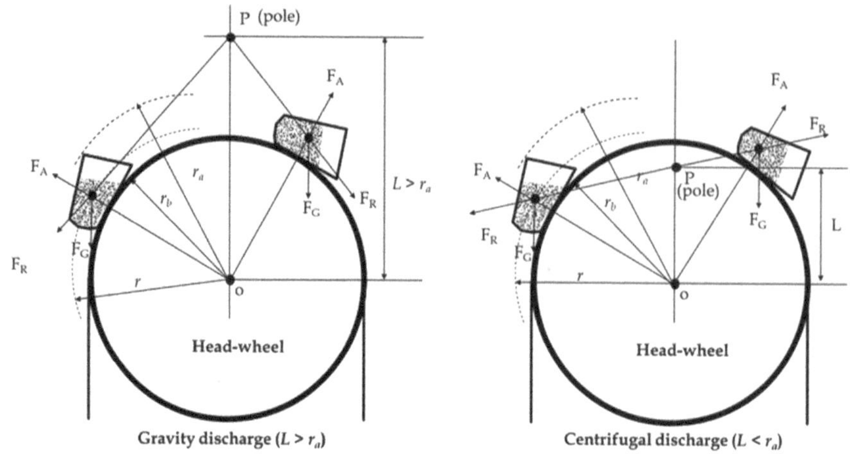

Figure 13.15 Representation of solid discharge conditions from the bucket. (a) Gravity discharge ($L > r_a$), (b) centrifugal discharge $L < r_b$.

- If $L < r_b$, the elevator is classed as centrifugal discharge (Figure 13.15b).
- If $r_b < L < r_a$, the discharge will involve a combination of gravity and centrifugal effects.

13.6.3 Discharge Position

- The point of discharge can be specified as when the radial component of the gravity force is equal to the centrifugal force (Figure 13.16), i.e.,

$$F_G \cos\theta = F_A \Rightarrow mg\cos\theta = \frac{mv^2}{r} \tag{13.65}$$

- The position of the bucket at the point of discharge is given by

$$\theta = \theta_L = \cos^{-1}\left(\frac{v^2}{gr}\right) \tag{13.66}$$

13.6.4 Capacity

- The actual volumetric capacity (or flow rate) may be expressed as

$$\dot{V}_s = \eta_b V_b \frac{v}{\lambda} \tag{13.67}$$

where V_b = the volume of each bucket, η_b = the bucket filling efficiency, v = belt velocity, λ = bucket pitch (2–3 times of bucket height).

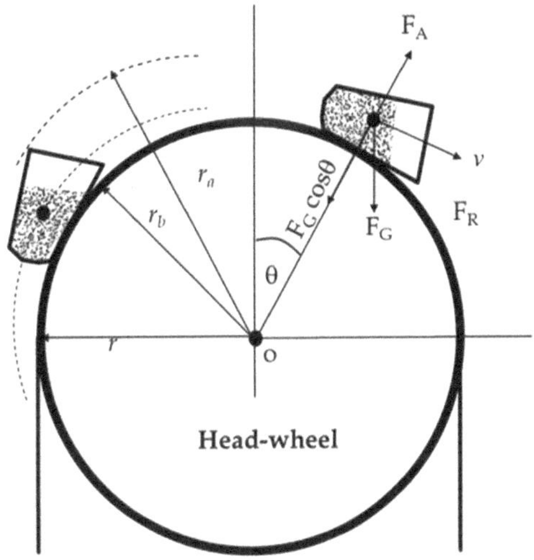

Figure 13.16 Components of force acting on the solid in the bucket at a certain radial position.

- The mass throughput is then given by

$$\dot{m}_s = \rho_b \dot{V}_s = \eta_b \rho_b V_b \frac{v}{\lambda} \tag{13.68}$$

where ρ_b is the bulk density of the material in the buckets. The efficiency η_b lies within 0.65–0.75.

13.6.5 Motor Power

- The motor power can be calculated by

$$P_{\text{mot}} = \frac{F_u v}{\eta_0} \tag{13.69}$$

- The combined (gravity and drag) resistive force (F_u),

$$F_u = F_H \left(1 + \frac{H_f}{H}\right) \tag{13.70}$$

where
H is the difference in height between the feed and discharge points.
F_H is the gravity force opposite to the motion of the elevator given by

$$F_H = \eta_b \rho_b g V_b \frac{H}{\lambda} \tag{13.71}$$

- H_f is the loading factor, as shown in Table 13.5.

Table 13.5 Loading factor based on material characteristics (Woodcock and Mason, 1987)

Type of Elevator	*Material Type*	*Loading Factor H_f (m)*
Continuous bucket	$\rho_b < 1600\,kg/m^3$	3.1
	$\rho_b > 1600\,kg/m^3$	4.6
Spaced bucket	Free-flowing grains	9.2
	Other materials	12.3

Example 13.5

A continuous chain bucket elevator is used to transport the cement of bulk density 1440 kg/m³ at an elevated height of 30 m at a speed of the elevator 2.5 m/s. The pitch of the buckets is two times of bucket height. The transport efficiency of the bucket is 0.75. Find the capacity of the bucket elevator and the motor power required to transport the material by the bucket elevator. The bucket dimensions (Length×projection×depth): 10 inch×6 inch×6.24 inch. Motor efficiency is 0.95.

Solution:

Volume of the bucket (V_b) = $6.14\times10^{-3}\,m^3$.

Bulk density of material, $\rho_b = 1440$ kg/m³.

The pitch of the bucket = 2×6.24 inch = 0.317 m

Speed of the bucket elevator = 2.50 m/s

Capacity of the elevator = $\dot{m}_s = \rho_b \dot{V}_s = \eta_b \rho_b V_b \dfrac{v}{\lambda} = (0.75)(1440)$

$$\times(6.14\times10^{-3})\frac{2.50}{0.317} = 52.26\text{ kg/s} = 188.13\text{ t/h}$$

F_H is the gravity force opposite to the motion of the elevator given by

$$F_H = \eta_b \rho_b g V_b \frac{H}{\lambda} = (0.75)(1440)(9.81)(6.14\times10^{-3})\frac{30}{0.317} = 6151.81\text{ N}$$

The loading factor (H_f) = 3.1 (as continuous bucket elevator and $\rho_b < 1600$ kg/m³).

The combined (gravity and drag) resistive force (F_u),

$$F_u = F_H\left(1+\frac{H_f}{H}\right) = (6151.81)\left(1+\frac{1.3}{30}\right) = 6418.385\text{ N}$$

Therefore, the motor power would be

$$P_{mot} = \frac{F_u v}{\eta_0} = \frac{(6418.385)(2.5)}{0.95} = 16890.49\text{ W} = 16.89\text{ kW}.$$

IMPORTANT POINT TO REMEMBER

- The belt conveyors consist of drums (driving and non-driving), belts, drive, idler rollers, belt tensioners, and belt cleaners.
- The main design components of the belt conveyors are belt width, speed, capacity, and power requirement.

- The temperature of the material to be transported should not be above 120°C.
- For most materials, the surcharge angle is 5°–15° less than the angle of repose.
- In practice, for a single drive, K_d should be 20% greater if the drum is bare and 30% greater if it is lagged, and for a tandem or dual drive, K_d should be about 25% greater for both bare and lagged drums.
- It is recommended that the return idler sets on a 3 m pitch, but on the carrying side, the idler spacing would typically be from 0.93 to 1.6 m.
- For most conveying applications, a single screw with continuous flighting and a constant pitch-to-diameter ratio of unity is the standard geometry used.
- For very easy-flowing materials, a pitch equal to 1.5 times the screw diameter may be sufficient to move the bulk material effectively.
- The total power for the screw conveyor operating is calculated by considering the mass transportation power (P_{mat}) and power required for frictional resistance (P_{frict}).
- Vibratory conveyors use both vibration and gravity to move material.
- The centrifugal action of this type of elevator requires relatively high operating speeds of about 1–3.0 m/s.
- Chain elevators are generally limited to a speed of about 1.3 m/s, whereas belt elevators are operated satisfactorily at speeds up to 2.5 m/s.

SELF-TESTING QUESTIONS

I. Multiple-Choice Questions

Q1. The choice of a suitable screw involves the consideration of____________.

(a) The overall diameter of the screw
(b) The type and pitch of helical light
(c) The diameter of the shaft
(d) All of these

Q2. The capacity of belt conveyor depends upon________________.

(a) Speed
(b) Particle volume
(c) Width of belt
(d) Bulk density

Q3. The _________ moves upwards at a sinusoidal angle in the direction of transport.

(a) Vibratory conveyor
(b) Bucket conveyor
(c) Motor power
(d) All of these

Q4. In screw conveyor power, the total power is the sum of power required to transport the ______ at a specified rate and power required to overcome __________ moving parts of conveyor.
(a) Shear material, frictional resistance
(b) Bulk material, frictional resistance
(c) Shear material, bulk resistance
(d) None of these

Q5. Ribbon fighting pattern of screw conveyor is used for conveying substances that are ______.
(a) Hard substances
(b) Sticky substances
(c) Gummy substances
(d) Viscous substances

Q6. Which of the following application is a belt conveyor used for?
(a) Material transportation over long distances.
(b) Material transportation within premises.
(c) Material transportation for processing.
(d) All of the above.

Q7. Statement 1: Belt conveyors cannot be used in the inclined position.
Statement 2: Screw conveyors can carry more amount of load compared to belt conveyors.
(a) True, True
(b) True, False
(c) False, True
(d) False, False

Q8. The capacity of belt conveyor depends upon ______.
(a) speed
(b) width of the belt
(c) bulk density and particle size
(d) all of these

Q9. Which pattern of screw conveyor is used for conveying substances that are sticky, gummy, or viscous?
(a) Helicoid Flighting
(b) Cut screw Flighting
(c) Ribbon Flighting
(d) Double flight conveyor screw

Q10. Which conveyor moves upwards at a sinusoidal angle in the direction of transport?
(a) Bucket conveyor
(b) Screw conveyor
(c) Vibratory conveyor
(d) None of the above

Q11. Which of the following options is true regarding belt and screw conveyors?
  (a) Belt conveyors cannot be used in the inclined position
  (b) Screw conveyors can carry more amount of load compared to belt conveyors
  (c) Belt conveyors can carry more amount of load compared to belt conveyors
  (d) None of the above

Q12. On which of the following factor(s) does the effectiveness of the conveyor drive depend(s)?
  (a) The difference in tension between the 'tight side'
  (b) The "slack side" of the belt, the friction between the belt and the drive drum
  (c) The angle of wrap, or arc of contact, of the belt to the drum
  (d) All of the above

Q13. Which of the following statement(s) is(are) false for screw conveyors?
  (a) Feeding lumpy materials, the screw diameter is suitable as tapered, usually from a pitch-to-diameter ratio of 1/3 to 1.0.
  (b) For most conveying applications, a single screw with a continuous flighting and a constant pitch-to-diameter ratio of unity is the standard geometry used.
  (c) For very easy-flowing materials a pitch equal to 1.5 times the screw diameter may be sufficient to move the bulk material effectively.
  (d) For vertical applications and for screw feeders, the pitch-to-diameter ratio is usually 2/3 to 1/2.

Q14. Which of the following is an incorrect bucket selection?
  (a) Abrasive products – Front lip of bucket strengthened to reduce wear damage when digging to material.
  (b) Very watery products and light free-flowing products – Low front lip to bucket to provide maximum capacity.
  (c) Products susceptible to aeration – Bucket drilled with air holes to help product to settle.
  (d) Cohesive or sticky products – Shallow rounded buckets to reduce the tendency for material to lodge in corners.

Q15. Which of the following factors should be considered for choosing a suitable screw size for screw conveyors?

I. the overall diameter of the screw, the diameter of the shaft,
II. the radial clearance between the shaft,
III. the containing trough (typically 12–15 mm), and
IV. the type and pitch of the helical flight:

(a) Only I, II, and IV
(b) Only II, III, and IV
(c) I, II, III, and IV
(d) Only I and III

Q16. Which of the following expresses the carrying capacity of conveyor belt as per parabolic surface of material on the belt?
(a) $\dot{m}_s = \rho_b k_s A v$
(b) $\dot{m}_s = \rho_b k_s^2 A v$
(c) $\dot{m}_s = \rho_b k_s A^2 v$
(d) $\dot{m}_s = \rho_b k_s A v^2$

Q17. Select the correct option for screw conveyors.
(a) Transportation is achieved by rotation of a helix conveyor
(b) Suitable for sticky materials
(c) Both a and c
(d) None of the above

Q18. Which of the following application is a belt conveyor used for?
(a) Material transportation over long distances
(b) Material transportation within premises
(c) Material transportation for processing
(d) All of the mentioned

Q19. A screw conveyor has____________________.
(a) Triangular trough
(b) U-shape trough
(c) Trapezoidal shape trough
(d) None of these

Q20. The capacity of the bucket elevator mainly depends on ________________.
(a) Bucket design, size, and spacing
(b) Number of buckets and characteristics of the material
(c) Way of loading and unloading of material
(d) All are correct

Answers to the multiple-choice questions

Q1. Ans. (d)	Q8. Ans. (d)	Q15. Ans. (c)
Q2. Ans. (a, c, d)	Q9. Ans. (c)	Q16. Ans. (a)
Q3. Ans. (a)	Q10. Ans. (c)	Q17. Ans. (c)
Q4. Ans. (b)	Q11. Ans. (c)	Q18. Ans. (d)
Q5. Ans. (b, c, d)	Q12. Ans. (d)	Q19. Ans. (b)
Q6. Ans. (d)	Q13. Ans. (a)	Q20. Ans. (d)
Q7. Ans. (d)	Q14. Ans. (b)	

II. Objective Type Questions

1. Explain the working principle of gravity chute.
2. Derive the exit velocity at which particle transport in the gravity chute.
3. Explain the working principle of belt conveyors.
4. What are the different types of belt conveyor used for material transport?
5. What are the basic needs for power input to the belt conveyor?
6. Mention the advantages and disadvantages of belt conveyor.
7. What factor governs the capacity of screw conveyor?
8. What are the two components of resistances of power consumption of screw conveyor.
9. Explain the different patterns of screw conveyor.
10. What are the operating ranges of vibratory conveyors? How do vibratory conveyors work?
11. What are the design features of bucket elevator?
12. What are the main selection parameters for bucket conveyors?
13. Describe the discharge characteristics of a bucket elevator.
14. How motor power is related to the resistive force in case of bucket elevator.
15. Give some examples of applications of bucket elevator.

III. Numerical Problems

Q1. An inclined belt conveyor is used to load goods. If the speed of the belt having a minimum of 500 mm width is 10 m/s, then what is the volumetric capacity of the conveyor? For 21°, $k_s=2.35\times10^{-4}$. [Ans. 3.76×10^{-4} m^3/s]

Q2. The mass capacity of a conveyor is 500 ton/h, if the speed of the belt is 4 m/s, then what is the minimum width of a horizontal flat belt conveyor carrying the load? (Surcharge factor=0.060 & density of solid is 1000 kg/m^3) [Ans. 0.114 m]

Q3. A belt is running over a pulley of 2 m diameter. The angle of contact is 120°, and the coefficient of friction is 0.40. If the maximum tension in the belt is 500 N, and the power transmitted by the belt is 4 KW. Find the speed of the driver in rpm. [Ans. 134.66 rpm]

Q4. Find the tension of the belt running over a pulley of 500 mm diameter at 200 rpm. The coefficient of friction between the pulleys is 0.3; the angle of lap 150° and maximum tension in the belt is 3 kN. [Ans. 1367.81 N]

Q5. Find the Power acting on belt running over a pulley of 500 mm diameter at 200 rpm. The coefficient of friction between the pulleys is 0.3; angle of lap 150° and maximum tension in the belt is 3 kN. [Ans. 8.55 kW]

Q6. The mass capacity of a conveyor is 400 ton/h, if the speed of the belt is 2 m/s, then what is the minimum width of a horizontal flat belt conveyor carrying the load? (Surcharge factor=0.080 & ρ=1000 kg/m^3) [Ans. 0.981 m]

Q7. An inclined belt conveyor is used to load goods. If the speed of the belt is 1000 mm width and is 4 m/s, what is the conveyor's volumetric capacity? (For 21°, k= 2.35×10^{-4}) [Ans. 6.799×10^{-4} m^3/s]

Q8. Determine the mass flow rate of the screw conveyor if the bulk density of the material is 1500 kg/m^3, the loading factor K is 0.7. Trough and Shaft diameter are 15 and 10 cm, respectively. (Given-Pitch of the screw is 5mm and the rotational speed is 200 rev/s). [Ans. 10.31 kg/s]

Q9. A conveyor belt is used to carry industrial load. The bulk density of the conveyed material is 3800 kg/m^3. The speed of the belt is 1.5 m/s, and the slope factor is 0.85. The width of the load stream is 1.25 m, and the surcharge angle is 4.5°. What is the belt carrying capacity? [Ans. 99.3 kg/s]

Q10. A screw conveyor is used to carry industrial load. The bulk density of the conveyed material is 3400 kg/m^3. The screw diameter is 0.08 m, and the pitch is 1.5 times the screw diameter. The rotational speed of the conveyor is 2 rev/s. The loading factor is 0.45. The trough diameter is 1 m, and the shaft diameter is 0.6 m. What is the carrying capacity of the screw conveyor? [184.6 kg/s]

Q11. A vibratory conveyor is used to carry an industrial load. The bulk density of the conveyed material is 2800 kg/m^3. The trough is oscillating with a frequency of 0.25 Hz and an amplitude of 0.5 m. The line of motion of the trough makes an angle of 22.5° with the horizontal. What is the maximum velocity of the trough in the horizontal direction? [Ans. 0.726 m/s]

Q12. Bucket Elevators are used in an industry to carry loads. The bulk density of the transported material is 2800 kg/m^3. The bucket volume is 65 L, and the pitch is 0.7 m. The belt velocity is 1.75 m/s, and the filling efficiency is 0.7. What is the capacity of the Bucket Elevator? [Ans. 293.5 kg/s]

Q13. Bucket Elevators are used in an industry to carry loads. The bulk density of the transported material is 1500 kg/m^3. The bucket's volume is 50 L, and the pitch is 0.75 m. The belt velocity is 1.25 m/s, and the height difference between the feed and discharge points is 1.5 m. The filling efficiency is 0.75. What is the motor power if the motor efficiency is 0.8? (The loading factor is to be taken as 3.1.) [Ans. 5288.203 kW]

Q14. An inclined belt conveyor is used to load goods. If the speed of the belt has a 1500 mm^2 area and is 3 m/s, what is the capacity of the conveyor? k_s=2.35×10^{-4} and bulk density of goods=2500 kg/m^3. [Ans. 2.6×10^{-6} kg/s]

Q15. Find the volumetric throughput of a screw conveyor if the area of the moving bed is 10 m^2, the pitch is 5 m, and the rational speed is 50 rev/s. [Ans. 2500 m^3/s]

Q16. A material of density 2600 kg/m^3 is transported by a screw conveyor having a shaft diameter is 1.5 cm, trough diameter is 3 cm, and pitch is 3 cm. If the screw conveyor rotates at a speed of 170 rev/s, what would be the mass flow rate of the material transported by the screw conveyor. The loading factor k=0.3. [Ans. 2.11 kg/s]

Q17. Calculate volumetric throughput for the bucket elevator if the bucket volume is 2 L, belt velocity is 5.5 m/s, pitch is 3.5 m. (The bucket filling efficiency η_b=0.7). [Ans. 2.2 lps]

Q18. A belt conveyor with an inclination of 15° to the horizontal will transport iron ore from the mine to the washing plant. The iron ore particles are 10–30 mm and of bulk density 2600 kg/m^3. Production at the mine is 2150 tons/h. Materials with around 10% moisture may be considered medium-flowable. The average cross-sectional area of the material stacked on the horizontal moving belt depends on the belt width, which is related as follows:

$$A_b = K_a C_i b^2$$

where K_a=a constant whose value depends on the flowability of the material and the angle of inclination of side rollers. C_i=correction factor for inclination, b=width of the belt carrying the load. For a belt speed of 1.6 m/s, calculate the belt width for the above purpose. (Data: K_a=0.067, C_i=0.95) [Ans. 1.5 m]

NOMENCLATURE

A	average cross-sectional area of the material stacked [m^2]
a	acceleration [m/s^2]
A_b	cross-sectional area of the moving bed of particulate material [m^2]
b	width of the load stream on the belt [m]
B	belt width [mm] as defined in Table 12.2.1
B_{min}	minimum width of the belt [m]
CF_f	flight capacity factor [-]
CF_m	mixing paddle capacity factor [-]
$C_{r,sc}$	radial clearance of the screw conveyor [mm]
D_{sc}	diameter of the screw conveyor [mm]
F	force acting in x-direction [N]
f	trough oscillating at frequency [-]
F_0	overload factor [-]
F_{fb}	belt friction resistance [N]

F_{fL}	load friction resistance [N]
F_H	gravity force opposite to the motion of the elevator [N]
F_H	main resistance [N]
F_m	coefficient to account for the nature of the conveyed material [-]
F_N	Load acceleration resistance [N]
F_s	coefficient to account for differences in the types of screw [-]
F_{st}	load slope resistance [N]
F_u	combined (gravity and drag) resistive force [N]
H	difference in height between the feed and discharge points [m]
H	net change in vertical elevation [m]
H_f	loading factor defined in Equation (13.70) [-]
K	conveyor loading factor [-]
K_d	drive factor or wrap factor [-]
K_s	sag factor [-]
K_s	slope factor as defined in Equation (13.6) [-]
K_{SR}	secondary resistance coefficient [-]
L	overall length of the conveyor (center-to-center) [m]
L	conveying distance [m]
L_{ic}	idler pitch on the carrying side [m]
L_{ir}	idler pitch on the return side [m]
m	mass of particle [kg]
m_b	mass of the belt per unit length overall [kg]
m_b	mass per unit length of the belt [kg]
M_c	total mass of moving parts [kg]
m_{ic}	mass of the rotating parts of the idlers per unit length of belt on the carry side [kg]
m_{ir}	mass of the rotating parts of the idlers per unit length of belt on the return side [kg]
m_L	mass per unit length of the conveyed material on the belt
m_s	carrying capacity of a belt conveyor [kg/s]
N	rotational speed [rev/s]
N	rotational speed of the head wheel [rps] defined in Equation [13.64]
N	normal force perpendicular to x-direction [N]
N_{max}	maximum operating speed [rev/min]
n_{mp}	number of mixing paddles per pitch [-]
P_{frict}	power required for frictional resistance [W]
P_m	motor power [W]
P_{mat}	mass transportation power [W]
r	radial distance of the center of mass C as defined in Equation (13.60)
S_T	The line of motion of the trough [m]
t	time [s]
T_e	effective tension [N]
v	Linear velocity of belt [m/s]
v	belt speed [m/s]

V_b volume of each bucket [m^3]
V_e exit velocity of the particle [m/s]
V_i initial velocity [m/s]
W weight of particle [kg]
x maximum lump size [mm]

GREEK LETTERS

α angle of elevation [degree]
ϑ angle of inclination [rad.]
ϑ angle of wrap [degree]
λ bucket pitch [m]
μ dynamic coefficients of friction [-]
η efficiency of the motor of the motor/drum transmission [%]
λ pitch of the screw [m]
δ surcharge angle [degree]
α angle of inclination of the belt [degree]
Γ material coefficient or throw factor [-]
λ trough oscillating amplitude [m]
β troughing angle [degree]
β angle of oscillation' or 'drive angle' as defined in Figure 13.12
η_b bucket filling efficiency [-]
ρ_b bulk density of the conveyed material [kg/m^3]
μ_r belt friction coefficient [-]

SUBSCRIPTS

b bed, bucket
c carrying
d drive
e exit, effective
f factor
frict frictional
H elevation
i initial
ic idler carrying side
ir idler return side
L length
m motor, material
mat mass transportation
max maximum
Min minimum

mp	mixing paddle
O	overload
s	solid, slope, screw
Sc	screw conveyor
SR	secondary resistance
T	trough

REFERENCE

Woodcock, C. R. and J. S. Mason, *Bulk Solids Handling: An Introduction to the Practice and Technology*, Springer, 1987.

Chapter 14

Storage of Solid

Hopper

I've never seen a job being done by a five-hundred-person engineering team that couldn't be done better by fifty people.

– C. Gordon Bell

14.1 STORAGE OF MATERIAL

- Large quantities of solids in the industry are usually stored in bins, hoppers, and silos, as shown in Figure 14.1.
- Silos are larger and more expensive than bins and hoppers. They are usually used for longer-term storage and are often grouped together with a common conveying system for loading and unloading.
- Many solid handling processes are required to meter the rate at which a bulk solid is fed from a storage bin or hopper. For example, when filling bags or supplying products to a processing plant. For this purpose, a feeder is used whenever there is a need to transfer solids at a controlled rate from the bin, silo, or hopper.

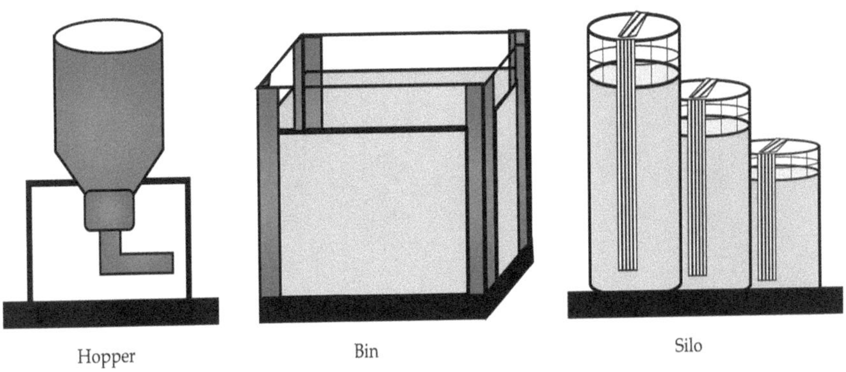

Figure 14.1 Storage bin or hopper or silo.

DOI: 10.1201/9781003571025-19

- There are several different types of feeders shown as follows:
  - **Apron feeders:** They can withstand extreme impact loading. Used for both small and large capacities.
  - **Belt feeders:** Smaller-sized material is fed. Used with smaller hoppers in secondary or tertiary applications.
  - **Drag and reclaim feeders:** Ideal for free-flowing materials. Does not require an operator.
  - **Reciprocating/plate feeders:** Suitable for small and large capacities. Usually located under a bin. Ideal for cyclic feed rates.
  - **Scalper feeders:** Appropriate to treat large size materials and they rough out the drier feeding material in low humidity sites.
  - **Wobbler feeders:** Ideal for small and medium capacities for pre-sorting before a crusher. Used for sticky types of materials to scrub the fines off.
  - **Vibrating pan feeders:** Ideal for feeding primary crushers and grizzly feeders. Accepts a variety of aggregate, mining, and recycled materials.
  - **Vibrating grizzly feeders:** Used for small and medium capacities. Ideal where pre-sorting is required before a crusher.
- The flow of material from the hopper must be "throttled" to the required rate by some "feeder."
- Once an appropriate feeder has been selected, it must be ensured that the product flows continuously from the hopper to the feeder at an adequate rate.
- Different geometries of the hopper are shown in Figure 14.2.

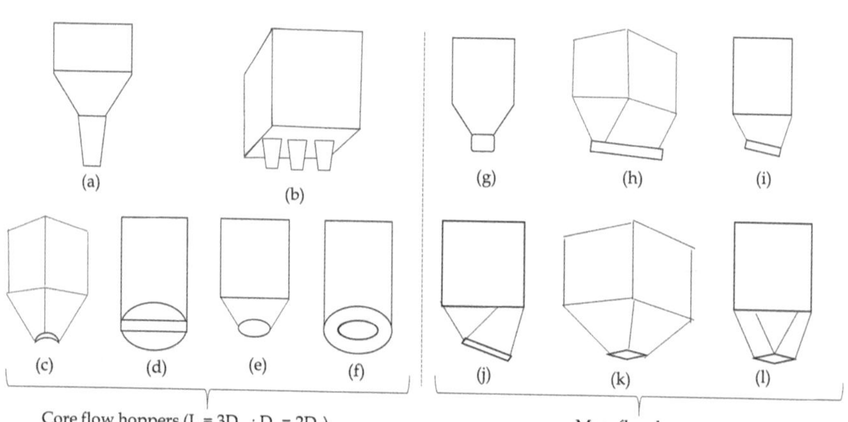

Figure 14.2 Different geometries of the hopper. (a) Basic expanded bin, (b) Multiple outlet bin, (c) Pyramid square opening bin, (d) Cylindrical flat-bottomed slot opening bin, (e) Conical bin, (f) Cylindrical flat-bottomed circular opening bin, (g) Conical hopper, (h) Wedge, plane-flow hopper, (i) Transition hopper, (j) Chisel, plane-flow hopper, (k) Pyramid hopper, (l) Square opening hopper.

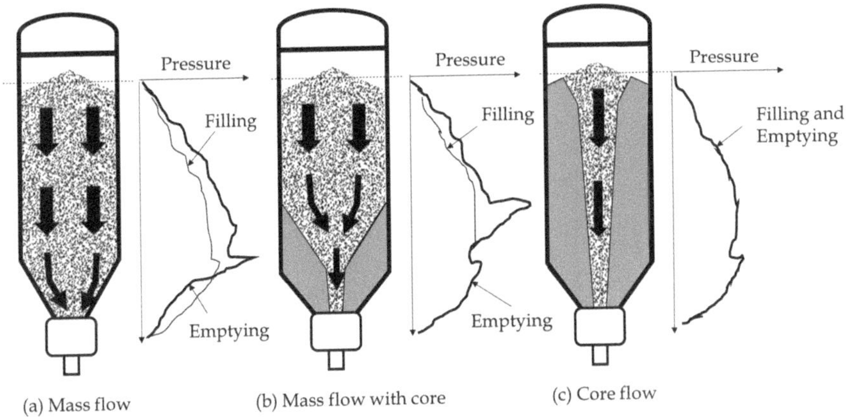

Figure 14.3 Mass and core flow in the hopper.

14.2 MASS AND CORE FLOWS IN HOPPERS

- The mass flow is considered to occur when the motion of the powder is uniform through the cross-section and steady-state whenever the material is drawn from the outlet of the hopper, as shown in Figure 14.3 (Shamlou, 1988; Woodcock and Mason, 1987).
- Core flow is said to happen when the powder flows toward the outlet of a silo in a channel formed within the powder itself.
- In the case of mass flow, the bulk density of the discharged powder is constant and practically independent of the height in the silo.
- Mass flow stresses are generally low throughout the mass of solids, giving low powder compaction.
- There are no stagnant regions in the mass-flow hopper.
- In core flow, the powder lower down in the hopper is stagnant until the hopper is almost empty. The inclined surface of the powder gives rise to size segregation.

14.3 EFFECTIVE STORAGE VOLUME OF A HOPPER

- The stored volume of the bin is divided into three parts, as shown in Figure 14.4 which is expressed as:

$$\begin{aligned}\text{Stored volume} &= \text{volume of conical section}\,(\text{Vol. 1})\\ &\quad+\text{volume of parallel section}\,(\text{Vol. 2})\\ &\quad+\text{volume resulting from surcharge on filling}\,(\text{Vol. 3})\end{aligned} \tag{14.1}$$

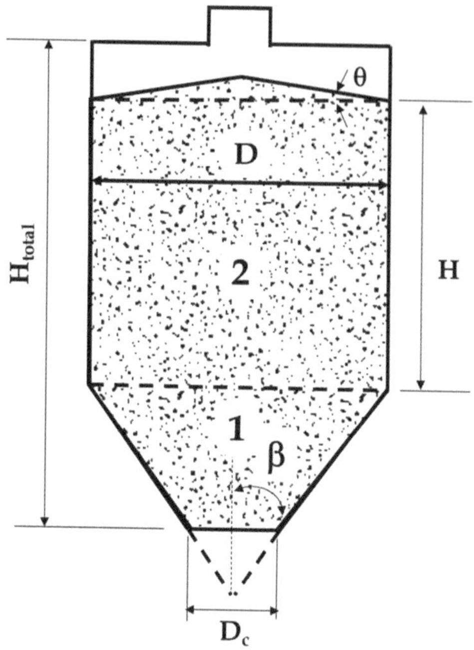

Figure 14.4 Hopper with geometrical dimensions.

- It can be mathematically expressed as:

$$V = \underbrace{\frac{\pi}{24\tan\beta}\left(D^2 - D_c^2\right)}_{\text{Vol.1}} + \underbrace{\frac{\pi D^2}{4}H}_{\text{Vol.2}} + \underbrace{\frac{\pi D^3}{24}\tan\theta}_{\text{Vol.3}}, \tag{14.2}$$

where D=the diameter of the bin, D_c=the diameter of the outlet, H=the height of the cylindrical section, β=the half-angle of the conical section, and ϑ=the surcharge angle of the "free surface" of the material in the hopper (which corresponds to the poured angle of repose).

- The total height of the bin is

$$H_{\text{total}} = H + 1.15D = \frac{1.274V}{D^2} + 0.765D \quad \text{for } \theta, \quad \beta < 30°, \tag{14.3}$$

where

$$H = \frac{1.274V}{D^2} - 0.385D \tag{14.4}$$

and

$$V = \frac{1.274m}{D^2 \rho_b} - 0.385D, \quad (14.5)$$

where m = mass of the solid and ρ_b is the bulk density.

- For a cylindrical mass-flow hopper, $D < H < 4D$. Note that for a rectangular hopper $H < 3D_{min}$.

Example 14.1

A conical bin of diameter 2.4 m and diameter of the outlet section 1.25 m is designed to store sand for use in a chemical process. Find the stored volume and total height of the bin. The height of the cylindrical section is 3.0 m. The half-angle of the conical section is 15°. The surcharge angle of the "free surface" of material in the hopper (which corresponds to the poured angle of repose) is 30°.

Solution:

D = the diameter of the bin = 2.4 m, D_c = the diameter of the outlet = 1.25 m, H = the height of the cylindrical section = 3.0 m, β = the half-angle of the conical section = 15°, and θ = the surcharge angle of the "free surface" of material in the hopper (which corresponds to the poured angle of repose) = 30°.

Therefore, after substituting the values, the stored volume of the bin is

$$V = \underbrace{\frac{\pi}{24 \tan \beta}\left(D^2 - D_c^2\right)}_{\text{Vol.1}} + \underbrace{\frac{\pi D^2}{4} H}_{\text{Vol.2}} + \underbrace{\frac{\pi D^3}{24} \tan \theta}_{\text{Vol.3}}$$

$$= \underbrace{\frac{\pi}{24 \tan(15°)}\left(2.4^2 - 1.25^2\right)}_{\text{Vol.1}} + \underbrace{\frac{\pi(2.5^2)}{4}(3.0)}_{\text{Vol.2}} + \underbrace{\frac{\pi(2.4^3)}{24} \tan(20°)}_{\text{Vol.3}} = 16.28 \text{ m}^3$$

The total height of the bin is

$$H_{total} = H + 1.15D = 3.0 + 1.15(2.4) = 5.76 \text{ m}$$

14.4 JENIKE'S "FLOW-NO FLOW" CRITERION IN HOPPER

- A conical mass-flow hopper in which a cohesive arch is formed at some point in the converging section, as shown in Figure 14.5 (Shamlou, 1988).
- The magnitude of the stress σ_a (arch stress) set up in such a stable arch is proportional to the span D_a of the arch.

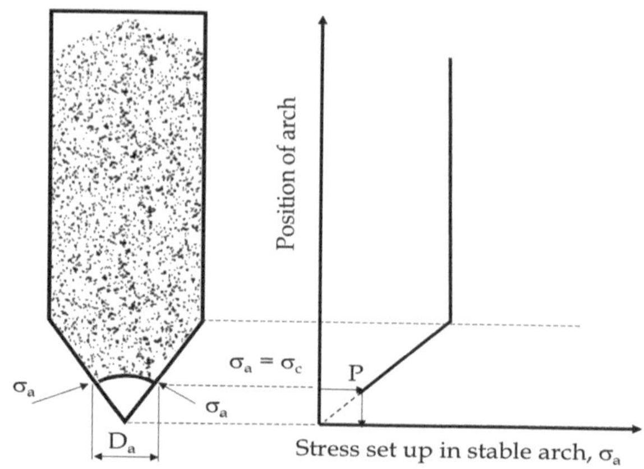

Figure 14.5 Cohesive arch in the converging section (Woodcock and Mason, 1987).

- The bulk solid must be capable of sustaining the stress σ_a without failing.
- The material of the free surface has an unconfined yield stress σ_c, which represents the strength of the material at the free surface.
- At the position where $\sigma_a = \sigma_c$ as shown in Figure 14.5, a stable arch is established.
- Jenike's "flow-no flow" criterion states "a bulk solid will flow if the strength, σ_c develops less than the stress, σ_a, which would exist in a stable obstruction to flow" (Woodcock and Mason, 1987).
- **Graphical representation of the "flow-no-flow" criterion in a hopper:**
  - The unconfined yield stress (σ_c) is generated by the consolidating (normal) stress (σ_1).
  - In the diagram, as shown in Figure 14.6a, the stresses (σ_a, σ_c, σ_1) are plotted that would exist in a stable arch, and it is evident that the point of intersection (*P*) of these lines represents a critical condition between "**flow**" **and "no flow."**
  - Above this point, *P*, ($\sigma_a > \sigma_c$), the flow condition is satisfied, and the product will flow.
  - Below *P*, however, ($\sigma_a > \sigma_c$), indicating that the bulk solid has enough strength to support a stable arch and, therefore, may not flow without some form of discharge aid for a rectangular hopper having $L < 3D_{min}$.
- The stresses σ_a and σ_1 approach zero at the virtual apex of the conical section.
- The ratio of σ_a to σ_1 is constant for a given hopper, called **flow factor** (*ff*) as shown in Figure 14.6b. This parameter characterizes how a given hopper discharges its contents.
- If the consolidating stress (σ_1) increases, the materials acquire greater arch strength (σ_a). The relation is called **flow function** (*FF*), as shown in Figure 14.6b.

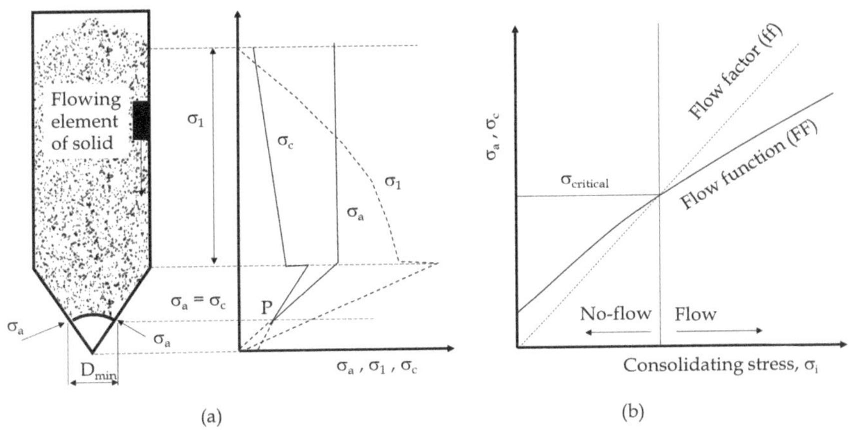

Figure 14.6 Representation of the "flow-no-flow" criterion and stress relation. (a) Representation of the "flow-no-flow" criterion in a hopper (b) Stress in the stable arch vs. consolidating (normal) stress (Woodcock and Mason, 1987).

14.5 CRITICAL OUTLET DIMENSION OF HOPPER

- The minimum value of the hopper outlet dimension (D_{min}) is calculated by

$$D_{\text{min}} = \frac{\sigma_{\text{crit}} J}{\rho_b g}, \quad (14.6)$$

where ρ_b is the bulk density of the stored material, g is the gravitational acceleration, J is an empirical dimensionless factor, and the stress σ_{crit} is the critical stress at flow-no flow conditions.
- The empirical dimensionless factor (which depends upon the shape of the hopper and the angle of the converging section in degree) can be calculated from the correlations as follows for different types of sections.
  a. For conical section

$$J = 2 + \left(\frac{\beta}{60}\right) \quad (14.7)$$

  b. For circular section

$$J = 2 + 0.9315\left(\frac{\beta}{60}\right) \quad (14.8)$$

  c. For square section

$$J = 1.822 + 0.7712\left(\frac{\beta}{60}\right) \quad (14.9)$$

d. For rectangular sections where the length is greater than three times of width of the outlet

$$J = 1.008 + 0.3240\left(\frac{\beta}{60}\right) \tag{14.10}$$

- The moisture content of the material affects its flow behavior.
- Free moisture of a solid can result in surface tension effects. This effect inhibits their movement relative to one another, resulting in the bulk being more cohesive and affecting free-flowing. Hence, a hopper requires a higher value of minimum outlet dimension D_{min}.

Example 14.2

In a rectangular hopper at flow-no flow condition, the material (calcium carbonate) experiences the critical arch stress of 3.5 kN/m² with the half-angle of the conical section of 20°. The bulk density of the calcium carbonate powder is 2,710 kg/m³. What would be the minimum value of the hopper outlet dimension to flow the powder through the bin rectangular outlet? The length of the cross-sectional region is greater than three times of width of the outlet.

Solution:

The critical stress, $\sigma_{crit} = 3.5\ \text{kN/m}^2$, bulk density, $\rho_b = 2710\ \text{kg/m}^3$.

For rectangular sections where the length is greater than three times of width of the outlet

$$J = 1.008 + 0.3240\left(\frac{\beta}{60}\right)$$

Therefore, the minimum value of the hopper outlet dimension would be

$$D_{min} = \frac{\sigma_{crit} J}{\rho_b g} = \frac{3.5 \times 1000 \times 1.116}{(2710)(9.81)} = 0.147\ \text{m}$$

14.6 SOLID DISCHARGE RATE FROM THE HOPPER

- The theoretical mass flow rate ($\dot{M}_p$) can be expressed as

$$\text{Mass flowrate, } \dot{M}_p = \frac{\pi}{4} d_o^2 u \rho_b, \tag{14.11}$$

where d_o=orifice diameter, ρ_b=bulk density. For cohesionless coarse particles free falling over a distance H, their velocity, neglecting drag and interaction, is

$$u = \sqrt{2gH}, \tag{14.12}$$

where H=free falling distance.

- The solid discharge rate (mass flow rate ($\dot{M}_p$)) depends on the orifice diameter of the outlet of the hopper. (Shamlou, 1988) introduce a correction factor kd_p to account for the empty annulus effect in the hopper during flow where k is a constant based on the shape of the particles and d_p is the particle size. As per (Shamlou, 1988), the correlation for mass flow rate is

$$\dot{M}_p = 0.58\rho_b g^{0.5}(d_o - kd_p)^{2.5} \tag{14.13}$$

The constant k can be estimated from the linear relationship of horizontal stress (σ_h) to vertical stress (σ_v) by the solid in the hopper as $\sigma_h = k\sigma_v$.

- Baserinia and Sinka (2018) proposed a correlation for the solid discharge rate of fine and cohesive powders under differential air pressure. The proposed model is:

$$\frac{\dot{M}_p}{\rho_b g^{0.5}(d_o - kd_p)^{2.5}} = c + a\left(\frac{\Delta P D_h^2}{\rho_b g H d_o^2}\right)^n, \tag{14.14}$$

where c, a, and n are the empirical parameters that depend on the type of material. D_h is the hopper diameter and d_p is the particle diameter.

For example, microcrystalline cellulose (MCC), Avicel (PH101, PH102, PH200, and PH302 manufactured by FMC BioPolymer), calcium phosphate (named as ATAB), and mannitol (manufactured by Roquette), the values of c, a, n are given in Table 14.1. For spherical particles, the typical value of k is 1.5, where $\Delta P/H$ is the differential pressure gradient near the orifice.

Example 14.3

A calcium phosphate powder of bulk density 3,150 kg/m³ and 180 μm particle size discharges from a flat-base cylindrical hopper (diameter

Table 14.1 The values of the coefficients used to predict the mass flow rate (Baserinia and Sinka, 2018)

Material	*Coefficients of Equation (14.14)*		
	c	*a*	*n*
PH101	0.32	0.099	0.5
PH102	0.32	0.086	0.5
PH302	0.40	0.085	0.5
PH200	0.42	0.074	0.5
ATAB	0.48	0.070	0.5
Mannitol	0.42	0.073	0.5

PH stands for "potential of hydrogen"

68 mm, height 70 mm) whose orifice diameter at the outlet is 10 mm under a pressure of 1,400 Pa. The linear relationship of horizontal to vertical stress by solid in the hopper can be expressed as $\sigma_h = 1.5\sigma_v$. Find the discharge rate of the material from the hopper (a) as per Baserinia and Sinka's (2018) proposed correlation, (b) as per Shamlou's (1988) proposed correlation. Consider that there is an empty annulus effect in the hopper during flow.

Solution:

Bulk density of the material $(\rho_b) = 3{,}150\ \text{kg/m}^3$.

Orifice diameter $(d_o) = 0.010\ \text{m}$.

The proportionality constant for the linear relationship $(k) = 1.5$.

Particle diameter $= 180\ \mu\text{m}$.

The differential pressure gradient near the orifice $(\Delta P/H) = 20{,}000$ Pa/m.

The hopper diameter $(D_c) = 0.068\ \text{m}$.

As per Baserinia and Sinka's (2018) proposed correlation, for the material calcium phosphate powder, $c = 0.48$, $a = 0.070$, and $n = 0.5$.

Therefore, the discharge rate of material as per Baserinia and Sinka's (2018) proposed correlation,

$$\frac{\dot{M}_p}{\rho_b g^{0.5}(d_o - kd_p)^{2.5}} = c + a\left(\frac{\Delta P D_c^2}{\rho_b g H d_o^2}\right)^n \quad \text{implies}$$

$$\dot{M}_p = \left[(3150)(9.81)^{0.5}(0.010 - 1.5 \times 180 \times 10^{-6})^{2.5}\right]$$

$$\times\left[0.48 + 0.070\left(\frac{0.068^2}{(3150)(9.81)(0.010)^2} \times 20000\right)^{0.5}\right] = 0.080\ \text{kg/s}$$

The discharge rate of material as per Shamlou's (1988) proposed correlation is

$$\dot{M}_p = 0.58\rho_b g^{0.5}(d_o - kd_p)^{2.5} = 0.58(3150)(9.81)^{0.5}(0.010 - 1.5 \times 180 \times 10^{-6})^{2.5}$$

$$= 0.053\ \text{kg/s}.$$

IMPORTANT POINTS TO REMEMBER

- Silos are larger and more expensive than bins and hoppers.
- The mass flow is considered to occur when the motion of the powder is uniform through the cross-section and steady-state whenever the material is drawn from the outlet of the hopper.
- Core flow is said to happen when the powder flows toward the outlet of a silo in a channel formed within the powder itself.

- There are no stagnant regions in the mass-flow hopper.
- For a cylindrical mass-flow hopper, $D < H < 4D$. Note that for a rectangular hopper $H < 3D_{min}$.
- Jenike's "flow-no flow" criterion states "a bulk solid will flow if the strength σ_c develops less than the stress σ_a which would exist in a stable obstruction to flow."
- The solid discharge rate depends on the orifice diameter of the outlet of the hopper.

SELF-TESTING QUESTIONS

I. Multiple-Choice Questions

Q1. The mass flow is considered to occur when ____________________.
- (a) the motion of the powder is nonuniform through the cross-section and steady-state
- (b) the motion of the powder is uniform through the cross-section and unsteady state
- (c) the motion of the powder is uniform through the cross-section and steady-state
- (d) the motion of the powder is nonuniform through the cross-section and unsteady state

Q2. Core flow is said to happen when the powder flows toward ____________________.
- (a) the inlet of a silo in a channel formed within the powder itself
- (b) the outlet of a silo in a channel formed within the powder itself
- (c) the outlet of a silo in a channel formed outside the powder itself
- (d) the inlet of a silo in a channel formed outside the powder itself

Q3. In which of the following cases is the bulk density of the discharged powder constant and practically independent of the height in the silo?
- (a) mass flow
- (b) mass flow with core
- (c) core flow
- (d) none

Q4. In which of the following cases is the powder low down in the hopper stagnant until the hopper is almost empty?
- (a) mass flow
- (b) mass flow with core
- (c) core flow
- (d) none

Q5. In which of the following cases there are no stagnant regions?
- (a) in the mass-flow hopper
- (b) in the mass-flow hopper with core

(c) in the core flow hopper
(d) none

Q6. In which of the following cases are stresses generally low throughout the mass of solids, giving low powder compaction?
(a) mass flow
(b) mass flow with core
(c) core flow
(d) none

Q7. Which one of the following options is correct for a cylindrical mass-flow hopper?
(a) the height of the cylindrical section should be less than four times of the diameter of the bin
(b) the height of the cylindrical section should be greater than four times of the diameter of the bin
(c) the height of the cylindrical section should be less than half of the diameter of the bin
(d) the height of the cylindrical section should be equal to the diameter of the bin

Q8. A conical mass-flow hopper in which a cohesive arch is formed at some point in ____________.
(a) the top section
(b) the bottom section
(c) the converging section
(d) the diverging section

Q9. In a conical mass-flow hopper, the magnitude of the stress (arch strength) set up in a stable arch is proportional to ____________.
(a) the span D_a of the arch
(b) the square of the span D_a of the arch
(c) the square root of the span D_a of the arch
(d) the inverse of the span D_a of the arch

Q10. Jenike's "flow-no flow" criterion states that____________.
(a) "a bulk solid will flow if the strength σ_c develops is greater than the stress σ_a which would exist in a stable obstruction to flow"
(b) "a bulk solid will flow if the strength σ_c develops is equal to the stress σ_a which would exist in a stable obstruction to flow"
(c) "a bulk solid will flow if the strength σ_c develops is half of the stress σ_a which would exist in a stable obstruction to flow"
(d) "a bulk solid will flow if the strength σ_c develops is less than the stress σ_a which would exist in a stable obstruction to flow"

Q11. The ratio of arch stress to the consolidating stress is constant for a given hopper which is called ____________.
(a) flow factor (*ff*)
(b) flow function (*FF*)
(c) flow-no flow criterion
(d) none

Q12. If the consolidating stress (σ_1) is increased, the materials acquire greater arch strength (σ_a). The relation is called ________________.
(a) flow factor (ff)
(b) flow function (FF)
(c) flow-no flow criterion
(d) none

Q13. More moisture content of the material to flow requires ________________.
(a) a higher value of minimum outlet dimension
(b) a lower value of minimum outlet dimension
(c) no change in the value of the minimum outlet dimension
(d) none of the above

Q14. The horizontal stress is related to vertical stress by solid in the hopper as________________.
(a) the horizontal stress is linearly related to vertical stress by solids
(b) the horizontal stress is exponentially related to vertical stress by solids
(c) the horizontal stress is inversely related to vertical stress by solid
(d) the horizontal stress is the square of the vertical stress by solid

Q15. For cohesionless coarse particles free falling over a distance h in a hopper. Under such conditions, the theoretical mass flow rate is proportional to________________.
(a) $\sqrt{d_o}$
(b) d_o^2
(c) d_o^3
(d) $1/d_o^2$

Q16. For cohesionless coarse particles free falling over a distance h in a hopper. Under such conditions, the theoretical mass flow rate is proportional to________________.
(a) $\sqrt{h}$
(b) h^2
(c) $h^{2.5}$
(d) $1/h^{0.5}$

Q17. For a rectangular hopper, which of the following options is correct?
(a) $H=3D_{min}$
(b) $H>3\,D_{min}$
(c) $H<3\,D_{min}$
(d) $3H<D_{min}$

Q18. For a cylindrical mass-flow hopper, which of the following relations is correct?
(a) $D<H<4D$
(b) $D^{0.5}>H>4D$
(c) $D<H<4D^{0.5}$
(d) $H=4D$

Q19. There are no stagnant regions in the mass-flow hopper. Is the statement true or false?
(a) true
(b) false

Q20. For spherical particles, the typical value of k is __________ where $\Delta P/H$ is the differential pressure gradient near the orifice.
(a) 0.01
(b) 0.15
(c) 1.50
(d) 5.10

Answers to the multiple-choice questions.

Q1. Ans. (c)	Q8. Ans. (c)	Q15. Ans. (b)
Q2. Ans. (b)	Q9. Ans. (a)	Q16. Ans. (a)
Q3. Ans. (a)	Q10. Ans. (d)	Q17. Ans. (c)
Q4. Ans. (c)	Q11. Ans. (a)	Q18. Ans. (a)
Q5. Ans. (a)	Q12. Ans. (b)	Q19. Ans. (a)
Q6. Ans. (a)	Q13. Ans. (a)	Q20. Ans. (c)
Q7. Ans. (a)	Q14. Ans. (a)	

II. Objective Type Questions

1. Describe the different types of feeders. Why feeders are important?
2. Distinguish mass and core flows in hoppers.
3. What are the different geometries of hoppers?
4. Explain the mathematical expression of the stored volume of a hopper based on its different parts.
5. Explain Jenike's "flow-no flow" criterion in hopper. Represent it graphically.
6. What are flow factors and flow functions?
7. How can you estimate the critical outlet dimension of the hopper?
8. How is mass flow related to the orifice of fixed diameter?
9. Why does a hopper require a higher value of minimum outlet dimension?
10. Explain the solid discharge from the apertures (x) in a flat-based cylindrical hopper.

III. Numerical Problems

Q1. A conical bin is designed to store catalyst powder in hydrocarbon conversion in a refinery where the bin diameter is 1.978 m, and the diameter of the outlet section is 0.75 m. The height of the cylindrical section is 2.5 m. The half-angle of the conical section is 14.1°. The surcharge angle of the "free surface" of material in the hopper (which corresponds to the poured angle of repose) is 30°. Find the stored volume of the bin. [Ans. 10.01 m^3]

Q2. A conical bin is designed to store catalyst powder in hydrocarbon conversion in a refinery where the bin diameter is 1.978 m, and the diameter of the outlet section is 0.75 m. The height of the cylindrical section is 2.5 m. The half-angle of the conical section is 14.1°. The surcharge angle of the "free surface" of material in the hopper (which corresponds to the poured angle of repose) is 30°. Find the total height of the bin. [Ans. 4.77 m]

Q3. In a conical hopper at flow-no flow condition, the material (iron oxide powder) experiences the arch stress of 1.8 kN/m^2 with the half-angle of the conical section of 30°. The bulk density of the powder is 4.8 g/cc. What would be the minimum value of the hopper outlet dimension to flow the powder through the bin outlet? [Ans. 0.0963 m]

Q4. A solid powder of bulk density 2,650 kg/m^3 discharges from the aperture of 0.007 m of a flat-base cylindrical hopper whose orifice diameter at the outlet is 0.005 m. The linear relationship of horizontal to vertical stress by solid in the hopper can be expressed as $\sigma_h = 0.1\sigma_v$. Find the discharge rate of the material from the hopper. [Ans. 5.83×10^{-3} kg/s]

NOMENCLATURE

d_o	orifice diameter [m]
D	diameter of the bin [m]
D_a	span of the arch [m]
D_c	diameter of the outlet of bin [m]
D_h	hopper diameter defined in Equation (14.14) [m]
$D_{\min}$	minimum value of the hopper outlet dimension [m]
d_p	particle diameter [m]
H	free falling distance defined in Equation (14.12)
H	height of the cylindrical section [m]
k	ratio of horizontal stress (σ_h) to vertical stress (σ_v) [-] defined in Equation (14.13)
m	mass of the solid [kg]

$\dot{M}_p$ theoretical mass flow rate [kg/s]
u velocity [m/s]
V volume [m^3]

GREEK LETTERS

σ_a arch stress [N/m^2]
ρ_b bulk density [kg/m^3]
σ_{crit} critical stress at flow-no flow conditions [N/m^2]
σ_1 consolidating (normal) stress [N/m^2]
σ_c unconfined yield stress [N/m^2]
β half-angle of the conical section [degree]
ϑ surcharge angle of the "free surface" of material in the hopper [degree]

SUBSCRIPTS

a arch
b bulk
c outlet
crit critical
h hopper
min minimum
o orifice
p particle

REFERENCES

Baserinia, R. and I. C. Sinka, Mass flow rate of fine and cohesive powders under differential air pressure, *Powder Technol.*, 334, 173–1821, 2018.
Shamlou, P. A., *Handling of Bulk Solids: Theory and Practice*, Butterworths, 1988.
Woodcock, C. R. and J. S. Mason, *Bulk Solids Handling: An Introduction to the Practice and Technology*, Springer, 1987.

Index

Note: **Bold** page numbers refer to tables and *italic* page numbers refer to figures.

FSC
www.fsc.org
MIX
Papier | Fördert gute Waldnutzung
FSC® C083411

Kolibri 360 GmbH
Ferdinand-Jühlke-Straße 7
99095 Erfurt, Deutschland
produktsicherheit@kolibri360.de